Operators and Representation Theory

Canonical Models for Algebras of Operators Arising in Quantum Mechanics

Palle E. T. Jorgensen
Department of Mathematics
The University of Iowa

Foreword by William Klink
With a New Preface to the Dover Edition by the Author

Dover Publications, Inc.
Mineola, New York

Copyright

Bibliographical Note

This Dover edition, first published in 2008, is an unabridged republication of the work originally published as Volume 147 in the North-Holland Mathematics Studies series by North-Holland Publishing Company, Amsterdam and New York, in 1988. The present volume adds a new Foreword written by William Klink and a new Preface to the Dover Edition written by the author.

Library of Congress Cataloging-in-Publication Data

Jørgensen, Palle E. T., 1947–
Operators and representation theory : canonical models for algebras of operators arising in quantum mechanics / Palle E. T. Jorgensen ; foreword by William Klink ; with a new preface to the Dover edition by the author. — Dover ed.
p. cm.
Originally published: Amsterdam ; New York : North-Holland Pub., 1988; in series: North-Holland mathematics studies series ; v. 147.
Includes bibliographical references and index.
ISBN-13: 978-0-486-46665-1
ISBN-10: 0-486-46665-5
1. Operator algebras. 2. Representations of algebras. I. Title.

QA326.J67 2008
512'.556—dc22

2007053037

Manufactured in the United States of America
Dover Publications, Inc., 31 East 2nd Street, Mineola, N.Y. 11501

FOREWORD

Einstein was one of the first to realize the importance of symmetry in grounding physical theories. His analysis of the contrasting symmetry structures of Maxwell's equations for electromagnetism and Newton's mechanics led directly to the special theory of relativity.

Symmetry has played an even more important role in the development of quantum theory. For example, early on in the development of quantum theory it was realized that the quantization of spin and orbital angular momentum was a direct consequence of rotational symmetry. Representations of the rotation group are used routinely to couple angular momentum for many-electron systems. Other compact groups such as the unitary groups have played an important role in nuclear physics, and with so called internal symmetries dealing with quantum numbers such as charge, baryon number and strangeness.

Although the earliest formulations of quantum theory did not explicitly draw on group theory, it was soon realized that many of the grounding principles of quantum theory could naturally be expressed in group theoretical language. This was made particularly clear when trying to combine the symmetry structure of the special theory of relativity with quantum theory. Here the underlying group is the Poincaré group, the group of transformations leaving the structure of space-time invariant. Wigner's analysis of the representations of the Poincaré group form today the basis of all versions of relativistic quantum theory, including quantum field theory. What is particularly striking is that the invariant operators of the Poincaré group are mass and spin operators, exactly the observables that characterize elementary particles such as electrons and protons.

Surprisingly the symmetry properties of what is now called nonrelativistic quantum theory, in which the transformations in space-time called collectively the Galilei transformations, those that leave the form of Newton's equations unchanged, were not investigated until after the work by Wigner on the Poincaré group. Bargmann and Wigner, among others, investigated the representation structure of the Galilei group and found that many of the grounding principles of the earliest formulations of quantum theory could be understood in group theoretical language. In particular the necessity of infinite dimensional Hilbert spaces for describing even single particle systems was seen to be a direct consequence of unitary irreducible representations of a noncompact group, the Galilei group.

Along with developments in atomic and nuclear physics, it also became clear in the 1930's that for processes involving the creation and annihilation of particles such as the photon, the quantum of light, that larger groups and Lie algebras than the compact and locally compact groups and Lie algebras were needed. The earliest such examples included the bosonic and fermionic algebras generated by the commutation and anticommutation relations of creation and annihilation operators. Large classes of representations of these algebras were discovered and work with these and other infinite dimensional algebras continues to be an area of active research today, in both the mathematics and mathematical physics communities.

The mathematical expression of symmetry is group theory in all of its ramifications, including in particular Lie algebras and representations of groups and Lie algebras on Hilbert spaces. There are many ways in which group theory is used in quantum theory, going well beyond the applications mentioned above. It is therefore of interest to have a mathematician develop in modern mathematical language some of the newer developments occurring in the intersection between group theory (in its most generalized sense) and quantum theory. In this monograph Palle Jorgensen begins with some of the mathematical machinery needed to understand modern developments. He introduces the notions of projective representations and central extensions and finishes part I with the imprimitivity theorem, which grounds in more mathematical language the work of Wigner on representations of the Poincaré and Galilei groups.

Part II of the monograph deals with algebras of operators on Hilbert spaces, broadening the mathematics used in earlier versions of quantum theory. There are many examples where the Hamiltonian, the operator that translates a quantum system in time, can be written as a polynomial in elements of an underlying Lie algebra. This part of the monograph deals with properties of such operators, for example their spectral properties.

Concrete examples, including the behavior of charged particles in curved magnetic fields, make the connection between the abstract mathematics and interesting physical systems.

Part III deals with infinite dimensional Lie algebras, a topic of increasing importance in a variety of different approaches to quantum theory, including string theory, theories of gravitation and gauge quantum field theories. The structures of interest here include the gauge (map) groups, Diff groups and infinite dimensional groups such as the infinite unitary groups. New mathematical machinery is needed to deal with these structures, and this is done by looking at several different examples; in particular the Virasoro algebra is analyzed in some detail.

This monograph is an excellent example of the interplay between mathematics and mathematical physics. Most of the mathematical topics are motivated by physical systems, and then developed in their own right. Anyone interested in the mathematical formulation of modern quantum theory will benefit from this monograph.

2007

WILLIAM KLINK
Department of Physics
The University of Iowa

PREFACE TO THE DOVER EDITION

This book grew out of seminars and courses we taught covering a core of mathematics used in Quantum Mechanics, and in other applications.

Motivated by quantum theory, we begin with Hilbert space, and with the corresponding linear transformations (linear operators). A central theme is the use of representations of groups and algebras by operators in Hilbert space: differential operators, wave operators, Schroedinger operators, etc. The questions we ask, both in the context of mathematics and physics, relate to these operators: Are they selfadjoint? What is the spectrum; the spectral resolution? Where does the Hilbert space come from? How do we use symmetries in computations of spectrum? What is the appropriate notion of equivalence in infinite dimensions?

In the teaching of mathematics and physics, there is a core of ideas with a good amount of permanence: principles that are more timeless than others. For quantum mechanics, on the mathematical side, they center around our use of Hilbert space $\mathcal{H}$, and of transformations (linear operators) transforming the vectors in $\mathcal{H}$. This is the language used in asking questions about spectral lines, scattering and the like.

This amounts to a generalization of linear analysis to infinite dimensions (operator theory). It entails an infinite-dimensional extension of the fundamentals from undergraduate courses in linear algebra: matrix, transformation, vectors, eigenvector, eigenvalue, eigenspace, basis, subspace, and the spectral theorem; i.e., the radical idea of asking for a diagonalized version of a given normal linear transformation.

At its inception in the 1920's, Quantum Theory seemed perhaps radical: Furthermore, at the time, it appeared somewhat artificial to select axioms and constructs from mathematics with view to making precise sense of such physical "realities" as, say, state, observable, measurement, spectrum, transition probability, symmetry, energy, momentum, position, electron, photon, elementary particle, entanglement, quantum fields, quantum communication, etc. But in the seven decades following the pioneering discoveries of Heisenberg, Born, Dirac, Schrödinger, Pauli; the mathematical framework has now become common place; and a permanent fixture.

Despite this state of affairs, there is still a divide between how students are taught the subject in math and in physics. In mathematics and physics department the emphasis differs; and what is more, the use of lingo, terminology, and definitions continues to diverge. Interdisciplinary communication has not become easier. Rather the other way around. Some talk of a divorce between math and physics. Take a look, for example, at such a basic notion as "state" (as in state of a particle, ground state, excited state,). Comparing physics and math texts, you will find that in fact this fundamental notion of a "state" is defined differently depending on which side of the divide you are.

True, the authors of the two texts might both have been motivated by Paul Dirac's vision, but the product in current textbooks came out differently; or at least so it would seem.

The present book aims to organize this material on a common ground: A recurrent theme is the use of mathematics in understanding symmetry, and the use of groups and algebras in computations of spectra. This framework is part of what has become known as operator theory, the use of linear operators in Hilbert space in computations of mathematical physics. Here "spectrum" refers to some underlying selfadjoint operator in Hilbert space. And the Spectral Theorem for selfadjoint operators allows us to use operator theory in creating a precise mathematical description of measurements in quantum mechanics; in giving an answer to questions such as these: "Given some interval J on the real axis, and some specific observable A; what is the probability of measuring values of the observable A in the interval J when an experiment is prepared in a specific state? And if there are two states, what is the corresponding transition probability?"

Since the work of J. von Neumann, E. Wigner, V. Bargmann, and H. Weyl, we now take for granted that groups and algebras allow us to use symmetry in computations, but the distinction between the groups on the one hand, and their representations on the other is occasionally blurred in textbooks. While the groups G and their isomorphism classes may be easy entities, based only on a small number of coordinates, their representations (for non-compact G) are typically infinite-dimensional; i.e., their irreducible "pieces" cannot be broken down into finite-dimensional constituents. Since the irreducible representations of the appropriate symmetry group G correspond to elementary particles from physics, infinite-dimensional Hilbert space is inescapable. Starting with infinite-dimensional Hilbert space $\mathcal{H}$, we get raising and lowering operators, particle number, groups of unitary operators acting on $\mathcal{H}$; and much more. It isn't just a generalization for generalizations sake, dreamed up by some crazy mathematician.

On occasion, in the time period since the first edition, I was asked to explain one point. It is about the following two facts, and choices I made:

(1) The variety of types and classes of groups that are needed in physics applications make up for a vast and diverse gallery: Continuous vs discrete, finite-dimensional vs infinite-dimensional, abelian vs non-abelian, central extensions vs more general extensions, nilpotent vs semisimple, .

(2) For each of the possibilities for the group G under consideration, we are interested in understanding the associated unitary representations of G, as well as covariant systems of operators in Hilbert space.

And for each class of the groups and their representations, the literature discussing analysis of the unitary representations and the equivalence classes of irreducibles is vast. In fact, for a number of groups, the answers to natural questions about classification are not even within reach.

Because of (1) we have adopted a terminology that is a bit more "expansive" that is perhaps customary; for example in our use of the term "short exact sequence." Because of (2), our list of References is more extensive than would be the case in a more specialized monograph.

A more detailed overview of the book: The idea of symmetry is central in physics, especially in Quantum Physics, where symmetries manifest themselves in subtle ways. Quantization of symmetries is fundamental in Relativistic Quantum Physics via superselection sectors, anomalies, and spontaneous symmetry breaking.

The mathematics of symmetries was pioneered in the work of Herman Weyl, Eugene Wigner, George Mackey, Valentin Bargmann, Harish-Chandra, Edward Nelson... - Some of the mathematical subtleties in the subject include a transition from ray (projective) representations to unitary representations (in Hilbert space.)

Powerful tool for probing quantum manifestations of symmetry are induced representations, Mackey's systems of imprimitivity, central extensions of Lie algebras, representations of Lie Algebras by unbounded operators with dense domain in a Hilbert space; coupled with their associated spectral theory (energy levels, spectral type, continuous vs discrete). Stressing physical motivation, this is presented with an emphasis on Hilbert space geometry, and relying on basics in functional analysis, homological algebra, Lie groups and Lie algebras, and in especially representation theory.

We probe cutting-edge topics such as representations of Virasoro algebras (conformal field theory, string theory) and non-commutative differential-geometry, e.g., non-commutative tori in the form of specific C^*-algebras.

The reader should ideally have some acquaintance with basic functional analysis (Banach and Hilbert space, measure theory), Lie groups, and some idea of Quantum Mechanics. Yet, the required preliminaries are relatively minor, and are taught in most beginning courses.

In writing the book, the author had in mind both students in math courses and physics students (e.g., quantum theory courses). One isn't favored over the other.

2007

PALLE E. T. JORGENSEN
Iowa City

GENERAL COMMENTS TO THE DOVER EDITION

The Preface includes comments explaining terminology used in the book. Since the first edition, from time to time readers wrote me asking for clarifications of terminology.

So in the Dover Preface, I took the opportunity to flesh out explanations in a little more systematical fashion. For example, a few comments on our use of the term "short exact sequence" and semidirect sum (see e.g., pages 139, 225, 251, and 282). And we explain some of the items in References.

Throughout, we use two ways of representing physical operators in Hilbert space: (1) As operators in function spaces; or alternatively (2) as infinite-by-infinite matrices. The first is motivated by Schrödinger's formulation of quantum mechanics, and the second by Heisenberg's.

The reader will notice that the first edition was from the PreTex Era.

On page 134, Theorem 7.1.2, in the general case, more than L^1 is required of the function under the integral in (7.1.16), but this is automatic in the application to the heat kernel. See the cited reference, page 132, and the discussion following the theorem.

On pages 219 and 224, the tilde over a particular group G refers to the simply connected universal covering group; and the canonical homomorphism from the covering group to G itself.

On pages 59, 214, and elsewhere we discuss matrix groups, and their Lie algebras. Our convention is to use Capital letters for the group in question and lower case for the associated Lie algebra, for example U(n, 1) and u(n, 1).

PALLE E. T. JORGENSEN

PREFACE

Professor Marshall Stone told me, some time ago, that when he began the work on his book [St 3], he had in mind unitary representations of groups on Hilbert space. Of course, at the time, Quantum Mechanics, operator theory, and representation theory were emerging at a rapid pace and taking separate forms. As we now know, it was left to others to carry out what Stone may have had in mind (with regard to group representations).

Such books have indeed been written, and covering different time periods, reflecting the different stages in the development of the subjects.

In the intervening time, since Stone's book, the number of subspecialities in this area of mathematics and mathematical physics has grown (see, for example, the list of references below), and the diversity of subjects has increased as well. In the meantime, connections to developments in operator algebras have come to play an important role.

During the Eighties, there seems to have been tendencies for some of the related subjects to develop along separate paths, and the potential for fruitful interaction may not have been fully realized and utilized. It is the aim of the present monograph to improve on this state of affairs. We have picked certain subjects from the theory of operator algebras, and from representation theory, and showed that they may be developed starting with Lie algebras, extensions, and projective representations. In some cases, this Lie algebra approach to familiar problems in C^*-algebra theory is new. Furthermore, we have picked problems from the theory of representations of infinite-dimensional Lie algebras, and demonstrated that C^*-algebraic methods may be used in nontraditional ways with some success.

The book is addressed to graduate students who wish to get an introduction to some of the more recent developments in these interacting subject areas. It is also addressed to researchers with specialized knowledge in some, but not all, of the research fields. Connections to mathematical physics have been stressed throughout, and the book is written also with the mathematical physics community in mind. Our group theoretic treatment of curved magnetic field Hamiltonians is a case in point.

ACKNOWLEDGEMENTS

A portion of the research reported in the present monograph has been done in various collaborative projects. My co-authors in the joint research projects include (in alphabetical order): Ola Bratteli, Joachim Cuntz, George A. Elliott, David E. Evans, Frederick M. Goodman, Akitaka Kishimoto, William H. Klink, Robert T. Moore, Paul S. Muhly, Steen Pedersen, Costel Peligrad, Geoffrey L. Price, and Derek W. Robinson. I am pleased to acknowledge their contributions. When the picture emerges, it is difficult, and often impossible, to say who did exactly what. The pleasant fact about the collaborations is that the outcome is better, and larger, than the sum of the individual components.

I am also pleased to thank colleagues W.H. Klink, S. Pedersen, and my student H. Prado for help with proofreading.

Special thanks are due to Mrs. Ada Burns for the typing (and production) of the final manuscript. Her artistic skill is pleasantly visible, and greatly appreciated.

The research reported in the present monograph was supported in part by the National Science Foundation.

CONTENTS

CHAPTER 1. INTRODUCTION AND OVERVIEW

1. This book is about algebras of operators. For lack of a better name, we shall refer to these algebras as *algebras of unbounded operators*. This is because most people think about C^*-algebras and von Neumann algebras when the shorter title "operator algebras" appears, and our algebras will typically not be C^*-algebras, but they are utilized in the study of C^*-algebras as we shall show in Part III of the book. They are used especially in that part of noncommutative differential geometry which is concerned with the study of differentiable structures on simple C^*-algebras. (This subdiscipline has not really become a theory yet because only a few classes of simple C^*-algebras are understood up to now.)

The algebras which we consider arise frequently in the study of representations of Lie groups (both finite-dimensional, and infinite-dimensional ones). The theory is best developed for finite-dimensional groups, but the infinite-dimensional groups are playing an increasingly important role in the exciting recent work on loop groups.

In some cases the problem of understanding differentiable structures for C^*-algebras boils down to finding smooth Lie group actions on the C^*-algebra in question. It will be assumed that the given C^*-algebra has an atlas of "coordinate charts," and we shall see in examples how to identify this. Then the smoothness is a relative notion and refers to a given, more or less canonical, group action defined naturally in terms of the generators for the given C^*-algebra.

The more classical and familiar side of the subject is the study of unitary representations of Lie groups (on Hilbert space $\mathcal{H}$). Such a representation has an enveloping algebra of unbounded operators on $\mathcal{H}$, and it is obtained by differentiation of the given unitary representation along the Lie algebra. The resulting algebra is a $*$-algebra, i.e., comes with a natural involutory anti-automorphism or, equivalently, is given by a Hermitian representation.

The enveloping algebra of a unitary representation is useful because the elements in it are operators which play a central role in Quantum Mechanics. Examples of such algebras include the Weyl algebra of partial differential operators with polynomial coefficients. We shall also show how Hamiltonians for magnetic fields can be completely spectrally analyzed using unitary representations of certain nilpotent Lie groups. Finally we shall show how the representation theory of $SL_2(\mathbb{R})$ can be used for constructing models for operators in Hilbert space.

The converse problem is to decide if a given $*$-algebra of unbounded operators can be integrated (exponentiated) to a unitary representation. This is the more interesting and difficult part, and it too has its roots in Quantum Mechanics since the $*$-algebra is typically given from the physical context, and it is generated by raising and lowering operators (the notion Verma-module is now popular in representation theory). The problem is then to find the Hilbert space, and to "reconstruct" the unitary representation.

If the problem is not "unitarizable" then this procedure fails, and one tries instead to get a representation of the Lie group by bounded operators in some Banach space. The types of techniques which are used in solving these problems include Nelson's analytic vector method, the Arveson-Powers method based on complete positivity, C^∞-vector methods, and perturbation methods. The latter two types of methods were developed recently by R. T. Moore and the author in [JM].

Recently, heat equation methods have been developed by O. Bratteli, F.M. Goodman, D.W. Robinson and the author [BGJR], and they were shown to apply to the most general instance of the integrability problem in the setting of Banach spaces.

In this book we shall include a systematic treatment of the methods which have been developed more recently, and we shall apply them to some classical problems in Quantum Mechanics, and to some recent problems in C^*-algebras, and general operator theory.

CHAPTER 2. DEFINITIONS AND TERMINOLOGY

2.1 Notation

Let $\mathfrak{D}$ be a vector space over $\mathbb{R}$ or $\mathbb{C}$. Let $\mathfrak{g}$ be a Lie algebra over $\mathbb{R}$, and let $\mathfrak{A}$ be an associative algebra over $\mathbb{C}$. (As an example, take $\mathfrak{A}$ to be the associative universal enveloping algebra over $\mathfrak{g}$. It is denoted $\mathfrak{U}_{\mathbb{C}}(\mathfrak{g})$. Finally let End $\mathfrak{D}$ be the algebra of linear endomorphisms of $\mathfrak{D}$.

2.2 Brackets

Recall that End $\mathfrak{D}$ acquires the structure of a Lie algebra when equipped with the *commutator bracket*

$$[X,Y] = XY - YX.$$

The *anti-commutator* will be denoted

$$\{X,Y\} = XY + YX.$$

2.3 Representations

A *representation* of $\mathfrak{g}$ is a homomorphism, $\rho : \mathfrak{g} \longrightarrow \text{End } \mathfrak{D}$ of Lie algebras, and a representation of $\mathfrak{A}$ is a homomorphism, $\rho : \mathfrak{A} \longrightarrow \text{End } \mathfrak{D}$ of associative algebras. From the definition of the universal enveloping algebra $\mathfrak{U}_{\mathbb{C}}(\mathfrak{g})$, it follows that every representation ρ of $\mathfrak{g}$ extends uniquely to a representation of $\tilde{\rho}$ of $\mathfrak{U}_{\mathbb{C}}(\mathfrak{g})$.

2.4 Units

We shall denote the *identity element* in $\mathfrak{U}_{\mathbb{C}}(\mathfrak{g})$ by 1, and that of End $\mathfrak{D}$ by I. Recall that I is simply the identity operator of the linear space $\mathfrak{D}$. If $\mathfrak{A}$ has a unit, and $\rho(1) = I$, then we shall say that ρ is unital.

2.5 Projective Representations

One of the defining conditions of a representation, $\rho : \mathfrak{g} \longrightarrow \text{End } \mathfrak{D}$, is the property

$$\rho([x,y]) = [\rho(x),\rho(y)], \qquad x,y \in \mathfrak{g}. \tag{2.5.1}$$

If there is a bilinear function

$$B : \mathfrak{g} \times \mathfrak{g} \longrightarrow \mathbb{C}$$

such that

$$(2.5.2) \qquad \rho([x,y]) = [\rho(x),\rho(y)] - B(x,y)I,$$

then we say that ρ is a *projective representation*. In this case we shall always assume that B satisfies the *cocycle* property

$$(2.5.3) \qquad B(x,[y,z])+B(y,[z,x])+B(z,[x,y]) = 0, \qquad x,y,z \in \mathfrak{g}.$$

If $\mathfrak{g}$ is given, and if B is a real valued cocycle, then we may define a *central extension* as follows. The central extension $\mathfrak{g}_B$ is spanned by $\mathfrak{g}$ and some additional element, say b, satisfying

$$(2.5.4) \qquad [x,b]_B = 0, \qquad x \in \mathfrak{g}$$

(where the subscript refers to the Lie bracket of $\mathfrak{g}_B$), and

$$(2.5.5) \qquad [x,y]_B = B(x,y)b + [x,y], \qquad x \in \mathfrak{g}.$$

It can be checked that there is a natural one-to-one correspondence between, on the one hand,

(i) projective representations ρ of $\mathfrak{g}$ with co-cycle B,

and, on the other hand,

(ii) representations ρ_B of the central extension Lie algebra $\mathfrak{g}_B$.

If ρ is as in (i) satisfying (2.5.2) above, then ρ_B is defined on $\mathfrak{g}_B$ via

$$(2.5.6) \qquad \rho_B(x+\lambda b) = \rho(x) + \lambda I, \qquad x \in \mathfrak{g}, \quad \lambda \in \mathbb{R}.$$

If, conversely, ρ_B is as in (ii) then ρ is defined by

$$(2.5.7) \qquad \rho(x) = \rho_B(x+0\cdot b), \qquad x \in \mathfrak{g}.$$

A particular case of this construction, when $\mathfrak{g}$ is Abelian, is the Heisenberg Lie algebra. Other important examples arise in string theory and super Lie algebras [PS], and will be recalled in Chapter 8.5 below.

2.6 Central Extensions

The *short exact sequence*

$$0 \longrightarrow g \longrightarrow g_B \longrightarrow \mathbb{C}b \longrightarrow 0$$

motivates the terminology, central extension for g_B.

2.7 Modules

We include a comment on the familiar multiplicity of notation in representation theory: A representation may be thought of as

(i) a *homomorphism* from an algebraic object into operators on some vector space $\mathfrak{D}$;

(ii) $\mathfrak{D}$ may be viewed as a *module* over this object, e.g., a g-module, $\mathfrak{X}$-module, $\mathfrak{U}_{\mathbb{C}}(g)$-module, g_B-module, etc.;

finally,

(iii) the operators in the image of the representation may be thought of as an *algebra of operators*, or, in the case of group-representations, as a group of operators. In the case of Lie algebras, and associative algebras, the corresponding operator algebras will generally consist of *unbounded operators*.

2.8 Norm and Unboundedness

The unboundedness refers to some preordained (i.e., given) norm on $\mathfrak{D}$. In applications, this norm may be defined relative to some measure, e.g., some L^p-norm; it may be a C^*-*norm*, i.e., satisfying

$$\|a^*a\| = \|a\|^2, \qquad a \in \mathfrak{D}; \tag{2.8.1}$$

or it may be defined relative to some *sesquilinear form* on $\mathfrak{D}$. We shall use the notation $\langle\cdot,\cdot\rangle$ for such a form. The familiar assumptions on $\langle\cdot,\cdot\rangle$ include:

$$\langle a,a\rangle \ge 0, \qquad a \in \mathfrak{D}, \tag{2.8.2}$$

By Cauchy-Schwarz, we then have

(2.8.3) $$|\langle a,b\rangle|^2 \le \langle a,a\rangle\langle b,b\rangle, \qquad a,b \in \mathfrak{D}.$$

If the subspace $\mathfrak{N} \subset \mathfrak{D}$, defined by

(2.8.4) $$\mathfrak{N} = \{a \in \mathfrak{D} : \langle a,a\rangle = 0\},$$

can be checked to be *compatible* with the representation, in a sense which will be made explicit below, then the representation passes to the *quotient* $\mathfrak{D}/\mathfrak{N}$, or, equivalently, $\mathfrak{D}/\mathfrak{N}$ turns into a module relative to the given algebraic object: Lie algebra, algebra, group, etc. On $\mathfrak{D}/\mathfrak{N}$, we may define a *Hilbert-norm*,

$$\|a+\mathfrak{N}\|^2 := \langle a,a\rangle.$$

The Hilbert space, which results upon completion of $\mathfrak{D}/\mathfrak{N}$ in this norm, will be denoted $\mathfrak{H}$, and we shall refer to algebras of operators on $\mathfrak{H}$. Frequently these operators will be unbounded, and only defined on the dense domain $\mathfrak{D}/\mathfrak{N}$. If some of the operators, T say, are bounded, i.e.,

(2.8.5) $$\|Ta\| \le \text{const.}\|a\|, \qquad a \in \mathfrak{D},$$

then we shall extend T by closure to be defined on all of $\mathfrak{H}$. But in general operators cannot be extended by closure to be defined everywhere on the norm-completion. In fact, in some applications, the operators in question may not even have any closed extensions at all.

2.9 C^*-norms

A second example of norm-completion arises for C^*-algebras which are generated by a given set of elements and relations. The space $\mathfrak{D}$ will then be the linear span of all monomials in the generators, modulo the ideal generated by the given relation(s). A C^*-norm on $\mathfrak{D}$ may be constructed using linear algebra and inductive limits, as is the case in the familiar construction of the C^*-algebra over the canonical anticommutation relations (CAR). For the latter construction, the reader is referred to [Ar 1], [BR 2]. [CRu], [Kad 3], and references therein.

2.10 Groups and Operators

We conclude with a remark on some routine terminology used throughout. The bold face letters $\mathbb{C}$, $\mathbb{N}$, $\mathbb{Q}$, $\mathbb{R}$, and $\mathbb{Z}$ will denote the complex numbers, the natural numbers $\{1,2,\cdots\}$, the rational numbers, the reals, and the integers, respectivly.

A given Lie algebra will be assumed, unless otherwise stated, to be finite-dimensional, and the ground field will be $\mathbb{R}$. Gothic letters $\mathfrak{g}$, $\mathfrak{h}$, etc., will be used for Lie algebras. The Lie bracket will be denoted $[\cdot,\cdot]$. Recall the Jacobi identity

$$[x,[y,z]] + [y,[z,x]] + [z,[x,y]] = 0, \qquad x,y,z \in \mathfrak{g}, \tag{2.10.1}$$

which is the principal defining property.

We shall use without mention the basic properties of the universal enveloping algebra $\mathfrak{U}_{\mathbb{C}}(\mathfrak{g})$. It is an associative algebra over $\mathbb{C}$ with unit 1, and containing $\mathfrak{g}$. We have, in particular, the familiar one-to-one correspondence between representations of $\mathfrak{g}$, and representations of $\mathfrak{U}_{\mathbb{C}}(\mathfrak{g})$.

The exponential mapping of Lie theory will be denoted $\exp : \mathfrak{g} \longrightarrow G$ where G is the Lie group. We shall use that the image $\{\exp x : x \in \mathfrak{g}\}$ generates the connected component of the identity element e in G. The adjoint representation of G will be denoted Ad, and the automorphism $\mathrm{Ad}_g \in \mathrm{Aut}(\mathfrak{g})$ is defined by

$$\exp(\mathrm{Ad}_g(x)) = g(\exp x)g^{-1}, \qquad g \in G, \quad x \in \mathfrak{g}.$$

The derived representation $(\mathrm{Ad})_*$ will be denoted ad; it is a representation of $\mathfrak{g}$, and $\mathrm{ad}\, x$ is the derivation of $\mathfrak{g}$ given by

$$\begin{aligned} \mathrm{ad}\, x(y) &= \frac{d}{dt}\Big|_{t=0} \mathrm{Ad}(\exp tx)(y) \\ &= [x,y], \qquad x,y \in \mathfrak{g}. \end{aligned} \tag{2.10.2}$$

The Killing form of $\mathfrak{g}$ will be denoted

$$B(x,y) = \mathrm{trace}(\mathrm{ad}\, x\ \mathrm{ad}\, y),$$

and we have

$$B(\mathrm{Ad}_g\ x,\ \mathrm{Ad}_g\ y) = B(x,y), \qquad x,y \in \mathfrak{g},\ g \in G.$$

Differentiation along $g = \exp(tz)$, $z \in \mathfrak{g}$, leads to

$$B([z,x],y) + B(x,[z,y]) = 0, \qquad x,y,z \in \mathfrak{g}.$$

The basic result [Ser 1, VI, Thm. 2.1] regarding B which will be used is that the Lie algebra is semisimple if and only if the Killing form is nondegenerate. In that case, we may pick a pair of bases $x_1,\cdots,x_n$, $y_1,\cdots,y_n$ for $\mathfrak{g}$ such that $B(x_i,y_j) = \delta_{ij}$, and a simple argument shows that the element ω in $\mathfrak{U}_{\mathbb{C}}(\mathfrak{g})$, defined by

$$\omega = \sum_j x_j y_j, \tag{2.10.3}$$

is central. It is called the Casimir element, and it can be readily checked that ω is independent of the choice of bases.

Hilbert spaces will be assumed complex, and, unless specifically stated, infinite dimensional. The letter $\mathcal{H}$ will be reserved for Hilbert space, and $\langle\cdot,\cdot\rangle$ will denote the inner product. We will take it to be linear in the first variable. If G is a given Lie group, we shall denote by $\mathcal{H}_L$, resp., $\mathcal{H}_R$, the Hilbert space of all square integrable functions on G relative to the left, resp., the right, invariant Haar measure on G. We use that left, resp., right, Haar measure is unique up to scaling. As a result, we get the modular function on G. If "the" left-invariant Haar measure is denoted dg, then the modular function Δ is determined by

$$\Delta(g')\int_G \varphi(gg')dg = \int_G \varphi(g)dg. \tag{2.10.4}$$

We shall occasionally work with Δ for *locally compact* groups which are not necessarily Lie groups. It is defined in this generality. If $H \subset G$ is a closed subgroup, the homogeneous space G/H of cosets $\{g\cdot H : g \in G\}$ carries an action of G by left-translation. But G/H generally does not carry a nontrivial invariant measure, but only a quasi-invariant one. Let δ denote the modular function of H, and define $\rho(h) = \delta(h)\Delta(h)^{-1}$, $h \in H$. Then there is an *induced* invariant measure on G/H if and only if $\rho(h) = 1$, $h \in H$. If $\rho \not\equiv 1$, then the induced measure on the quotient is only quasi-invariant; it is defined as follows.

Let $\tau : C_c(G) \longrightarrow C_c(G/H)$ denote the *conditional expectation* defined by

$$(2.10.5) \qquad (\tau\varphi)(g) = \int_H \varphi(gh)dh$$

where C_c denotes the space of compactly supported functions on the respective spaces, and dh denotes the left-invariant Haar measure on H. It can readily be checked that the expectation τ maps onto $C_c(G/H)$. It therefore uniquely induces a measure μ on G/H determined by the identity:

$$(2.10.6) \qquad \int_{G/H} \tau\varphi \, d\mu = \int_G \varphi(g)\rho(g)dg, \qquad \varphi \in C_c(G).$$

It can be checked that μ is quasi-invariant. It scales (with a certain function closely related to ρ) under left-translation.

The reader is referred to [BRac], [HR] and [Ma 9] for further details on measures on quotients.

We shall consider linear operators on Hilbert space. Linear means complex linear. The algebra of all bounded linear operators on $\mathcal{H}$ will be denoted $\mathcal{B}(\mathcal{H})$. Vectors in $\mathcal{H}$ will be denoted by lower case letters $a, b, \cdots$, etc. An operator X, defined on $\mathcal{H}$ will be said to be bounded if the number

$$\sup\{\|Xa\| : a \in \mathcal{H}, \ \|a\| \leq 1\}$$

is finite. In that case, the number is denoted $\|X\|$ and is called the norm of X. The adjoint operator X^* is defined by the quadratic form identity

$$\langle Xa, b\rangle = \langle a, X^*b\rangle, \qquad a, b \in \mathcal{H},$$

and we have the familiar formula

$$\|X^*X\| = \|X\|^2, \qquad x \in \mathcal{B}(\mathcal{H}). \tag{2.10.7}$$

The latter formula is also the principal defining property for a C^*-algebra.

Let $\mathcal{H}$ be a given Hilbert space, and let the Cartesian product $\mathcal{H} \times \mathcal{H}$ be equipped with the "product" inner product given by

$$\left\langle \begin{bmatrix} a_1 \\ a_2 \end{bmatrix}, \begin{bmatrix} b_1 \\ b_2 \end{bmatrix} \right\rangle := \langle a_1, b_1 \rangle + \langle a_2, b_2 \rangle. \tag{2.10.8}$$

Then $\mathcal{H} \times \mathcal{H}$ is a Hilbert space. Operators on $\mathcal{H} \times \mathcal{H}$ will be identified with 2×2 matrices

$$\tilde{X} = \begin{bmatrix} X_{11} & X_{12} \\ X_{21} & X_{22} \end{bmatrix}$$

with operator entries X_{ij}. $1 \leq i,j \leq 2$. We shall need familiar facts on operator matrices, such as the formula for the adjoint

$$\tilde{X}^* = \begin{bmatrix} X_{11}^* & X_{21}^* \\ X_{12}^* & X_{22}^* \end{bmatrix}.$$

In particular, we shall use that $\tilde{X}$ is an orthogonal projection if and only if it is of the form $\begin{bmatrix} A_1 & B \\ B^* & A_2 \end{bmatrix}$ where the diagonal entries are selfadjoint, and

$$A_1 = A_1^2 + BB^*,$$

$$B = A_1B + BA_2,$$

and

$$A_2 = A_2^2 + B^*B.$$

CHAPTER 3. OPERATORS IN HILBERT SPACE

In this chapter we recall some concepts and results from the theory of operators in Hilbert space. We shall restrict attention only to those results which will be used in later chapters and are not readily available in textbooks, in precisely the form in which we shall use them.

Special attention is paid to unbounded linear operators with dense domain in a Hilbert space, and to commutation relations for such operators.

Our treatment of normal operators follows the beautiful (but generally overlooked) paper by M.H. Stone [St 4].

General references for this chapter include [DS], [RN], [RS] and [St 3].

3.1 Domain and Graph

Let $\mathcal{H}$ be a complex Hilbert space and let T be a *linear operator* which is defined on a subspace of $\mathcal{H}$. This subspace will be called the *domain*, $D(T) \subset \mathcal{H}$. The *graph* of T is a subspace $G(T) \subset \mathcal{H} \times \mathcal{H}$ (Cartesian product) consisting of pairs of vectors a, Ta with $a \in D(T)$. We shall say that T is closed if $G(T)$ is a closed subspace of $\mathcal{H} \times \mathcal{H}$, and we shall say that T is closable if there is some closed operator S such that

$$(3.1.1) \qquad G(T) \subset G(S).$$

If T is closable, then $G(T)^{-}$ is the graph of some operator, and this operator will be denoted $\overline{T}$. It is the smallest closed operator containing T.

The inclusion (3.1.1) will be summarized in the terminology

$$(3.1.2) \qquad T \subset S,$$

and the inclusion holds precisely when $D(T) \subset D(S)$, and $Ta = Sa$ for all $a \in D(T)$.

If T and S are given operators satisfying (3.1.2), then we shall say that T is a *restriction* of S, or equivalently that S is an *extension* of T. The following notation for restrictions will be used,

$$T = S|_{D(T)}. \tag{3.1.3}$$

If S is a given operator, and $\mathscr{D} \subset D(S)$ is a subspace of the domain, then $S|_{\mathscr{D}}$ is a restriction operator. If $\mathscr{D}$ is dense in $D(S)$ relative to the graph norm given by

$$a \longrightarrow \|Sa\| + \|a\|, \tag{3.1.4}$$

then we say that $\mathscr{D}$ is a *core* for S. In this case, we have

$$G(S|_{\mathscr{D}})^{-} \supset G(S).$$

If S is also closed, then this inclusion is an identity, and we shall write

$$(S|_{\mathscr{D}})^{-} = S.$$

3.2 The Adjoint Operator

If T is a given operator with $D(T)$ dense in $\mathscr{H}$, then the *adjoint* T^* is a well-defined operator. Its domain $D(T^*)$ is defined as follows. If $b \in \mathscr{H}$ is given, consider the linear form, $a \longrightarrow \langle Ta,b\rangle$, defined on $D(T)$. If this form is *bounded*, there is by Riesz' representation theorem [RN] a unique vector $c \in \mathscr{H}$ such that

$$\langle Ta,b\rangle = \langle a,c\rangle. \tag{3.2.1}$$

The domain $D(T^*)$ consists precisely of those vectors b such that the form is bounded, and we define

$$T^*b = c. \tag{3.2.2}$$

It can be checked that T^* is a closed operator.

3.3 Selfadjoint Operators

The operator T is said to be *Hermitian*, resp., *skew-Hermitian*, if

$$T \subset T^*, \tag{3.3.1}$$

resp.,

$$T \subset -T^*. \tag{3.3.2}$$

If the inclusion (3.3.1) is an identity, then we say that T is *self-adjoint*. Similarly, if (3.3.2) is an identity, i.e., $T = -T^*$, then T is said to be *skew-adjoint*. (We shall avoid the commonly used but self-contradictory term "skew-self-adjoint"!)

3.4 The Operator T^*T

An operator T is said to be *regular* if $D(T)$ is dense and T is closed. In this case we have the familiar:

Theorem 3.4.1 (von Neumann). Let T be a regular operator in a Hilbert space $\mathcal{H}$.

(i) Then the operator T^*T is selfadjoint; in particular, $D(T^*T)$ is dense in $\mathcal{H}$.

(ii) $D(T^*T)$ is a core for T.

Remark 3.4.2. We shall consider closed subspaces $\mathcal{V} \subset \mathcal{H}$ of a given Hilbert space, and make use of the projection theorem [RN] which gives a 1-1 correspondence between the lattice of closed subspaces of $\mathcal{H}$ and the lattice of all projections. (Only selfadjoint projections will be considered, and the self-adjointness is part of our definition.) Let the orthogonal complement be defined by

$$\mathcal{V}^{\perp} = \{b \in \mathcal{H} : \langle a,b\rangle = 0,\ a \in \mathcal{V}\}. \tag{3.4.1}$$

Then the projection theorem follows from the familiar orthogonal decomposition:

$$\mathcal{H} = \mathcal{V} \oplus \mathcal{V}^{\perp}. \tag{3.4.2}$$

Proof of Theorem 3.4.1. Recall that $\mathcal{H} \times \mathcal{H}$ is the orthogonal sum of $G(T)$ and the orthogonal complement $G(T)^{\perp}$ (use (3.4.2) on $\mathcal{H} \times \mathcal{H}$). Let χ be the operator in $\mathcal{H} \times \mathcal{H}$ given by $\chi(a,b) = (-b,a)$. Then $G(T)^{\perp} = \chi G(T^*)$. For given $c \in \mathcal{H}$, consider the decomposition,

$$(c,0) = (a,Ta) + (-T^*b,b)$$

with $a \in D(T)$, and $b \in D(T^*)$ uniquely determined. This follows from formula (3.4.2) above applied to $\mathcal{H} \times \mathcal{H}$. Solving the two resulting equations,

$$c = a - T^* b, \qquad 0 = Ta + b$$

yields $a \in D(T^*T)$ and $a + T^*Ta = c$. We have proved that the operator $I + T^*T$ maps $D(T^*T)$ onto $\mathcal{H}$. Since this operator is clearly 1-1, it is invertible with an everywhere defined bounded inverse $B = (I+T^*T)^{-1}$. It is immediate that $\|B\| \leq 1$, and $B^* = B$, i.e., B is selfadjoint. The selfadjointness of T^*T follows from this.

To prove that $D(T^*T)$ is a core for T we check that $D(T^*T)$ is dense in $D(T)$ relative to the graph-norm. We show that the orthogonal complement, relative to the graph-norm, is zero. Suppose $b \in D(T)$, and

$$\langle (a, Ta), (b, Tb) \rangle_{\mathcal{H} \times \mathcal{H}} = 0$$

for all $a \in D(T^*T)$. This translates into the identity,

$$\langle a + T^*Ta, b \rangle_{\mathcal{H}} = 0, \qquad a \in D(T^*T).$$

But we just saw that the operator $I + T^*T$ is onto, i.e., $R(I+T^*T) = \mathcal{H}$. It follows that $b = 0$, and the core property (ii) is proved.

3.5 The Polar Decomposition

The result is important since it allows us to state and prove the polar decomposition theorem for regular operators. First some definitions:

An operator E in $\mathcal{H}$ is said to be a *projection* if

$$E^2 = E = E^*,$$

i.e., it is a selfadjoint idempotent. An operator V in $\mathcal{H}$ is said to be a *partial isometry* if V^*V is a projection. It follows then that VV^* is also a projection. The two projections V^*V, resp., VV^*, are called the *initial*, resp., *final* projection of V.

For operators T we shall denote by $N(T)$ the kernel, and by $R(T)$ the range. We define the *initial space* of T to be

$N(T)^{\perp} \equiv \mathcal{H} \ominus N(T)$, and the *final space* to be the closure of the range, i.e., $R(T)^{-}$.

Theorem 3.5. Let T be a regular operator.

(a) Then there is a unique partial isometry V such that

$$\text{(i)} \qquad T = V(T^*T)^{\frac{1}{2}}$$

(ii) the projection onto the initial space of T is V^*V, and

(iii) the projection onto the final space of T is VV^*.

(b) The partial isometry V is also uniquely determined by properties (i′), (ii), and (iii), where

$$\text{(i}'\text{)} \qquad T = (TT^*)^{\frac{1}{2}} V$$

3.6 Remarks

We have used the selfadjointness of the two operators T^*T and TT^* to define the respective square-roots in (i) and (i′). If P denotes the projection valued (Borel) spectral measure of T^*T, then $T^*T = \int_0^\infty \lambda \, dP(\lambda)$, and the square-root is defined by

$$(T^*T)^{\frac{1}{2}} = \int_0^\infty \lambda^{\frac{1}{2}} \, dP(\lambda).$$

We shall omit the proof of the polar decomposition theorem, and only note the following corollary.

Restrict the operator T^*T to the complement of $N(T)$, and restrict TT^* to $R(T)^{-}$. Then the resulting two operators are *unitarily equivalent*. It follows that, aside from possibly the point 0, the two operators T^*T and TT^* have the same spectrum.

3.7 Normal Operators

An operator T in a Hilbert space $\mathcal{H}$ is said to be *normal* if it is regular, and if

$$(3.7.1) \qquad T^*T = TT^*.$$

It is said to be *formally normal* if the inclusion

$$T^*T \subset TT^* \tag{3.7.2}$$

is valid.

If T is closable, but not closed, and if the closure $\overline{T}$ is normal, then we shall say that T is *essentially normal*.

The normal operators include the selfadjoint ones, and they are important because they have spectral resolutions. The *Spectral Theorem* states that an operator T is normal if and only if there is a projection valued Borel measure P, in the complex plane, such that

$$T = \int \lambda \, dP(\lambda), \tag{3.7.3}$$

and, moreover, the support of this measure equals the spectrum of T.

We shall consider von Neumann algebras of operators on $\mathcal{H}$. If $\mathcal{M}$ is a set of bounded operators, the commutant $\mathcal{M}'$ is defined by

$$\mathcal{M}' = \{B \in \mathcal{B}(\mathcal{H}) \ : \ BX = XB, \ X \in \mathcal{M}\}.$$

The *von Neumann Double-Commutant Theorem* states that every von Neumann algebra satisfies the identity

$$\mathcal{M} = \mathcal{M}''. \tag{3.7.4}$$

If two operators are given, T and B with only B bounded, then we say that they *commute* if the inclusion

$$BT \subset TB \tag{3.7.5}$$

holds. We say that T is *affiliated* with a von Neumann algebra $\mathcal{M}$ if T commutes with every unitary operator in the commutant $\mathcal{M}'$.

Let T be a regular operator, and let Q be the projection onto the closed subspace $G(T) \subset \mathcal{H} \times \mathcal{H}$. It is called the *characteristic projection*, and it may be represented as a 2×2 matrix $(Q_{ij})_{i,j=1}^{2}$ with entries in $\mathcal{B}(\mathcal{H})$. We have the following:

Theorem 3.7 (M.H. Stone). Let T be a regular operator in a Hilbert space $\mathcal{H}$. Then the following four conditions are equivalent:

(i) T is affiliated with an Abelian von Neumann algebra;

(ii) T is affiliated with a maximal Abelian von Neumann algebra;

(iii) the four entries $Q_{ij} \in \mathcal{B}(\mathcal{H})$ of the characteristic matrix are mutually commuting operators on $\mathcal{H}$;

(iv) T is a normal operator.

Proof. The reader is referred to the beautiful (and generally overlooked) paper by Stone [St 4].

3.8 Functional Calculus

Let T be a normal operator on a given Hilbert space $\mathcal{H}$, and let P be the corresponding spectral resolution. Recall that P is an orthogonal measure defined on the sigma algebra of Borel subsets of the complex plane and taking values in the lattice of projections on $\mathcal{H}$. We know from the spectral theorem that P is supported on the spectrum of T. Since T may be unbounded, the spectrum may be an unbounded subset of $\mathbb{C}$ although closed.

We have the integral formulas

$$T = \int \lambda dP(\lambda), \tag{3.8.1}$$

$$\|Ta\|^2 = \int |\lambda|^2 \|dP(\lambda)a\|^2 = \int |\lambda|^2 \langle dP(\lambda)a, a\rangle, \quad \text{valid for } a \in D(T), \tag{3.8.2}$$

$$\|a\|^2 = \int \|dP(\lambda)a\|^2, \qquad a \in \mathcal{H}. \tag{3.8.3}$$

Moreover, if φ is a given Borel-function defined on $\mathbb{C}$, then the functional calculus $\varphi(T)$ is well defined. We have

$$\|\varphi(T)a\|^2 = \int |\varphi(\lambda)|^2 \|dP(\lambda)a\|^2, \quad \text{valid for } a \in D(\varphi(T)). \tag{3.8.4}$$

We wish to stress that the two formulas (3.8.2) and (3.8.4)

carry more information than is immediately apparent. Formula (3.8.2) states (implicitly) that the domain $D(T)$ consists of *precisely* those vectors a in $\mathcal{H}$ for which the integral $\int |\lambda|^2 \, \|dP(\lambda)a\|^2$ is finite. On such vectors, Ta is determined by formula (3.8.1).

Similarly, the domain $D(\varphi(T))$ consists precisely of vectors a such that the integral on the right hand side of (3.8.4) is finite. Then $\varphi(T)$ applied to a is given by

$$\varphi(T)a = \int \varphi(\lambda)dP(\lambda)a. \tag{3.8.5}$$

We shall now proceed to apply formula (3.8.5) to the following two functions,

$$\varphi_1(\lambda) = \operatorname{Re} \lambda , \quad \varphi_2(\lambda) = \operatorname{Im} \lambda , \qquad \lambda \in \mathbb{C}. \tag{3.8.6}$$

The resulting operators $\varphi_k(T) \equiv T_k$, $k = 1,2$, are called the *real part*, respectively the *imaginary part*, of the operator T. We note that T_1 and T_2 commute in the *strong sense* of having commuting spectral projections. If conversely, T_1 and T_2 are given selfadjoint operators on some fixed Hilbert space $\mathcal{H}$ having commuting spectral resolutions, then it follows from Theorem 3.4.1 above that the operator $T := T_1+iT_2$ is normal.

We now turn to the details of the spectral resolutions.

3.9 Real and Imaginary Parts

Theorem 3.9. Let T be a normal operator in a Hilbert space $\mathcal{H}$ and let T_1, resp., T_2, be the real, resp., imaginary, parts. Let P be the spectral resolution of T.

(a) Then the spectral resolutions P_k of the respective selfadjoint operators T_k, $k = 1,2$, are given as follows:

(i) For Borel subsets $B_k \subset \mathbb{R}$, $k = 1,2$, we have

$$P_1(B_1) = P(B_1 \times \mathbb{R})$$

and $$P_2(B_2) = P(\mathbb{R} \times B_2).$$

(ii) The two measures in (i) are mutually commuting, i.e., the listed operators mutually commute for all B_1, B_2.

(iii) Let $U_1(t) = e^{itT_1} = \int_{\mathbb{R}} e^{it\lambda}\, dP_1(\lambda)$ be the unitary one-parameter group generated by T_1. Then

$$U_1(t)P_2(B)U_1(t)^* = P_2(B) \tag{3.9.1}$$

for $t \in \mathbb{R}$, and all Borel subsets $B \subset \mathbb{R}$.

(b) Conversely, let T_k be a pair of selfadjoint operators with spectral resolutions P_k, $k = 1,2$. Then the following are equivalent:

(i) $$P_1(B_1)P_2(B_2) = P_2(B_2)P_1(B_1);$$

(ii) $$U_1(t_1)P_2(B_2)U_1(t_1)^* = P_2(B_2);$$

(iii) $$U_2(t_2)P_1(B_1)U_2(t_2)^* = P_1(B_1),$$

for $t_k \in \mathbb{R}$, $B_k \subset \mathbb{R}$(Borel), $k = 1,2$;

(iv) T_1 commutes with $P_2(B)$ for all Borel sets $B \subset \mathbb{R}$;

(v) T_2 commutes with $P_1(B)$ for all Borel sets $B \subset \mathbb{R}$;

(vi) T_1 is affiliated with the commutant $\{P_2(B) : B \subset \mathbb{R}, \text{Borel}\}'$;

(vii) T_2 is affiliated with the commutant $\{P_1(B) : B \subset \mathbb{R}, \text{Borel}\}'$; and

(viii) the operator

$$T = T_1 + iT_2 \tag{3.9.2}$$

is normal.

(c) If the conditions in (b) are satisfied, then the spectral resolution P of the (normal) operator T in (3.9.2) is the product measure $P_1 \times P_2$ on $\mathbb{R} \times \mathbb{R} \simeq \mathbb{C}$ with the usual Fubini type product measure construction.

Proof. The proof is based on formulas (3.8.1) through (3.8.5) from Section 3.8 above. We shall omit the complete details but rather refer the reader to the two excellent textbooks [Na 1] and [Put].

We now turn to a generalization of the covariance formula (3.9.1) from Theorem 3.9 above.

3.10 Positive Operators

We conclude with some definitions which will be needed. Let T be a regular operator in a Hilbert space $\mathcal{H}$. The spectrum $sp(T)$ is the set of points $\lambda \in \mathbb{C}$ such that the operator $\lambda I - T$ is not invertible, and the resolvent-set is the complement, $res(T) := \mathbb{C} \setminus sp(T)$. We shall say that T is positive if the spectrum $sp(T)$ is contained in the half-line $\{\lambda : 0 \leq \lambda \leq \infty\}$. If T is known to be bounded, i.e., $T \in \mathcal{B}(\mathcal{H})$ we then have the following

Lemma 3.10. The operator $T \in \mathcal{B}(\mathcal{H})$ is positive if and only if any one of the equivalent conditions holds:

(i) $T = S^2$ for some selfadjoint operator S.

(ii) $T = V^*V$ for some $V \in \mathcal{B}(\mathcal{H})$.

Proof. The proof is an interesting application of the Spectral Mapping Theorem concerning the functional calculus $\varphi(S)$ applied to selfadjoint operators S and continuous functions φ. It states that

$$sp\ \varphi(S) = \varphi(sp\ S) \tag{3.10.1}$$

and we refer to Theorems 4.1.6 and 4.2.2 in [KR] for detailed proofs of both (3.10.1) and the theorem.

CHAPTER 4. THE IMPRIMITIVITY THEOREM

In this chapter we prove the Imprimitivity Theorem of Mackey but in its modern formulation due to Poulsen and Ørsted [Ør 3], and we record the classical application to the Stone-von Neumann uniqueness theorem. Both results will be stated and proved in their maximum generality (in the setting of nonseparable Hilbert space, and locally compact, generally nonunimodular groups) since this is the context of the applications later in the book.

There are some technical points regarding strongly measurable functions with values in (generally nonseparable) Hilbert spaces which we shall only touch lightly without in-depth discussion. The precise definition is recorded in [Ør 3], and the reader is referred to [DSI] for additional details on this technical point.

Our treatment of the Imprimitivity Theorem follows [Ør 3] closely. A refinement and correction is added. Our version of the Stone-von Neumann uniqueness theorem is that which is well suited to the theory of C^*-algebras where it will be applied.

4.1 Integration and Unitary Equivalence

Let G be a locally compact group and H a given closed subgroup. Let the corresponding right invariant Haar measures be denoted dg, resp., dh. Recall that the right invariant Haar measure on a locally compact group is unique up to normalization. Let $X = H\backslash G$ be the homogeneous space of left-cosets equipped with the quotient topology. Then X carries a canonical transitive action of G which is given by right G-translation of cosets, i.e., the mapping

$$G \times X \longrightarrow X$$

is defined as follows: If $x = H\cdot a$ for $a \in G$, then $x\cdot g := H\cdot(ag)$, $g \in G$. Let $\pi : G \longrightarrow X$ be the quotient mapping, $\pi(g) = H\cdot g$. We shall recall a natural measure on X which is quasi-invariant under the G-action on X.

Let $C_c(X)$, resp., $C_c(G)$, denote the space of compactly supported continuous functions on X, resp., on G, and let

$$\tau : C_c(G) \longrightarrow C_c(X) \tag{4.1.1}$$

be the conditional expectation mapping obtained by integration over the subgroup H, viz.,

$$(\tau\varphi)(\pi(g)) = \int_H \varphi(hg)dh, \tag{4.1.2}$$

for $\varphi \in C_c(G)$, and $g \in G$. Note that the integral on the right hand side of (4.1.2) exists since φ is assumed to be continuous of compact support. It can be checked that τ maps *onto* $C_c(X)$.

We shall consider *systems of imprimitivity based on the transformation group* (G,X) and we proceed with the definition.

A *unitary representation* of G on a Hilbert space $\mathscr{H}$ is a strongly continuous homomorphism from G into the group of unitaries in $\mathscr{B}(\mathscr{H})$. A linear mapping

$$P : C_c(X) \longrightarrow \mathscr{B}(\mathscr{H}) \tag{4.1.3}$$

is said to be *positive* if it takes positive functions to positive operators.

We say that the mapping P from (4.1.3) is *nondegenerate* if the set of vectors

$$P(C_c(X))\mathscr{H} = \{P(\varphi)a : \varphi \in C_c(X),\ a \in \mathscr{H}\}$$

spans a dense subspace of $\mathscr{H}$.

Recall that the G-action lifts to a representation of G on $C_c(X)$. We shall denote this representation by R and it is given by

$$(R_g\varphi)(x) = \varphi(x\cdot g), \qquad x \in X, \quad g \in G. \tag{4.1.4}$$

A *system of imprimitivity* is a triple $(U,P,\mathscr{H})$ as above satisfying the *covariance condition*

$$U_gP(\varphi)U_g^* = P(R_g\varphi). \qquad \varphi \in C_c(X),\ g \in G. \tag{4.1.5}$$

Note that (4.1.5) generalizes (3.9.1) from the previous chapter.

Two systems $(U, P, \mathscr{H})$ and $(U', P', \mathscr{H}')$ are said to be *unitarily equivalent* if there exists a unitary isomorphism

$$W : \mathscr{H} \longrightarrow \mathscr{H}'$$

satisfying the two *intertwining relations*:

$$U'_g W = W U_g, \qquad g \in G, \tag{4.1.6}$$

and

$$P'(\varphi) W = W P(\varphi), \qquad \varphi \in C_c(X). \tag{4.1.7}$$

We now give an example of a system of imprimitivity which is constructed naturally from an *induced representation*.

We shall start with a unitary representation λ of the subgroup, i.e., $\lambda : H \longrightarrow \mathfrak{B}(\mathscr{V})$ is a strongly continuous homomorphism from the subgroup H into the group of unitary operators on a given Hilbert space $\mathscr{V}$.

We proceed to construct the triple $(U^{\lambda}, P^{\lambda}, \mathscr{H}^{\lambda})$ associated to this representation. As for U^{λ}, it is the representation of G which is induced from the representation λ of the subgroup, and the Hilbert space $\mathscr{H}^{\lambda}$ is the representation space of the induced representation.

This space is constructed by the procedure which we outlined in Section 2.8, and the reader is referred to [Se 2] and [KR] for additional details. Suffice it to say that $\mathscr{H}^{\lambda}$ is obtained by a completion of a space of vector valued functions

$$f : G \longrightarrow \mathscr{V}$$

which are *strongly measurable*, with $\|f(\cdot)\|^2$ locally integrable, and satisfying

$$f(hg) = \rho(h)^{\frac{1}{2}} \lambda(h)(f(g)), \qquad h \in H, \quad g \in G. \tag{4.1.8}$$

The function ρ in (4.1.8) is determined by the choice of Haar measures on G, resp., H. Let Δ, resp., δ, be the modular function of dg on G, resp., dh on H. Then ρ is defined by

$$\rho(h) = \delta(h)\Delta(h)^{-1}.$$

On the space $\mathcal{D}$ of functions satisfying the above conditions (4.1.8), we now define a sesquilinear form $\langle f_1, f_2\rangle$, and we let $\mathcal{H}^\lambda$ be the completion of $\mathcal{D}/\mathcal{N}$ where $\mathcal{N} = \{f \in \mathcal{D} : \langle f,f\rangle = 0\}$.

Let $f_1, f_2 \in \mathcal{D}$, and consider the associated linear functional on $C_c(X)$ which is defined by:

$$\tau\varphi \longrightarrow \int_G \langle f_1(g), f_2(g)\rangle_{\mathcal{V}}\ \varphi(g)dg. \tag{4.1.9}$$

It can be checked to be a complex Radon measure, see [Ør 3, Lemma 2] for details. Let μ_{f_1,f_2} be the measure given in (4.1.9) and define

$$\langle f_1, f_2\rangle := \mu_{f_1,f_2}(X). \tag{4.1.10}$$

Then this is the sesquilinear form from which $\mathcal{H}^\lambda$ is constructed with the familiar quotient-completion procedure, cf. above.

We define the operator U_g^λ on $\mathcal{D}$ to be right translation, and it is immediate from (4.1.8) that $\mathcal{D}$ is invariant under this translation to the right on G. For $\varphi \in C_c(X)$, we define $P^\lambda(\varphi)$, on $\mathcal{D}$, to be multiplication by the scalar valued function φ. We shall identify φ with the corresponding H-periodic function, i.e.,

$$\varphi(h\cdot g) = \varphi(g), \qquad h \in H, \quad g \in G,$$

and it follows again from (4.1.8) that $\mathcal{D}$ is invariant under this multiplication operator. The pair (U^λ, P^λ) is discussed in more detail below.

Theorem 4.1.1 (The Imprimitivity Theorem). Let $(U,P,\mathcal{H})$ be an imprimitivity system based on the transformation group (G,X), and assume that P is normalized and nondegenerate.

(a) Then there is a unitary representation λ of the subgroup H such that the system $(U,P,\mathcal{H})$ is unitarily equivalent to a subsystem of $(U^\lambda, P^\lambda, \mathcal{H}^\lambda)$.

(b) The representation λ is determined uniquely up to unitary equivalence.

(c) The subsystem in (a) is $(U^\lambda, P^\lambda, \mathscr{H}^\lambda)$ if and only if P is multiplicative. In general, $P = W^* P^\lambda W$ where $W : \mathscr{H} \longrightarrow \mathscr{H}^\lambda$ is isometric.

Remark 4.1.2. Part (b) is the uniqueness part, and is a special case of an *isomorphism theorem for intertwining operators*. The latter result is needed in applications, and we include it.

Theorem 4.1.3. Let λ, λ' be unitary representations of H on Hilbert spaces $\mathscr{V}$, resp., $\mathscr{V}'$, and let $I(\lambda, \lambda')$ be the set of unitary isomorphisms $w : \mathscr{V} \longrightarrow \mathscr{V}'$ satisfying

$$\lambda'(h)w = w\lambda(h), \qquad h \in H. \tag{4.1.11}$$

Similarly, let $I(U^\lambda, P^\lambda; U^{\lambda'}, P^{\lambda'})$ be the set of unitary isomorphisms

$$W : \mathscr{H}^\lambda \longrightarrow \mathscr{H}^{\lambda'}$$

which intertwine the two systems $(U^\lambda, P^\lambda, \mathscr{H}^\lambda)$ and $(U^{\lambda'}, P^{\lambda'}, \mathscr{H}^{\lambda'})$.

Then there is a *natural bijection* between the two sets $I(\lambda, \lambda')$ and $I(U^\lambda, P^\lambda; U^{\lambda'}, P^{\lambda'})$.

Proof of Theorem 4.1.3 (sketch). Since $\mathscr{H}^\lambda$ is obtained by completion of a space $\mathscr{D}$ of functions from G to $\mathscr{V}$ satisfying (4.1.8), and similarly $\mathscr{H}^{\lambda'}$ is obtained by completion of the space $\mathscr{D}'$ of functions from G to $\mathscr{V}'$ satisfying (4.1.8) with λ replaced by λ', the proof reduces to show that every $w : \mathscr{V} \longrightarrow \mathscr{V}'$, $w \in I(\lambda, \lambda')$, extends to a unitary isomorphism

$$W : \mathscr{H}^\lambda \longrightarrow \mathscr{H}^{\lambda'}$$

which intertwines the two systems, cf. (4.1.6)-(4.1.7).

This is not hard, since we simply define W on $\mathscr{D}$ by letting it act pointwise via w, i.e.,

$$W(f)(g) = w(f(g)), \qquad g \in G. \tag{4.1.12}$$

To verify the two relations,

$$WU_g^{\lambda} = U_g^{\lambda'} W. \qquad g \in G, \tag{4.1.13}$$

and

$$WP^{\lambda}(\varphi) = P^{\lambda'}(\varphi)W, \qquad f \in C_c(X), \tag{4.1.14}$$

we must recall the formulas for the two operators U_g^{λ}, $g \in G$, and $P^{\lambda}(\varphi)$, $\varphi \in C_c(X)$, which define the imprimitivity systems $(U^{\lambda}, P^{\lambda}, \mathcal{H}^{\lambda})$. They are given by

$$(U_g^{\lambda} f)(g') = f(g'g), \qquad g, g' \in G, \quad f \in \mathfrak{D}, \tag{4.1.15}$$

and

$$(P^{\lambda}(\varphi)f)(g) = \varphi(g)f(g), \quad g \in G, \ \varphi \in C_c(H\backslash G), f \in \mathfrak{D}. \tag{4.1.16}$$

The verification of (4.1.6)-(4.1.7) is now immediate upon substitution of (4.1.11-12), and (4.1.15-16).

We leave to the reader the verification of invariance of the space $\mathfrak{D}$, given by (4.1.8), under the two operators U_g^{λ}, $g \in G$, and $P^{\lambda}(\varphi)$, $\varphi \in C_c(X)$; but we include below the detailed verification of formulas (4.1.6) and (4.1.7).

Let $f \in \mathfrak{D}$, $g, g' \in G$. We now evaluate both sides of (4.1.6) on the function $f(\cdot)$:

$$\begin{aligned}(WU_g^{\lambda} f)(g') &= w((U_g^{\lambda} f)(g')) \\ &= w(f(g'g)) \\ &= (Wf)(g'g) \\ &= (U_g^{\lambda'} W)f(g'),\end{aligned}$$

and (4.1.6) follows.

Similarly, for (4.1.7), we have:

$$\begin{aligned}(WP^{\lambda}(\varphi)f)(g) &= w(P^{\lambda}(\varphi)f(g)) \\ &= w(\varphi(g)f(g)) \\ &= \varphi(g)w(f(g)) \\ &= \varphi(g)(Wf)(g) \\ &= (P^{\lambda'}(\varphi)W)f(g).\end{aligned}$$

This completes the proof of (4.1.7).

It is clear that the correspondence, $w \longrightarrow W$, from $I(\lambda,\lambda')$ to $I(U^{\lambda},P^{\lambda};U^{\lambda'},P^{\lambda'})$ is natural, and it is not hard to show that it is a bijection as claimed in Theorem 4.1.3.

Proof of Theorem 4.1.1 (sketch). The measures μ_{f_1,f_2} in (4.1.10) above, $f_1 \neq f_2$ in $\mathfrak{D}$, may be reconstructed by polarization from the measures $\mu_{f,f}$, $f \in \mathfrak{D}$, which are given by the simpler positive linear functionals on $C_c(X)$:

$$\tau\varphi \longrightarrow \int_G \|f(g)\|^2_{\mathcal{V}} \varphi(g)dg,$$

or equivalently,

$$\int_X (\tau\varphi)(x)d\mu_{f,f}(x) = \int_G \|f(g)\|^2\varphi(g)dg, \qquad \varphi \in C_c(G).$$

Recall that $\mu_{f,f}$ is determined uniquely since $\tau(C_c(G)) = C_c(X)$.

We now do this construction, starting with a given imprimitivity system $(U,P,\mathcal{H})$ with a positive operator function which is nondegenerate on $\mathcal{H}$. For vectors a,b in the dense space

$$\mathcal{H}_0 = \text{span}\{U(\varphi)c : \varphi \in C_c(G),\ c \in \mathcal{H}\}$$

where

$$U(\varphi) = \int_G \varphi(g)U(g^{-1})dg, \tag{4.1.17}$$

we consider the Radon measure $d\mu_{a,b}$ given by

$$\varphi \longrightarrow \langle P(\tau\varphi)a,b\rangle. \tag{4.1.18}$$

Again the measure is positive when $a = b$. The important observation needed in the reconstruction of λ is that $d\mu_{a,b}$ is absolutely continuous relative to the given Haar measure on G.

To see this, consider the bilinear form:

$$\varphi_1,\varphi_2 \longrightarrow \langle P(\tau\varphi_1)U(\varphi_2)a,b\rangle.$$

It also defines a Radon measure, but now on $G \times G$, and we shall denote this measure by ν_2.

To check the absolute continuity claim for the measure $d\mu_{a,b}$, it is enough to consider the case where $a = U(\psi_1)a_1$ and $b = U(\psi_2)b_1$ with vectors a_1 and b_1 in $\mathcal{H}$. We now let ν_2 be the measure on $G \times G$ associated to the vector pair a,b. An easy calculation, using only Fubini's theorem then yields

$$\int_G \varphi(g)d\mu_{a,b}(g) = \int_G \varphi(g)F(g)dg$$

where $F(\cdot)$ is a continuous scalar function on G. It can be calculated explicitly as follows,

$$F(g) = \int_{G\times G} \overline{\psi_2(g_1^{-1}g)}\psi_1(g_2g_1^{-1}g)\Delta(g_1^{-1})d\nu_2(g_1,g_2),$$

and the continuity assertion follows by inspection. Naturally this function depends on the vector pair a,b given at the outset. We indicate this with the subscript notation $F_{a,b}$.

Now this function may be evaluated at the identity element e in G, and we get a sesquilinear form on $\mathcal{H}_0$ determined by

(4.1.19) $$\langle a,b\rangle := F_{a,b}(e),$$

and the Hilbert space $\mathcal{V}$ will be reconstructed from this form by the usual quotient-completion procedure, cf. Section 2.8.

For vectors a in $\mathcal{H}_0$ we denote by [a] the corresponding coset in

$$\mathcal{H}_0/\{b \in \mathcal{H}_0 : \langle b,b\rangle = 0\}.$$

We get a unitary representation λ of the subgroup H on this Hilbert space $\mathcal{V}$ as follows:

$$\lambda(h)[a] = [\rho(h)^{-\frac{1}{2}}U(h)a] \quad \text{defined for } h \in H,\ a \in \mathcal{H}_0.$$

Let $a \in \mathcal{H}_0$ and define the function $f_a : G \longrightarrow \mathcal{V}$ as follows:

$$f_a(g) = [U(g)a], \qquad g \in G.$$

It can be checked that f_a satisfies the periodicity condition (4.1.8). It is now only a matter of some formal verifications to check that the mapping:

$$W : a \longrightarrow f_a : \mathcal{H}_0 \longrightarrow \mathcal{H}^\lambda$$

extends to a unitary isomorphism from $\mathcal{H}$ to $\mathcal{H}^\lambda$, and that the transformation, so extended (also denoted by W), *intertwines* the two imprimitivity systems $(U\ P\ \mathcal{H})$ and $(U^\lambda, P^\lambda, \mathcal{H}^\lambda)$. The nondegeneracy assumption on P is needed in this verification.

4.2 Applications

As our first application of the imprimitivity theorem we record the *Stone-von Neumann Uniqueness Theorem* in its generalized form in the setting of locally compact Abelian groups.

Let Γ be a locally compact Abelian group, and let $\hat{\Gamma}$ be the dual group of continuous unitary characters on Γ, i.e., homomorphisms

$$\chi : \Gamma \longrightarrow \mathbb{T}^1$$

where $\mathbb{T}^1$ is the circle group

$$\mathbb{T}^1 = \{z \in \mathbb{C} : |z| = 1\}.$$

We introduce the notation

$$(\chi, \gamma) := \chi(\gamma), \qquad \gamma \in \Gamma, \quad \chi \in \hat{\Gamma}.$$

Recall that $\hat{\Gamma}$ is also a locally compact Abelian group (with the "compact-open" topology), and

$$(\chi_1\chi_2, \gamma) = (\chi_1, \gamma)(\chi_2, \gamma) \quad \text{for } \chi_1, \chi_2 \in \hat{\Gamma}, \quad \gamma \in \Gamma.$$

It follows that Γ embeds as a subgroup of $\hat{\hat{\Gamma}}$. But the *Pontryagin Duality Theorem* states that the embedding is in fact onto, so

$$\Gamma \cong \hat{\hat{\Gamma}}$$

and the pairing between Γ and $\hat{\Gamma}$ is in fact symmetric.

We shall consider the following general form of the *Canonical Commutation Relations*: Let V_1 be a unitary representation of Γ on the Hilbert space $\mathcal{H}$, and let V_2 be a unitary representation of $\hat{\Gamma}$ on the same Hilbert space, and assume that

$$(4.2.1) \qquad V_1(\gamma)V_2(\chi) = (\chi,\gamma)V_2(\chi)V_1(\gamma), \quad \gamma \in \Gamma, \quad \chi \in \hat{\Gamma}.$$

As an *example* of a system like that consider the Hilbert space $L^2(\Gamma)$ relative to "the" Haar measure on Γ, and define

$$(4.2.2) \qquad (U_1(\gamma)f)(\xi) = f(\xi+\gamma), \quad f \in L^2(\Gamma), \quad \gamma,\xi \in \Gamma,$$

and

$$(4.2.3) \qquad (U_2(\chi)f)(\xi) = (\chi,\xi)f(\xi), \quad f \in L^2(\Gamma), \quad \chi \in \hat{\Gamma}, \quad \xi \in \Gamma.$$

It is immediate that this is a pair of unitary representations. The strong continuity is an implicit condition which can be checked to hold. We shall omit details, and refer the reader to [Ru2] for details. An inspection reveals that the commutation relation (4.2.1) is satisfied for this pair.

We also have the

Lemma 4.2.1 (Irreducibility). The pair of representations U_1, U_2 on $L^2(\Gamma)$ is irreducible, i.e., there is no closed subspace of $L^2(\Gamma)$ which is invariant under both of the groups $\{U_1(\gamma) : \gamma \in \Gamma\}$ and $\{U_2(\chi) : \chi \in \hat{\Gamma}\}$.

Proof. Suppose that some closed subspace of $L^2(\Gamma)$ is invariant under the two groups. Then let Q be the projection onto this subspace. It follows that Q commutes with the two groups. Since Q commutes with U_2, it follows by Fourier analysis that

$$(4.2.4) \qquad Q(\varphi f) = \varphi Q(f), \quad \varphi \in C_0(\Gamma), \quad f \in L^2(\Gamma),$$

where $C_0(\Gamma)$ is the algebra of continuous functions on Γ vanishing at ∞. Now take a sequence $f_n \in L^2(\Gamma)$ which converges to the constant function 1 in the weak* topology. Apply this to (4.2.4), and it follows that there is a weak*-limit for the sequence $Q(f_n)$. If ψ is that limit $\psi \in L^\infty(\Gamma)$ then

$$Q(\varphi) = \psi\varphi, \qquad \varphi \in C_c(\Gamma).$$

Since $C_c(\Gamma)$ is dense in $L^2(\Gamma)$, it follows that the operator Q is just multiplication by ψ. But Q also commutes with U_1, and it follows that ψ is constant. Since $Q = Q^2 = Q^*$ the constant must be 0 or 1. Hence, the only closed invariant subspaces are the two trivial extreme subspaces. This concludes the proof of irreducibility.

The representation given by U_1 and U_2 from (4.2.2)-(4.2.3) is called a *generalized Schrödinger representation*.

In the special case when $\Gamma = \mathbb{R}$, and $\hat{\Gamma} = \hat{\mathbb{R}} \simeq \mathbb{R}$, this is the familiar Schrödinger representation.

We have

Theorem 4.2.1. Let $(V_1, V_2, \mathcal{H})$ be a representation system based on a locally compact Abelian group Γ and satisfying the canonical commutation relations (4.2.1).

Then this system is unitarily equivalent to an orthogonal direct sum of identical copies of the generalized Schrödinger system $(U_1, U_2, L^2(\Gamma))$.

We also have the more familiar result concerning representations of the *Heisenberg group* G_3 of upper triangular 3×3 real matrices with ones in the diagonal

$$g = \begin{bmatrix} 1 & u & w \\ 0 & 1 & v \\ 0 & 0 & 1 \end{bmatrix}, \qquad u, v, w \in \mathbb{R}.$$

For every $\lambda \in \mathbb{R}\setminus\{0\}$ the irreducible representation U_λ is defined by

$$U_\lambda(g)f(x) = e^{i\lambda(xv+w)}\, f(x+u), \quad f \in L^2(\mathbb{R}), \quad x \in \mathbb{R}, \quad g \in G_3. \tag{4.2.5}$$

It is called the *Schrödinger representation*, and λ is related to Planck's constant h. Except for the trivial representations, this is the list of all the irreducible unitary representations of G_3. (The latter result follows from Mackey's semidirect product "machine", and the reader is referrred to [Ma 9] for additional details on this point.)

We are now ready to state:

Corollary 4.2.2. Let U be a unitary representation of G_3 on some Hilbert space $\mathcal{H}$ such that the restriction of U to the center of G_3 is a scalar, i.e., $U(0,0,w) = e^{i\lambda w}I$, $\lambda \in \mathbb{R}\setminus\{0\}$.

Then it follows that U is unitarily equivalent to a direct sum of copies of the Schrödinnger representation with Planck's constant λ, i.e., $\mathcal{H}$ is the direct sum of copies of $L^2(\mathbb{R})$ with the restriction of U to each copy unitarily equivalent to the Schrödinger representation U_λ.

Proof of Theorem 4.2.1. Consider the Fourier transform

$$\hat{f}(\chi) = \int_\Gamma \overline{(\chi,\gamma)} f(\gamma)\, d\gamma, \tag{4.2.6}$$

defined for $f \in L^2(\Gamma)$, with Plancherel Fourier-inversion

$$f(\gamma) = \int_{\hat{\Gamma}} (\chi,\gamma)\hat{f}(\chi)\, d\chi, \tag{4.2.7}$$

and suitably normalized Haar measures on Γ, respectively, $\hat{\Gamma}$.

Let P_0 be the multiplication representation given by

$$P_0(\varphi)f := \varphi f, \qquad \varphi \in C_c(\Gamma), \quad f \in L^2(\Gamma). \tag{4.2.8}$$

In view of (4.2.7) we have

$$P_0(\varphi) = \int_{\hat{\Gamma}} \hat{\varphi}(\chi)U_2(\chi)d\chi. \tag{4.2.9}$$

Since the relation (4.2.1) is satisfied by the representation pair (U_1, U_2), it follows by simple substitution of (4.2.9) that

$$U_1(\gamma)P_0(\varphi)U_1(\gamma)^* = P_0(R_\gamma\varphi), \quad \text{for} \quad \gamma \in \Gamma, \quad \varphi \in C_c(\Gamma), \tag{4.2.10}$$

where

$$(R_\gamma\varphi)(\xi) = \varphi(\xi+\gamma), \qquad \xi, \gamma \in \Gamma. \tag{4.2.11}$$

Indeed,

$$\begin{aligned} U_1(\gamma)P_0(\varphi)U_1(\gamma)^* &= \int_{\hat{\Gamma}} \hat{\varphi}(\chi)U_1(\gamma)U_2(\chi)U_1(\gamma)^*d\chi \\ &= \int_{\hat{\Gamma}} \hat{\varphi}(\chi)(\chi,\gamma)U_2(\chi)d\chi \\ &= \int_{\hat{\Gamma}} (R_\gamma\varphi)\hat{}(\chi)U_2(\chi)d\chi \\ &= P_0(R_\gamma\varphi) \end{aligned}$$

with formula (4.2.9) being applied twice. We also used the familiar formula

$$(R_\gamma\varphi)\hat{}(\chi) = (\chi,\gamma)\hat{\varphi}(\chi), \qquad \gamma \in \Gamma, \quad \chi \in \hat{\Gamma},$$

from Fourier analysis.

The proof yields the following stronger conclusion: Let (V_1, V_2) be a pair of unitary representations, on some Hilbert space $\mathcal{H}$, satisfying the commutation relations (4.2.1), and define the operator function

$$P(\varphi) := \int_{\hat{\Gamma}} \hat{\varphi}(\chi)V_2(\chi)d\chi, \qquad \varphi \in C_c(\Gamma). \tag{4.2.12}$$

Then the covariance relation

$$V_1(\gamma)P(\varphi)V_1(\gamma)^* = P(R_\gamma\varphi), \quad \gamma \in \Gamma, \quad \varphi \in C_c(\Gamma), \tag{4.2.13}$$

is satisfied.

Moreover, $\varphi \longrightarrow P(\varphi)$ is positive and nondegenerate.

The positivity follows from the stronger set of properties

(i) $$P(\varphi)^* = P(\varphi^*), \quad \text{where } \varphi^*(\xi) = \overline{\varphi(-\xi)}, \quad \xi \in \Gamma,$$

and

(ii) $$P(\varphi\psi) = P(\varphi)P(\psi), \qquad \varphi,\psi \in C_c(\Gamma).$$

We leave the proof of (i) to the reader and check (ii). We have

$$\begin{aligned} P(\varphi\psi) &= \int_{\hat{\Gamma}} (\varphi\psi)^{\wedge}(\chi)V_2(\chi)d\chi \\ &= \int_{\hat{\Gamma}} (\hat{\varphi}*\hat{\psi})(\chi)V_2(\chi)d\xi \\ &= P(\varphi)P(\psi) \end{aligned}$$

which proves (ii). In this calculation, familiar properties of the convolution product were used.

To check that P is nondegenerate, pick some vector $b \in \mathscr{H}$ satisfying

$$\langle P(\varphi)a,b\rangle = 0$$

for all $\varphi \in C_c(\Gamma)$ and $a \in \mathscr{H}$. It follows that

(4.2.14) $$\int_{\hat{\Gamma}} \hat{\varphi}(\chi)\langle V_2(\chi)a,b\rangle d\chi = 0.$$

We may pick $a = b$, and a net of functions φ such that $\hat{\varphi} \longrightarrow$ (the delta function on $\hat{\Gamma}$). It then follows from (4.2.14), and strong continuity of V_2, that $\langle b,b\rangle = \|b\|^2 = 0$. Hence $P(C_c(\Gamma))\mathscr{H}$ spans a dense subspace of $\mathscr{H}$, which is the desired nondegeneracy condition.

We now apply Theorem 4.1.1 to the imprimitivity system $(V_1,P,\mathscr{H})$. This is a system on the group Γ, acting on itself by translation. We shall consider just the trivial subgroup with the single element 0. The representation of this subgroup is trivial, of course. It follows that there is a Hilbert space $\mathscr{V}$ such that the system $(V_1,P,\mathscr{H})$ is unitarily equivalent to the following one: $(U_1, P_0, L^2(\Gamma, \mathscr{V}))$ where U_1, resp., P_0, are given in formulas (4.2.2), resp. (4.2.8), above and $L^2(\Gamma, \mathscr{V})$ is the Hilbert space of square-integrable functions on Γ with values in the Hilbert space $\mathscr{V}$.

Strictly speaking, formulas (4.2.2) and (4.2.8) are just specified for scalar valued functions, but they make sense also for functions $f \in L^2(\Gamma, \mathscr{V})$.

Finally, let $\{v_\alpha\}_{\alpha \in I}$ be an orthonormal basis for the space $\mathscr{V}$, with some index set I. This set may be taken to be countable, of course, if $\mathscr{H}$, and therefore $\mathscr{V}$, is known to be separable.

Now every $f \in L^2(\Gamma, \mathscr{V})$ has an orthogonal expansion

$$f = \sum_\alpha^\oplus f_\alpha$$

with $f_\alpha \in L^2(\Gamma)$ given by

$$f_\alpha(\xi) = \langle f(\xi), v_\alpha \rangle_{\mathscr{V}}, \qquad \xi \in \Gamma, \quad \alpha \in I.$$

It follows that the system $(U_1, U_2, L^2(\Gamma, \mathscr{V}))$ is unitarily equivalent to a direct sum of copies of the system $(U_1, U_2, L^2(\Gamma))$. But then the original system $(V_1, V_2, \mathscr{H})$ is unitarily equivalent to the direct sum system as asserted.

This concludes the proof of the Theorem.

We shall not elaborate on the proof of Corollary 4.2.2, as it is immediate from the theorem.

Remark. The assumption in Corollary 4.2.2 that $U(0,0,w) = e^{iw\lambda} I$ cannot be omitted as is evidenced by the following unitary representation U of G_3 on $L^2(\mathbb{R}^2)$:

$$U(u,v,w)f(x,y) = f(x+u, y+xv+w).$$

This representation is reducible, but it has no minimal invariant subspaces, and is therefore *not* equivalent to a direct sum of Schrödinger representations.

Additional References for Part I

Chapter 1: [Ad], [Ado], [AG], [Arv 4], [BRac], [Bou 3], [BR 2], [Ch], [Co], [Dix 8], [DS], [Goo 9], [GℓJa], [Ham], [HR], [HP], [JM], [Loo 1], [Sak 6-7].

Chapter 2: [GGV], [KR], [Ka 6-7], [Ma 10], [St 3], [Yo], [Heℓ 1,3], [Kn], [Vo].

Chapter 3: [Arv 4], [DS], [GGV], [Na], [Put], [RN], [SK], [St 4], [Va].

Chapter 4: [Blat 1-3], [Loo 2], [Ma 13], [Ørs 3], [Ri 1-3].

CHAPTER 5. DOMAINS OF REPRESENTATIONS

In this chapter we introduce representations of Lie groups and Lie algebras. Starting with a representation of a given Lie group we recall the definition of the corresponding derived representation obtained by differentiation along the Lie algebra. The derived representation is extended to the universal enveloping algebra.

In case the Lie representation is given to be unitary we show that the corresponding derived representation of the enveloping algebra is a selfadjoint representation of the enveloping algebra in the sense of R.T. Powers [Pow 1]. The main result of the present chapter concerns the converse problem. Starting with a given selfadjoint representation of the enveloping algebra, when is it exact, i.e., when is it the derived representation of some unitary representation of the corresponding Lie group? The answer is known to be negative for certain representations associated to Abelian, and nilpotent, Lie algebras. But we show that the answer is affirmative for a very large class of selfadjoint representations of the enveloping algebra of $s\ell_2(\mathbb{R})$.

5.1 Introduction

Let $\mathfrak{A}$ be an algebra over the complex numbers $\mathbb{C}$, and assume that $\mathfrak{A}$ carries a **-operation*, i.e., a period 2, conjugate linear, antiautomorphism. Specifically, the mapping $x \longrightarrow x^*$ of $\mathfrak{A}$ into itself is conjugate linear, and satisfies:

$$x^{**} = x, \qquad x \in \mathfrak{A},$$

$$(xy)^* = y^* x^* \qquad x,y \in \mathfrak{A}.$$

Let ρ be a representation of $\mathfrak{A}$ and let $\mathfrak{D}(\rho)$ be the domain of this representation. This means that each operator $\rho(x)$, $x \in \mathfrak{A}$, is *defined* on $\mathfrak{D}(\rho)$ and that $\mathfrak{D}(\rho)$ is *invariant*.

We shall assume that $\mathfrak{D}(\rho)$ is a dense linear subspace of some given Hilbert space $\mathcal{H}$, and we shall develop the general notions from Chapter 3, following R.T. Powers [Pow 1-2], in this algebraic setting. The special case when $\mathfrak{A}$ is Abelian with a single generator, $x = x^*$, corresponds to the study of Hermitian operators with dense invariant domain in a Hilbert space $\mathcal{H}$.

Similarly, the case where $\mathfrak{A}$ is generated by two commuting elements $x_k = x_k^*$, $k = 1,2$, corresponds to the study of *formally* normal operators T in $\mathcal{H}$ with the property that $D(T)$ is dense and invariant under both T and T^*.

As a generalization, we shall also consider the case where $\mathfrak{A}$ is generated by two elements x_k, $k = 1,2$, satisfying (a) and (b) below:

(a) $$x_k^* = x_k, \qquad k = 1,2,$$

(b) the Lie subalgebra of $\mathfrak{A}$ which is generated by x_1, x_2, and iterated commutators of x_1, x_2, along with linear combinations over $\mathbb{C}$, is assumed *finite-dimensional*.

Let $\langle \cdot , \cdot \rangle$ denote the inner product of $\mathcal{H}$, and let ρ be an arbitrary representation with domain $\mathfrak{D}(\rho)$. We say that ρ is *Hermitian* if

(5.1.1) $$\langle \rho(x)a, b \rangle = \langle a, \rho(x^*)b \rangle, \qquad x \in \mathfrak{A}, \quad a,b \in \mathfrak{D}(\rho).$$

Lemma 5.1.1. Let ρ be a representation of $\mathfrak{A}$ with domain $\mathfrak{D}(\rho)$ dense in a given Hilbert space $\mathcal{H}$. Then the following two conditions (a) and (b) are equivalent:

(a) The intersection

(5.1.2) $$\cap \{ D(\rho(x)^*) : x \in \mathfrak{A} \} := \mathfrak{D}(\rho^*)$$

is dense in $\mathcal{H}$.

(b) The mapping ρ^* on $\mathfrak{A}$ defined by

(5.1.3) $$\rho^*(x) := \rho(x^*)^*, \qquad x \in \mathfrak{A},$$

is a representation.

Remark. Note that (a) is satisfied whenever

$$\mathfrak{D}(\rho) \subset \mathfrak{D}(\rho^*)$$

but there are examples when $\mathfrak{D}(\rho^*)$ contains only the zero vector.

Proof. If ρ^* is assumed to be a representation, then the density of $\mathfrak{D}(\rho^*)$ is part of the definition. Assume now that $\mathfrak{D}(\rho^*)$ is dense in $\mathcal{H}$. We claim that $\rho^*(x)$ maps $\mathfrak{D}(\rho^*)$ into itself. In view of (5.1.2), we must show that $\rho(x)^*$ maps $\mathfrak{D}(\rho^*)$ into itself for all $x \in \mathfrak{A}$. Pick some vector $a \in \mathfrak{D}(\rho^*)$. To show that $\rho(x)^*a \in \mathfrak{D}(\rho^*)$, we must verify that $\rho(x)^*a \in D(\rho(y)^*)$ for all $y \in \mathfrak{A}$. This amounts to checking that the linear form given by

$$b \longrightarrow \langle \rho(y)b, \rho(x)^*a \rangle,$$

is bounded on $\mathfrak{D}(\rho)$. We use the definitions from Section 3.2. But if $b \in \mathfrak{D}(\rho)$, then

$$\begin{aligned} \langle \rho(y)b, \rho(x)^*a \rangle &= \langle \rho(x)\rho(y)b, a \rangle \quad (\text{since } \rho(y)b \in \mathfrak{D}(\rho)) \\ &= \langle \rho(xy)b, a \rangle \quad (\text{since } \rho \text{ is a representation}) \\ &= \langle b, \rho(xy)^*a \rangle \quad (\text{since } a \in D(\rho(xy)^*)). \end{aligned}$$

It follows that

$$\rho(x)^*a \in D(\rho(y)^*), \text{ and}$$

$$\rho(y)^*\rho(x)^*a = \rho(xy)^*a. \tag{5.1.4}$$

Since the elements x and y were chosen arbitrarily in $\mathfrak{A}$, we conclude that $\rho(x)^*a \in \mathfrak{D}(\rho^*)$, and, by virtue of (5.1.4), that ρ^* is a representation.

Let ρ and π be two representations of $\mathfrak{A}$ by operators in the same Hilbert space $\mathcal{H}$. We shall say that

$$\rho \subset \pi \qquad (\textit{containment})$$

if

$$\rho(x) \subset \pi(x), \qquad x \in \mathfrak{A},$$

where the operator inclusion is defined in Section 3.1 above.

It follows that a representation ρ is Hermitian if and only if

$$\rho \subset \rho^*. \tag{5.1.5}$$

With the latter inclusion (5.1.5) we have the implicit assumption that $\mathfrak{D}(\rho) \subset \mathfrak{D}(\rho^*)$, and ρ^* is then automatically a representation.

Lemma 5.1.2. Let ρ be a Hermitian representation of $\mathfrak{A}$. Then the operator $\rho(x)$ is closable for all $x \in \mathfrak{A}$, i.e., it has a closed extension.

Proof. Since ρ is Hermitian, we have noted that

$$\rho(x) \subset \rho(x^*)^*.$$

But the operator $\rho(x^*)$ is densely defined, so its closed adjoint $\rho(x^*)^*$ is well defined. It follows that $\rho(x)$ has a closed extension as asserted.

Lemma 5.1.3. Let ρ be a Hermitian representation of $\mathfrak{A}$. Then ρ has a least closed extension.

Proof. Following [Pow 1] we say that a representation π is closed if each operator $\pi(x)$, $x \in \mathfrak{A}$, is closable and

$$\mathfrak{D}(\pi) = \cap\{D(\overline{\pi(x)}) : x \in \mathfrak{A}\}.$$

Let ρ be a given Hermitian representation. By virtue of (5.1.5) and Lemma 5.1.1, we have that ρ^* is also a representation, although it is generally *not* Hermitian.

We now define

$$\mathfrak{D}(\overline{\rho}) := \cap\{D(\overline{\rho(x)}) : x \in \mathfrak{A}\} \tag{5.1.6}$$

and

$$\overline{\rho}(x) := \rho(x^*)^*\Big|_{\mathfrak{D}(\overline{\rho})}. \tag{5.1.7}$$

It is routine to verify the following four claims:

(i) ρ^* is a closed representation;

(ii) $\overline{\rho}$ is a closed Hermitian representation;

(iii) we have the two inclusions:

$$\rho \subset \overline{\rho} \subset \rho^*;$$

(iv) $\overline{\rho}$ is the smallest closed extension of ρ.

Details are left to the reader as an exercise.

5.2 Selfadjoint Representations

We first recall two definitions from [Pow 1] which are natural extensions of the corresponding definitions given in Section 3.3 above for single operators.

We say that a representation ρ of a $*$-algebra $\mathfrak{A}$ is *selfadjoint* if (it is closed), and

$$\rho = \rho^{*}. \tag{5.2.1}$$

We say that ρ is *essentially selfadjoint* if (it is Hermitian), and

$$\overline{\rho} = \rho^{*}.$$

The definitions are important for applications to the theory of unitary representations of Lie groups, and to quantum field theory. In Wightman's formulation, the smeared fields generate a $*$-representation, or equivalently a $*$-algebra of operators on the underlying Hilbert space.

The selfadjoint representations are important more generally in Quantum Mechanics since they are maximal in the following sense.

Lemma 5.2.1. Let ρ be a Hermitian representation.

(a) Then ρ is selfadjoint if and only if

$$\mathcal{D}(\rho) = \mathcal{D}(\rho^{*}).$$

(b) Every selfadjoint representation is maximal Hermitian.

Proof. We leave (a) to the reader. Suppose ρ is selfadjoint, and that π is a Hermitian representation satisfying $\rho \subset \pi$. Then it follows that $\pi^{*} \subset \rho$. But then $\rho \subset \pi \subset \pi^{*} \subset \rho$, and it follows that $\rho = \pi$.

In conclusion, selfadjoint means that the domain is maximal and no boundary conditions have been overlooked for the physical problem.

There are very few known conditions for checking selfadjointness for the most general examples of Hermitian representations. But we have results when $\mathfrak{A}$ is Abelian and when $\mathfrak{A}$ is the universal enveloping algebra of a finite-dimensional Lie algebra. Both of the latter examples are important for the theory of unitary representations of Lie groups.

Proposition 5.2.2 (Powers). Let $\mathfrak{A}$ be the free commutative $*$-algebra on a single generator, $x = x^*$, i.e., $\mathfrak{A}$ is simply the algebra of all polynomials in a single indeterminate.

Then a Hermitian representation ρ of $\mathfrak{A}$ is essentially selfadjoint if and only if the single operator $\rho(x)^n$ is essentially selfadjoint for all $n = 1,2,\cdots$.

Proof. Assume that ρ is an essentially selfadjoint representation. We shall show that the single Hermitian operator $T = \rho(x)$ is essentially selfadjoint on the domain $\mathfrak{D}(\rho)$. By von Neumann's theory of deficiency indices, this amounts to checking that the two closed subspaces $N(T^* \pm iI)$ are zero. Now let $Q_\pm$ denote the projections onto the respective deficiency spaces. We must show that $Q_\pm = 0$. The calculations will be done only for Q_-. Vectors a in the range of Q_- satisfy $Q_- a = a$ and $T^* a = ia$. If $Q_- \neq 0$, then there is a nonzero vector a in the range. We claim that $a \in \mathfrak{D}(\rho^*)$, and this follows from the observation that

$$\mathfrak{D}(\rho^*) = \cap\{D((T^n)^*) : n = 1,2,\cdots\}.$$

We shall prove, by induction, that $a \in D((T^n)^*)$ with $(T^n)^* a = i^n a$. The case $n = 1$ is immediate from the definition of Q_-. But, for $b \in \mathfrak{D}(\rho) = \mathfrak{D}(T)$, we have

$$\begin{aligned} \langle T^{n+1} b, a\rangle &= \langle TT^n b, a\rangle \\ &= \langle T^n b, T^* a\rangle \end{aligned}$$

since $T^n b \in D(T)$. Hence,

$$\begin{aligned}\langle T^{n+1}b, a\rangle &= \langle T^n b, ia\rangle \\ &= \langle b, i(T^n)^* a\rangle \\ &= \langle b, i^{n+1}\ a\rangle\end{aligned}$$

by the induction hypothesis. The conclusion $a \in \mathfrak{D}(\rho^*)$ follows from this. But

$$\mathfrak{D}(\rho^*) = \mathfrak{D}(\bar{\rho}) = \cap\{D((T^n)^-) : n = 1,2,\cdots\} \subset D(\bar{T}).$$

Since $\bar{T}$ is Hermitian, we have

$$\begin{aligned}i\|a\|^2 &= \langle ia, a\rangle = \langle T^* a, a\rangle \\ &= \langle a, \bar{T}a\rangle = \langle a, T^* a\rangle \\ &= \langle a, ia\rangle = -i\|a\|^2.\end{aligned}$$

It follows that $a = 0$.

The same argument works for Q_+, and we conclude that T is essentially selfadjoint.

We must show also that T^n is essentially selfadjoint on $\mathfrak{D}(\rho)$ when $n > 1$. But the same argument applies. We may consider the restriction of ρ to the subalgebra $\mathfrak{A}_0$ of $\mathfrak{A}$ which is generated by x^n. This restriction is clearly also a selfadjoint representation with the same domain $\mathfrak{D}(\rho)$. Since

$$\rho(x^n) = T^n,$$

the argument from above yields the essential selfadjointness of T^n.

Assume conversely that T^n is essentially selfadjoint on $\mathfrak{D}(\rho)$ for all n. The essential selfadjointness of ρ amounts to the identity

$$\mathfrak{D}(\rho^-) = \mathfrak{D}(\rho^*) \tag{5.2.2}$$

which translates into:

$$\cap\{D((T^n)^-) : n = 1,\cdots\} = \cap\{D((T^n)^*) : n = 1,\cdots\}.$$

Using the invariance $T\mathfrak{D}(\rho) \subset \mathfrak{D}(\rho)$ we show

$$D((T^n)^-) = D((T^n)^*), \qquad n = 1,\cdots,$$

and the desired conclusion (5.2.2) follows from this.

Remark. The higher dimensional analogue of the Proposition is false as demonstrated by examples of Nelson [Ne, p. 605], Powers [Pow 1, Section 5], and Jorgensen-Moore [JM, p. 292].

Surprisingly, the situation is better for non-Abelian algebras $\mathfrak{A}$, as we proceed to show.

We conclude this section with two results on selfadjoint representations. They will be of use to us in Sections 5.4-5.5 in the analysis of selfadjoint representations of the enveloping algebra of semisimple noncompact Lie algebras.

Lemma 5.2.3. Let ρ be a selfadjoint representation of a $*$-algebra $\mathfrak{A}$ in a Hilbert space $\mathcal{H}$, and let $\mathcal{M}$ be the commutant of $\rho(\mathfrak{A})$, i.e.,

$$\mathcal{M} := \{B \in \mathfrak{B}(\mathcal{H}) : \langle B\rho(x)a,b\rangle = \langle Ba,\rho(x^*)b\rangle, \quad x \in \mathfrak{A},\ a,b \in \mathfrak{D}(\rho)\}.$$

Then $\mathcal{M}$ is a von Neumann algebra.

Proof. In general, the commutant may be defined for any Hermitian representation, but it is usually *not* a von Neumann algebra. (It fails to be closed under product.)

It is immediate that $\mathcal{M}$ is weak*-closed in $\mathfrak{B}(\mathcal{H})$ so we need only to check the following nontrivial assertions:

(i) If $B \in \mathcal{M}$ then $B^* \in \mathcal{M}$, and

(ii) If $B_1,B_2 \in \mathcal{M}$ then $B_1B_2 \in \mathcal{M}$.

In fact, (i) holds in general, but not (ii).

Using the definition of the adjoint operator, it is easy to check that the defining property for $\mathcal{M}$ may be rewritten as follows: $B \in \mathcal{M}$, if and only if $B\mathfrak{D}(\rho) \subset \mathfrak{D}(\rho^*)$ and $B\rho(x)a = \rho^*(x)Ba$ for all $x \in \mathfrak{A}$, and $a \in \mathfrak{D}(\rho)$. Since ρ was assumed selfadjoint, we have $\mathfrak{D}(\rho) = \mathfrak{D}(\rho^*)$, and the implications (i) and (ii) follow.

Theorem 5.2.4 (R.T. Powers [Pow 1]). Let ρ be a selfadjoint representation of a $*$-algebra $\mathfrak{A}$ on a Hilbert space $\mathcal{H}$, and let $\mathcal{M}$ be the commutant.

Then there is a bijection between the lattice of all projections in $\mathcal{M}$ and the lattice of all reducing closed subspaces of $\mathcal{H}$.

Moreover, ρ restricts to a selfadjoint representation on each of its closed reducing subspaces.

Proof. The bijection is the simplest possible, and the most natural one. If $P \in \mathcal{M}$, $P = P^* = P^2$, then the reducing subspace is simply

$$P\mathcal{H} = \{Pa : a \in \mathcal{H}\} = \{a \in \mathcal{H} : Pa = a\}.$$

Conversely, if $\mathcal{H}_r \subset \mathcal{H}$ is a closed reducing subspace, then let P be the projection onto $\mathcal{H}_r$, and we check that $P \in \mathcal{M}$.

Since ρ is selfadjoint, every projection $P \in \mathcal{M}$ satisfies

$$P\mathcal{D}(\rho) \subset \mathcal{D}(\rho^*) = \mathcal{D}(\rho),$$

and

$$P\rho(x)a = \rho^*(x)Pa = \rho(x)Pa, \qquad a \in \mathcal{D}(\rho), x \in \mathfrak{X};$$

and the conclusions in the theorem follow from this: The representation ρ restricts to a representation ρ_P on $P\mathcal{H}$ with

$$\mathcal{D}(\rho_P) = P\mathcal{D}(\rho).$$

5.3 The Derived Representation

Let U be a representation of a Lie group G in a Banach space $\mathcal{B}$. (Strong continuity will always be an implicit assumption.) Let $\mathfrak{g}$ be the Lie algebra of G with exponential mapping $\exp : \mathfrak{g} \longrightarrow G$. Consider the one-parameter group

$$\{U(\exp tx) : t \in \mathbb{R}\}$$

which is a group of bounded operators for each $x \in \mathfrak{g}$. Its infinitesimal generator will be denoted $dU(x)$. Formally, $dU(x) = \frac{d}{dt} U(e^{tx})\Big|_{t=0}$.

We proceed to show that the family of closed operators $dU(x)$, $x \in \mathfrak{g}$, generates an algebra of operators with dense invariant domain in $\mathcal{B}$, and that dU may be viewed as a representation of the universal enveloping algebra $\mathfrak{U}_{\mathbb{C}}(\mathfrak{g})$. In the case where U is unitary, and $\mathfrak{B}$ is a Hilbert space, this representation is *Hermitian*.

We show that it is, in fact, *selfadjoint*. The converse is not true. There are examples of selfadjoint representations ρ of $\mathfrak{U}_{\mathbb{C}}(\mathfrak{g})$ which are not of the form dU for any unitary representation, in fact by no representation by bounded operators at all. The counterexamples are those mentioned at the end of Section 5.2.

To describe the representation dU in the general setting of Banach spaces, we recall the *Gårding space* for U, and the space of C^∞-vectors.

The Gårding space is defined to be the linear span of

$$\{U(\varphi)a \,:\, \varphi \in C_c^\infty(G),\ a \in \mathfrak{B}\}$$

where

$$U(\varphi)a = \int_G \varphi(g)U(g)a\, dg \tag{5.3.1}$$

(integration with respect to a *left-invariant* Haar measure on G), and

$$C_c^\infty(G) = \text{all compactly supported smooth functions on } G.$$

The C^∞-vectors for U are defined as those vectors $a \in \mathfrak{B}$ such that the orbit function $g \longrightarrow U(g)a$ is smooth from G to $\mathfrak{B}$.

The Gårding space is contained in the space of C^∞-vectors, and it was proved recently that the two spaces coincide. The latter result is due to Dixmier and Malliavin [DM].

One inclusion is straightforward, and we give the details below: We show that the function $g \longrightarrow U(g)U(\varphi)a$ is smooth from G to $\mathfrak{B}$ for all $\varphi \in C_c^\infty(G)$, and $a \in \mathfrak{B}$. Indeed, by invariance of Haar measure, we get

$$U(g)U(\varphi)a = \int_G \varphi(g^{-1}g')U(g')a\,dg', \tag{5.3.2}$$

and the integral depends smoothly on g since φ is smooth.

For every $x \in \mathfrak{g}$, we have vector fields

$$(5.3.3) \qquad (\tilde{x}\varphi)(g) := \frac{d}{dt}\,\varphi(\exp(-tx)g)\Big|_{t=0},$$

and formula (5.3.2) yields the important additional information that

$$(5.3.4) \qquad dU(x)U(\varphi)a = U(\tilde{x}\varphi)a.$$

In particular, the Gårding vectors are included in the domain of the operator $dU(x)$ for all $x \in \mathfrak{g}$, and the Gårding space is invariant.

Let $\mathfrak{D}$ denote the Gårding space. Then (5.3.4) yields the representation property

$$dU([x,y]) = [dU(x),dU(y)], \qquad \text{for } x,y \in \mathfrak{g},$$

i.e.,

$$dU : \mathfrak{g} \longrightarrow \mathrm{End}(\mathfrak{D})$$

is a representation, and we noted in Section 2.3 that dU extends naturally to a representation of the associative enveloping algebra $\mathfrak{U}_{\mathbb{C}}(\mathfrak{g})$.

In the rest of the chapter, we let $\mathfrak{U}$ be this enveloping algebra, and we shall restrict to the case when U is a *unitary* representation of G on some fixed Hilbert space $\mathfrak{H}$.

We prove

Theorem 5.3.1 (R.T. Powers [Pow 1]). Let U be a unitary representation of a Lie group G with Lie algebra $\mathfrak{g}$. Then dU is a selfadjoint representation of $\mathfrak{U}_{\mathbb{C}}(\mathfrak{g})$ with $\mathfrak{D}(dU)$ equal to the Gårding space.

Proof. First note that the associative algebra $\mathfrak{U} := \mathfrak{U}_{\mathbb{C}}(\mathfrak{g})$ is a $*$-algebra with a $*$-operation defined as follows:

$$(5.3.5) \qquad 1^* = 1, \quad \text{and} \quad x^* = -x \qquad \text{for } x \in \mathfrak{g}.$$

It now follows from the universal property of $\mathfrak{U}$ that the operation, so defined on the generators (5.3.5), extends to a period 2, conjugate linear antiautomorphism of $\mathfrak{U}$, i.e., to a $*$-operation.

Let $x \in \mathfrak{g}$. Recall that the operator $dU(x)$ is the infinitesimal generator of the unitary one-parameter group $\{U(e^{tx}) : t \in \mathbb{R}\}$. By Stone's theorem [RN, St 2], it is skew-adjoint. But then the restriction of $dU(x)$ to the Gårding space $\mathfrak{D}$ is clearly skew-Hermitian, i.e.,

$$\langle dU(x)a,b\rangle = -\langle a,dU(x)b\rangle, \qquad x \in \mathfrak{g}, \quad a,b \in \mathfrak{D}.$$

Since $x^* = -x$, we have the inclusion

$$dU(x) \subset dU(x^*)^* \tag{5.3.6}$$

satisfied for $x \in \mathfrak{g}$, and we define

$$dU(1) = I.$$

Using again the universal property of $\mathfrak{U}$, relative to the given Lie algebra $\mathfrak{g}$, we conclude that (5.3.6) is also satisfied for all $x \in \mathfrak{U}$, i.e., that dU is a Hermitian representation.

To prove that dU is selfadjoint, we need to verify the identity

$$\mathfrak{D}((dU)^*) = \mathfrak{D}$$

where $\mathfrak{D}$ is the domain of dU, i.e., the Gårding space. It is clear from the Hermitian property of dU that

$$\mathfrak{D} \subset \mathfrak{D}((dU)^*)$$

and we now proceed to prove the other inclusion.

Recall [DM] that $\mathfrak{D}$ equals the space of C^∞-vectors for the representation U, and to finish the proof we use the following characterization of the C^∞-vectors in terms of the Laplace operator. Let $x_1, \cdots, x_d$ be a basis for $\mathfrak{g}$, and let

$$\Delta = \sum_{i=1}^{d} x_i^2. \tag{5.3.7}$$

Recall now that, by virtue of formula (5.3.4), the elements in $\mathfrak{U}$ may be identified with left-invariant analytic partial differential operators on G. With that identification $\mathfrak{U}$ is, in fact, the algebra of *all* left-invariant partial differential operators on G, and $\mathfrak{g}$ is the Lie algebra of all left-invariant vector fields on G. In particular, the Laplace operator Δ is elliptic and left-invariant.

Let

$$L = \overline{dU(\Delta)}$$

where $\overline{}$ refers to operator closure.

We have

Lemma 5.3.2 (Poulsen [Po 3]).

$$\mathcal{D} = \bigcap_{n=1}^{\infty} D(L^n). \tag{5.3.8}$$

Proof. Poulsen's result was really about the space $C^\infty(U)$ of C^∞-vectors for U, but, by Dixmier-Malliavin, we have

$$C^\infty(U) = \mathcal{D}$$

so (5.3.8) is obtained by putting the two results together.

It is immediate that $C^\infty(L) \supset C^\infty(U)$ where we have used the notation $C^\infty(L)$ for the right hand side of (5.3.8).

We now prove the implication $C^\infty(L) \subset C^\infty(U)$. Let $a \in C^\infty(L)$, and let $b \in \mathcal{H}$, and consider the function f on G given by

$$f(g) = \langle U(g^{-1})a,b\rangle, \qquad g \in G.$$

It is clear that f is continuous, and we proceed to show that it is also infinitely often differentiable.

Consider also the functions f_n given by

$$f_n(g) = \langle U(g^{-1})L^n a,b\rangle, \quad g \in G,\ n = 1,2,\cdots.$$

Each f_n is continuous since a is in the domain of L^n for all n.

We claim that f is a weak solution to the elliptic system of partial differential equations

$$\Delta_{tr}^n f = f_n, \qquad n = 1,2,\cdots, \tag{5.3.9}$$

where Δ_{tr} denotes the transposed operator. More specifically, we have

$$\int_G (\Delta^n\varphi)(g)f(g)dg = \int_G \varphi(g)f_n(g)dg, \qquad \varphi \in C_c^\infty(G),\ n = 1,2,\cdots. \tag{5.3.10}$$

The proof of the latter formula (5.3.10) follows from the following related formula:

$$U(\varphi)L^n a = U(\Delta^n\varphi)a \tag{5.3.11}$$

which in turn follows from (5.3.4). In the present setting $U(\varphi)$ is defined in (4.1.17) and the integral on G is with respect to the right-invariant Haar measure.

Starting with the left-hand side of (5.3.10), and using (5.3.11), we have

$$\begin{aligned}
\int_G (\Delta^n\varphi)(g)f(g)dg &= \int_G (\Delta^n\varphi)(g)\langle U(g^{-1})a,b\rangle dg \\
&= \langle \int_G (\Delta^n\varphi)(g)U(g^{-1})adg, b\rangle \\
&= \langle U(\Delta^n\varphi)a, b\rangle \\
&= \langle U(\varphi)L^n a, b\rangle \\
&= \langle \int_G \varphi(g)U(g^{-1})L^n adg, b\rangle \\
&= \int_G \varphi(g)\langle U(g^{-1})L^n a, b\rangle dg \\
&= \int_G \varphi(g)f_n(g)dg,
\end{aligned}$$

concluding the proof of (5.3.10).

The transposed operator Δ^{tr} from (5.3.9) is defined relative to the right-invariant Haar measure dg, and Δ_{tr} is generally different from Δ since Δ is a left-invariant operator, rather than right-invariant.

Now there is a fairly simple formula for calculating Δ_{tr} in terms of the basis $x_1,\cdots,x_d$ and the adjoint representation Ad of G on $\mathfrak{g}$, and we refer to [BGJR] for details on this point. In any case, Δ_{tr} is elliptic. So f is a solution to a system of elliptic equations (5.3.9). The right-hand side of each equation is continuous, and the $n^{\underline{th}}$ equation in the system is of degree $2n$. It follows now by elliptic regularity, or rather the Sobolev embedding theorem, that f is a C^{∞}-function on G. Since $f(g) = \langle U(g^{-1})a,b\rangle$, and b is arbitrary in $\mathcal{H}$, it follows that

$$g \longrightarrow U(g^{-1})a \tag{5.3.12}$$

is weakly C^{∞}. It follows now from general theory [Po 3, Lemma 1.2] that the function (5.3.12) is also strongly C^{∞}, i.e., C^{∞} relative to the norm on $\mathcal{H}$.

This completes the proof of (5.3.8), and therefore Lemma 5.3.2.

We now return to the proof of the theorem.

Suppose $a \in \mathfrak{D}((dU)^{*})$. To show that $a \in \mathfrak{D}$, it is enough to check that a is in the domain of the operator L^{n} for all $n = 1,2,\cdots$.

But it was proved by Nelson and Stinespring [NS] that the operator L^{n} is essentially selfadjoint on $\mathfrak{D}$ for all n. Since

$$\mathfrak{D}((dU)^{*}) \subset \bigcap_{n=1}^{\infty} D((L^{n})^{*}),$$

it follows that $a \in D((L^{n})^{*})$ for all n. But $D((L^{n})^{*}) = D(L^{n})$ by the essential selfadjointness, and it follows that $a \in D(L^{n})$ for all n. We conclude from the lemma that $a \in \mathfrak{D}$, and the proof of Theorem 5.3.1 is completed.

5.4 Integrability of Selfadjoint Representation

Let $\mathfrak{g}$ be the Lie algebra $s\ell_{2}(\mathbb{R})$ of 2×2 real matrices $\begin{bmatrix} u & v \\ w & -u \end{bmatrix}$ of trace zero, $u,v,w \in \mathbb{R}$ and let $\mathfrak{U}$ be the universal enveloping algebra $\mathfrak{U}_{\mathbb{C}}(\mathfrak{g})$ equipped with the $*$-operation which we described in Section 5.3. We shall prove that every selfadjoint representation ρ of $\mathfrak{U}$ on some Hilbert space $\mathcal{H}$ is integrable. Let G be the simply connected universal covering group of the matrix group $SL_2(\mathbb{R})$. The Lie algebra of G is $s\ell_2(\mathbb{R})$, and we shall be interested in constructing unitary representations U of G on $\mathcal{H}$. We say that the representation ρ is *integrable* if a unitary representation U can be found such that $\rho = dU$. Alternatively, we say that the representation ρ *exponentiates*, or that it is *exact*. When the result of this section is combined with Theorem 5.3.1 above, we get

Theorem 5.4.1. Let ρ be a selfadjoint representation of the universal enveloping algebra of $s\ell_2(\mathbb{R})$ and let ω be the Casimir element. Then ρ is exact if and only if $\overline{\rho(\omega)}$ is affiliated with the commutant of ρ.

Proof. We already have one implication from Theorem 5.3.1, and we proceed to show that every restricted selfadjoint representation ρ is integrable. We noted in Section 5.3 that this implication is false for the cases when the Lie algebra $\mathfrak{g}$ is Abelian, or nilpotent. It is probably true for any noncompact semisimple $\mathfrak{g}$, but we have not carried out all the details in the general case, and we restrict the discussion here to $\mathfrak{g} = s\ell_2(\mathbb{R})$. Some remarks before the start of the proof:

Remarks. In a private communication, Professor K. Schmüdgen has reported the existence of a selfadjoint representation of $\mathfrak{U}_{\mathbb{C}}(s\ell_2(\mathbb{R}))$ which is *not* exact. It follows that the assumption in Theorem 5.4.1 on the Casimir operator cannot be omitted.

In general, the integrability is closely connected to the existence of K-finite vectors.

We should mention that nonintegrable Hermitian representations ρ have recently been constructed for the case $\mathfrak{U}_{\mathbb{C}}(\mathfrak{g})$ with $\mathfrak{g} = \mathfrak{su}(p,q)$, special restricted values of p and q, and $\mathcal{H}$ a certain Hilbert space of metrics, by Kostant and Ørsted [KØ]. In these representations there are no nonzero

K-finite vectors.

In general, the study of Hermitian representations ρ gets increasingly complex as the dimension of the semisimple Lie algebra $\mathfrak{g}$ increases.

We begin the proof with a general:

Lemma 5.4.2. Let ρ be a selfadjoint representation of a $*$-algebra $\mathfrak{A}$ on a Hilbert space $\mathcal{H}$, and let $\omega = \omega^*$ be an element in the center of $\mathfrak{A}$. Finally, let V be the Cayley transform of $\rho(\omega)$, and let $\mathcal{M} = \rho(\mathfrak{A})'$ be the commutant.

Then the following are equivalent:

(a) $V \in \mathcal{M}$.

(b) $\rho(\omega)$ is essentially selfadjoint on $\mathcal{D}(\rho)$.

(c) $\overline{\rho(\omega)}$ is affiliated with $\mathcal{M}$.

Proof. Examples show that the operator $\rho(\omega)$ need not be essentially selfadjoint if ρ is selfadjoint.

Suppose it is. Then set $\Omega = \overline{\rho(\omega)}$, and let P be the spectral resolution of Ω, i.e.,

$$\Omega = \int_{\mathbb{R}} \lambda \, dP(\lambda). \tag{5.4.1}$$

Then the projection $P(B)$, $B \subset \mathbb{R}$, Borel, is in the commutant $\mathcal{M} = \rho(\mathfrak{A})'$ for all B.

Recall that for a Hermitian representations ρ, $\mathcal{M}$ is generally not a von Neumann algebra (of bounded operators on $\mathcal{H}$). But if ρ is assumed selfadjoint, then it is by virtue of Lemma 5.2.3 above. We are asserting that, if $V \in \mathcal{M}$, then the self-adjoint operator Ω is affiliated with $\mathcal{M}$.

Let V be the Cayley transform of Ω, i.e.,

$$V := (\Omega - iI)(\Omega + iI)^{-1}.$$

Then V is a partial isometry, and the space of the initial projection V^*V is

$$(\Omega + iI)D(\Omega) = \overline{(\rho(\omega) + iI)\mathcal{D}(\rho)},$$

while the space of the final projection VV^* is

$$(\Omega - iI)D(\Omega) = \overline{(\rho(\omega) - iI)\mathcal{D}(\rho)}.$$

The projection onto the deficiency space $N(\rho(\omega)^* - iI)$ is $Q_- := I - V^*V$, and the projection onto the other deficiency space $N(\rho(\omega)^* + iI)$ is $Q_+ := I - VV^*$.

Assumption (a) states that

$$V \in \mathcal{M} = \rho(\mathfrak{X})' \tag{5.4.2}$$

Since $\mathcal{M}$ is a von Neumann algebra, by Lemma 5.2.3, we conclude that V^*, as well as V^*V and VV^*, are also in $\mathcal{M}$. Hence, $Q_\pm \in \mathcal{M}$. Since $\mathcal{D}(\rho) = \mathcal{D}(\rho^*)$, the projections $Q_\pm$ commute with $\overline{\rho(\omega)}$. We have $\rho(\omega) = \rho(\omega)^*\Big|_{\mathcal{D}(\rho)}$, so $\overline{\rho(\omega)}$ is a closed restriction of the operator $\rho(\omega)^*\Big|_{\mathcal{D}(\rho^*)}$, and the asserted commutativity follows from this.

For $a \in D(\Omega)$, we have $Q_\pm a \in D(\Omega)$, and

$$\Omega Q_\pm a = Q_\pm \Omega a \tag{5.4.3}$$

But we also have

$$\Omega Q_\pm a = \Omega^* Q_\pm a = \mp iQ_\pm a,$$

and therefore, the number

$$\begin{aligned} \langle \Omega Q_\pm a, Q_\pm a\rangle &= \langle \mp iQ_\pm a, Q_\pm a\rangle \\ &= \mp i\|Q_\pm a\|^2 \end{aligned}$$

is purely imaginary. Since Ω is Hermitian, it is also real. So it is zero. Hence, $\|Q_\pm a\|^2 = 0$, and $Q_\pm a = 0$. Since both operators $Q_\pm$ are bounded, and $D(\Omega)$ is dense in $\mathcal{H}$, we get the desired conclusions $Q_\pm = 0$, i.e., the deficiency indices are $(0,0)$. We proved that $Q_\pm \equiv 0$. But then

$$V^*V = I - Q_- = I,$$

and

$$VV^* = I - Q_+ = I$$

which is to say: the Cayley transform V of the operator Ω is unitary. But $V \in \mathcal{M}$, $\mathcal{M}$ is a von Neumann algebra, and the spectral projections of Ω from (5.4.1) may be written as functions of V by use of measurable functional calculus on the unitary V. Hence, Ω is affiliated with $\mathcal{M}$ as asserted in part (b) of the lemma.

We now examine more closely condition (5.4.2), or equivalently, the identity

$$(5.4.4) \qquad \langle \rho(x)a, V^* b\rangle = \langle Va, \rho(x^*)b\rangle, \qquad x \in \mathfrak{X},\ a,b \in \mathfrak{D}(\rho).$$

We decompose the two vectors as follows:

$$(5.4.5) \qquad a = (\Omega+i)a_1 + Q_- a$$

and

$$(5.4.6) \qquad b = (\Omega-i)b_1 + Q_+ b,$$

where the new vectors a_1 and b_1 in $D(\Omega)$ are uniquely determined. The vector $(\Omega+i)a_1$ will be approximated with a vector from $R(\rho(\omega)+i)$, that is, a vector of the form $(\rho(\omega)+i)a_2$ with $a_2 \in \mathfrak{D}(\rho)$; and, similarly, the vector $(\Omega-i)b_1$ will be approximated with $(\rho(\omega)-i)b_2$ for some $b_2 \in \mathfrak{D}(\rho)$. We have the additional information:

$$V^* Q_+ b = 0 = VQ_- a.$$

When the approximations for (5.4.5) and (5.4.6) are substituted into (5.4.4), we may subtract the following two equal terms on both sides of the equation:

$$\langle \rho(x)(\rho(\omega)+i)a_2, (\rho(\omega)+i)b_2\rangle = \langle (\rho(\omega)-i)a_2, \rho(x^*)(\rho(\omega)-i)b_2\rangle$$

where the assumptions are, $x \in \mathfrak{X}$, $a_2, b_2 \in \mathfrak{D}(\rho)$.

What remains of (5.4.4) is the identity:

$$\langle \rho(x)Q_- a, V^* b\rangle = \langle Va, \rho(x^*)Q_+ b\rangle$$

which may also be rewritten in the form:

(5.4.7) $\langle V\rho(x)Q\_a,b\rangle = \langle\overline{\rho(x)}Va,Q_+b\rangle, \quad x \in \mathfrak{X},\ a,b \in \mathfrak{D}(\rho),$

since ρ is selfadjoint and

$$Va \in \bigcap_x D(\rho(x^*)^*) = \mathfrak{D}(\rho^*) = \mathfrak{D}(\rho).$$

If $\overline{\rho(\omega)}$ is affiliated with $\mathcal{M}$, it follows that this last identity (5.4.7) is satisfied, which means that condition (a) follows.

The proof of the lemma is completed.

Two steps remain in the proof of Theorem 5.4.1. First we must check that when $\mathfrak{X} = \mathfrak{U}_{\mathbb{C}}(s\ell_2(\mathbb{R}))$, then the lemma applies to the Casimir element ω where

(5.4.8) $$\omega = -x_0^2 + x_1^2 + x_2^2$$

(5.4.9) $$x_0 = \frac{1}{2}\begin{bmatrix} 0 & -1 \\ 1 & 0 \end{bmatrix}, \quad x_1 = \frac{1}{2}\begin{bmatrix} 1 & 0 \\ 0 & -1 \end{bmatrix}, \quad x_2 = \frac{1}{2}\begin{bmatrix} 0 & 1 \\ 1 & 0 \end{bmatrix}$$

and ρ is an arbitrary selfadjoint representation. This amounts to showing that the operator $\rho(\omega)$ is an essentially selfadjoint operator on $\mathfrak{D}(\rho)$ if it is affiliated with the commutant of a selfadjoint representation.

The second step in the proof of Theorem 5.4.1 is a reduction to the special case when the operator Ω is a scalar, i.e., $\Omega = qI$ for $q \in \mathbb{R}$. To carry out the reduction, we combine the two lemmas, Lemma 5.2.3 and Lemma 5.4.2 above, and we use direct integral theory for the von Neumann algebra $\mathcal{M}$ relative to the spectral decomposition (5.4.1) for the selfadjoint operator Ω.

We shall not include here the technical details regarding the direct integral decomposition, but refer instead to the texts [Dix 14], and [Mau 1].

We turn, in the next section, to the proof of the special case of Theorem 5.4.1 when Ω is a scalar. If the original representation ρ has been decomposed into an integral of selfadjoint representations $\rho(\lambda)$, and if each $\rho(\lambda)$ is exact, then it follows that ρ is itself exact. For, if $U(\lambda)$ is a unitary representation of G satisfying $\rho(\lambda) = dU(\lambda)$

for each λ in the spectrum of Ω, then we may define a "grand" unitary representation U as the direct integral of the representations $U(\lambda)$ relative to the same direct integral decomposition, and then it is a simple matter to check that

$$\rho = dU. \tag{5.4.10}$$

The proof of Theorem 5.4.1 has then been completed.

In the verification of formula (5.4.10), we need to do the direct integral decomposition in the setting of C^∞-vectors, and, in effect, do the direct integral theory in the setting of the Fréchet spaces of C^∞-vectors, both the C^∞-vectors for U, and those for the component representations $U(\lambda)$. The technical lemmas needed for carrying out this program have been worked out recently by R. Penney, and the reader is referred to the excellent paper [Pe] for the details.

5.5 Scalar Casimir Operator: A Special Case

In this section we include a result from [Jo 25] which amounts to precisely the missing implication in the above discussion.

Let $\mathfrak{A}$ be the enveloping algebra $\mathfrak{U}_{\mathbb{C}}(\mathfrak{g})$, $\mathfrak{g} = s\ell_2(\mathbb{R})$, and let ω be the Casimir element, cf. formula (5.4.8).

Theorem 5.5.*1* (Jorgensen [Jo 25]). Let ρ be a selfadjoint representation of $\mathfrak{A}$ such that $\rho(\omega)$ is a scalar. Then it follows that ρ is exact.

Proof. In the course of the proof we shall use results on analytic domination and integrability of Lie algebras of operators. However this particular application is a very simple one and the steps where analytic domination is used can be verified directly with elementary induction arguments. A systematic treatment of analytic domination and integrability in a much more general setting will not be taken up until Section 6.3.

Let $\{x_k : k = 0,1,2\}$ be the basis for $\mathfrak{g} = s\ell_2(\mathbb{R})$ which is listed in (5.4.9), and let

$$X_k = \rho(x_k)$$

be the corresponding operators with dense invariant domain $\mathcal{D}(\rho)$. The underlying Hilbert space of the representation will be denoted $\mathcal{H}$, with inner product $\langle \cdot , \cdot \rangle$, and norm $\| \cdot \|$.

We introduce the Casimir operator

$$\Omega = \rho(\omega)$$

and raising, resp., lowering, operators

$$A_{\pm} := i(X_1 \pm iX_2). \tag{5.5.1}$$

The assumption in the theorem on Ω amounts to: $\Omega = qI$ for some real number q, and the Hermitian property of ρ amounts to:

$$\langle A_+ a, b\rangle = \langle a, A_- b\rangle, \quad a,b \in \mathcal{D}(\rho), \tag{5.5.2}$$

and

$$\langle X_0 a, b\rangle = -\langle a, X_0 b\rangle.$$

We introduce the Hermitian operator:

$$A_0 := iX_0 = i\rho(x_0),$$

and the commutation relations of $s\ell_2(\mathbb{R})$ now take the following form:

$$[A_0, A_{\pm}] = \pm A_{\pm} \tag{5.5.3}$$

and

$$[A_-, A_+] = 2A_0.$$

For the Casimir, we have the identity

$$A_0^2 - \frac{1}{2}(A_+A_- + A_-A_+) = -qI \tag{5.5.4}$$

on $\mathcal{D}(\rho)$. More specifically, for the commutator, and anticommutator we have

$$[A_+, A_-] = -2A_0$$

and

$$\{A_+, A_-\} = 2(qI + A_0^2).$$

When properties (5.5.2) and (5.5.4) are combined, we get for $a \in \mathfrak{D}(\rho)$:

$$q\|a\|^2 = \langle qa, a\rangle = \frac{1}{2}(\langle A_+A_-a, a\rangle + \langle A_-A_+a, a\rangle) - \langle A_0^2 a, a\rangle$$
$$= \frac{1}{2}(\|A_-a\|^2 + \|A_+a\|^2) - \|A_0 a\|^2,$$

and we shall use this in the form:

$$\|A_-a\|^2 + \|A_+a\|^2 = 2q\|a\|^2 + 2\|A_0a\|^2. \tag{5.5.5}$$

After substitution of (5.5.1) into (5.5.5), we arrive at the two estimates

$$\|X_k a\|^2 \le q\|a\|^2 + \|A_0 a\|^2, \quad a \in \mathfrak{D}(\rho),\ k = 1,2. \tag{5.5.6}$$

In fact, the following weaker inequality,

$$\|X_k a\| \le \sqrt{|q|}\ \|a\| + \|A_0 a\|, \quad a \in \mathfrak{D}(\rho),\ k = 1,2, \tag{5.5.7}$$

will suffice for our present purpose.

We now use the commutator identities (5.5.7), and an induction, to arrive at the following analytic power estimates:

Let $c = \max\{1, \sqrt{|q|}\}$

$$\|X_k^n a\| \le c \sum_{m=0}^{n} \binom{n}{m} c^{n-m} \|A_0^m a\|. \tag{5.5.8}$$

Using again standard estimates for the Hermitian operator A_0, one can show that (5.5.8) implies the estimate

$$\|X_k^n a\| \le c_1 \|a\| + c c_2^n \|A_0^n a\| \tag{5.5.9}$$

for suitable choices of the positive constants c_1, c_2. We shall return to the combinatorics of these types of estimates in Section 6.3 below. It was developed, in some cases, first by Nelson [Ne 3], and extended further in [Goo 2] and [GJ 2]. The special situation for $s\ell_2(\mathbb{R})$ was also treated in [JM, Ch. 12].

Both of the estimates (5.5.8) and (5.5.9) hold for $k = 1,2$, $a \in \mathcal{D}(\rho)$, and $n = 1,2,\cdots$. The symbol $\begin{bmatrix} n \\ m \end{bmatrix}$ designates the usual binomial coefficient

$$\begin{bmatrix} n \\ m \end{bmatrix} = \frac{n!}{m!(n-m)!}.$$

When two operators X_k and A_0 satisfy an estimate of the form (5.5.9), then we say that A_0 *analytically dominates* X_k. We shall need the following stronger version of the power estimates (5.5.9).

To state them, we recall the notion of *degree* in the enveloping algebra $\mathfrak{U} = \mathfrak{U}_{\mathbb{C}}(s\ell_2(\mathbb{R}))$. First, by the Birkhoff-Witt theorem [Ser 1, LA 3.4], we note that the elements

$$x_0^i x_1^j x_2^k, \quad i,j,k = 0,1,2,\cdots \tag{5.5.10}$$

form a basis for $\mathfrak{U}$ over $\mathbb{C}$, and we say that $i+j+k$ is the degree of the monomial (5.5.10). We say that an element $x \in \mathfrak{U}$ is of degree n if it is a (finite) linear combination over $\mathbb{C}$ of the monomials (5.5.10) with $i+j+k = n$.

We now state the generalized form of the estimate (5.5.9) which is also proved by induction: If $x \in \mathfrak{U}$ is given of degree n, then there are constants c_1 and c_2 such that

$$\|\rho(x)a\| \le c_1\|a\| + c_2^n\|A_0^n a\|, \quad a \in \mathcal{D}(\rho). \tag{5.5.11}$$

Moreover, if x is one of the monomials (5.5.10), then the constants c_1 and c_2 may be chosen to depend only on the Lie algebra structure constants for $s\ell_2(\mathbb{R})$.

It follows from (5.5.11) that

$$\cap\{D(\overline{\rho(x)}) : x \in \mathfrak{U}\} = \cap\{(A_o^n)^- : n = 1,2,\cdots\}. \tag{5.5.12}$$

More explicitly, let ρ_0 denote the restriction of ρ to the subalgebra $\mathfrak{U}_0$ of $\mathfrak{U}$ which is generated by the single element

x_0, i.e.,

(5.5.13) $$\rho_0(x) = \rho(x), \qquad x \in \mathfrak{X}_0.$$

Then the information (5.5.12) translates into the statement

$$\mathcal{D}(\overline{\rho}) = \mathcal{D}(\overline{\rho}_0).$$

But ρ is assumed selfadjoint so we have $\rho = \overline{\rho} = \rho^*$, or equivalently,

$$\mathcal{D}(\rho) = \mathcal{D}(\overline{\rho}) = \mathcal{D}(\rho^*).$$

It also follows from the selfadjointness that ρ_0 is Hermitian, i.e.,

$$\rho_0 \subset \rho_0^*,$$

but we shall establish the nontrivial conclusion that ρ_0 is, in fact, essentially selfadjoint.

For this purpose, we need estimates similar to (5.5.11) but for the adjoint operators.

We begin with the first order estimates, and, in fact, the simplest instance of these estimates. Consider the estimate

$$\|A_+a\|^2 \leq c^2(\|a\|^2+\|A_0a\|^2), \qquad a \in \mathcal{D}(\rho).$$

Since $\|ia+A_0a\|^2 = \|a\|^2+\|A_0a\|^2$, the operator $iI+A_0$ is invertible on $R(i+A_0)$, and

$$\|A_+a\| \leq c\|(iI+A_0)a\|.$$

It follows that $A_+(i+A_0)^{-1}$ is well defined on $R(i+A_0)$ and bounded. Let

$$B = A_+(i+A_0)^{-1}.$$

Then

(5.5.14) $$A_+ = B(i+A_0)$$

and B is determined uniquely on $R(i+A_0)$ by this equation. Define B to be zero on the complement $R(i+A_0)^\perp = N(A_0^*-iI)$.

Then B is bounded and everywhere defined, and we claim that B maps $\mathcal{D}(\rho)$ into itself. Let

$$a \in \mathcal{D}(\rho) = \bigcap_n D((A_0^n)^-) \subseteq \bigcap_n D((\bar{A}_0)^n),$$

and consider the orthogonal decomposition

$$a = (iI+\bar{A}_0)b + a^\perp \tag{5.5.15}$$

where the two vectors

$$b \in D(\bar{A}_0)$$

and

$$a^\perp \in R(i+A_0)^\perp$$

are unique. Because of the estimate (5.5.5) we have $D(\bar{A}_0) \subset D(\bar{A}_+)$. When the bounded operator B is applied to (5.5.15), we get

$$\begin{aligned} Ba &= B(iI+\bar{A}_0)b + Ba^\perp \\ &= \bar{A}_+ b. \end{aligned}$$

Recall that $\bar{A}_+$ maps $\mathcal{D}(\rho)$ into itself since ρ is a closed representation. If we show that $b \in \mathcal{D}(\rho)$, it follows that

$$Ba = \bar{A}_+ b \in \mathcal{D}(\rho) \tag{5.5.16}$$

which means that B leaves invariant the domain $\mathcal{D}(\rho)$. But it follows from (5.5.15) and induction that

$$b \in \bigcap D((A_0^n)^-) = \mathcal{D}(\rho),$$

and we have the desired conclusion (5.5.16).

Recall the commutation relation

$$[A_0, A_+] = A_+ \quad \text{on } \mathcal{D}(\rho). \tag{5.5.17}$$

We show below that B satisfies the analogous relation

$$[A_0, B] = B. \tag{5.5.18}$$

A substitution of (5.5.14) into (5.5.17) yields,

$$A_+ = [A_0, B(i+A_0)] = [A_0, B](i+A_0).$$

But (5.5.14) determines uniquely the operator B on $R(i+A_0)$. Hence, (5.5.18) follows from the uniqueness.

We now return to (5.5.14) and take the adjoint on both sides of the equation. Since B is bounded, we get

$$A_+^* = (A_0^* - i)B^*$$

where the operators on the two sides of the equation have *equal* domains.

Using (5.5.18) we shall prove the stronger conclusions

$$D(A_0^*) \subset D(A_+^*) \tag{5.5.19}$$

and

$$A_+^* a = B^*(A_0^* - i)a - B^* a \quad \text{for all} \quad a \in D(A_0^*). \tag{5.5.20}$$

It follows from (5.5.20) that

$$\|A_+^* a\| \leq c\|A_0^* a\| + 2c\|a\|, \qquad a \in D(A_0^*), \tag{5.5.21}$$

where $c = \|B^*\| = \|B\|$.

The proof of (5.5.19) follows below. Let $a \in D(A_0^*)$ be given and consider the linear form,

$$b \longrightarrow \langle A_+ b, a\rangle \qquad \text{on} \quad \mathfrak{D}(\rho).$$

We have

$$\begin{aligned} \langle A_+ b, a\rangle &= \langle B(A_0+i)b, a\rangle \\ &= \langle (A_0+i)Bb, a\rangle - \langle Bb, a\rangle \\ &= \langle Bb, A_0^* a - ia\rangle - \langle b, B^* a\rangle \end{aligned}$$

which is well defined since $a \in D(A_0^*)$, and bounded in b since B is a bounded operator. Using finally that $\mathfrak{D}(\rho)$ is dense in $\mathcal{H}$, the desired formula (5.5.19) results.

We now turn to the more general type of first order estimates, and then finally to the higher degree commutator formulae.

Since the two operators $X_k = \rho(x_k)$, $k = 1,2$, are complex linear combinations of the two operators $A_\pm$, cf. formula (5.5.1), it follows that there are bounded operators B_k, $k = 1,2$, such that

$$X_k = B_k(i+A_0), \qquad k = 1,2,$$

with B_k given by the corresponding linear combination of the two operators $B_\pm$ from

$$A_\pm = B_\pm(i+A_0).$$

Using the commutation relations for the operators $B_\pm$ (which have already been derived) we arrive at

$$\text{(5.5.22)} \qquad \begin{cases} [A_0, B_1] = iB_2 \\ [A_0, B_2] = -iB_1 \end{cases}.$$

We now prove the inclusions

$$D(A_0^*) \subset D(X_k^*), \qquad k = 1,2,$$

and the formulae,

$$X_1^*a = B_1^*(A_0^*-i)a + iB_2^*a, \qquad a \in D(A_0^*),$$

$$X_2^*a = B_2^*(A_0^*-i)a - iB_1^*a, \qquad a \in D(A_0^*),$$

by the argument which was used above for (5.5.19)-(5.5.20).

Indeed, let $a \in D(A_0^*)$ be given. We then have, for $b \in \mathscr{D}(\rho)$,

$$\begin{aligned} \langle X_1 b, a\rangle &= \langle B_1(A_0+i)b, a\rangle \\ &= \langle (A_0+i)B_1 b, a\rangle - \langle iB_2 b, a\rangle \\ &= \langle B_1 b, A_0^*a-ia\rangle + \langle b, iB_2^*a\rangle. \end{aligned}$$

Since B_1 is bounded, we conclude that $a \in D(X_1^*)$. This proves the inclusion $D(A_0^*) \subset D(X_1^*)$, and the first formula. The same argument applies to the second formula.

The estimates

$$\|X_k^* a\| \leq c\|A_0^* a\| + 2c\|a\|, \quad a \in D(A_0^*), \qquad k = 1,2,$$

follow immediately.

We now turn to the induction step. Let $x \in \mathfrak{U} = \mathfrak{U}_{\mathbb{C}}(g)$ be a monomial of degree n, cf. (5.5.10) above; then we prove the domain inclusion

$$\bigcap_{m=1}^{n} D((A_0^m)^*) \subset D(\rho(x)^*). \tag{5.5.23}$$

This will imply that the restricted representation ρ_0 is essentially selfadjoint since

$$\mathfrak{D}(\rho_0^*) = \bigcap_n D((A_0^n)^*)$$

while

$$\mathfrak{D}(\rho^*) = \bigcap_x D(\rho(x)^*).$$

We then get the chain of inclusions:

$$\mathfrak{D}(\rho) \subset \mathfrak{D}(\overline{\rho}_0) \subset \mathfrak{D}(\rho_0^*) \subset \mathfrak{D}(\rho^*) = \mathfrak{D}(\rho)$$

since (5.5.23), for all n, is the "connecting" link in the chain, namely,

$$\mathfrak{D}(\rho_0^*) \subset \mathfrak{D}(\rho^*).$$

Let $\mathfrak{B}(\rho)$ be the algebra of bounded operators in $\mathcal{H}$ which map $\mathfrak{D}(\rho)$ into itself, specifically,

$$\mathfrak{B}(\rho) := \{B \in \mathfrak{B}(\mathcal{H}) : B\mathfrak{D}(\rho) \subset \mathfrak{D}(\rho)\}.$$

Our proof of (5.5.23) will be based on the following claim which is proved in turn by induction.

Let $x \in \mathfrak{U}$ be a degree n monomial. Then there are elements B_j, $0 \leq j \leq n$, in $\mathfrak{B}(\rho)$ such that

$$\rho(x) = A_0^n B_n + \cdots + A_0 B_1 + B_0. \tag{5.5.24}$$

This indeed implies (5.5.23). For let

$$a \in \bigcap_{m=1}^{n} D((A_0^m)^*),$$

then $b \longrightarrow \langle \rho(x)b,a\rangle$ is continuous on $\mathfrak{D}(\rho)$. Indeed,

$$\langle \rho(x)b,a\rangle = \sum_{m=0}^{n} \langle A_0^m B_m b,a\rangle$$

$$= \sum_{m=0}^{n} \langle B_m b,(A_0^m)^* a\rangle$$

depends continuously on b in $\mathfrak{D}(\rho)$ since the operators B_m are bounded.

We now finish the proof of (5.5.24). Suppose it holds for monomials in $\mathfrak{X}$ of degree n. Let x be a monomial of degree $n+1$. We have

$$x = x_0^i x_1^j x_2^k$$

for $i+j+k = n+1$. If $i > 0$, then

$$\rho(x) = A_0 \rho(x_0^{i-1} x_1^j x_2^k)$$

and the induction hypothesis applies to the degree n monomial $x_0^{i-1} x_1^j x_2^k$.

We now use the commutation relations (5.5.22) which state that the complex linear span of B_1 and B_2 is invariant under the derivation

$$(\mathrm{ad}\, A_0)(B) = A_0 B - B A_0.$$

Let $\mathfrak{B}_2 = \mathrm{span}\{B_1, B_2\}$, and $B \in \mathfrak{B}_2$. We then have

$$BA_0^m = A_0^m B + \sum_{k=1}^{m} (-1)^k \begin{bmatrix} m \\ k \end{bmatrix} A_0^{m-k} (\mathrm{ad}\, A_0)^k (B)$$

and $(\mathrm{ad}\, A_0)^k (B) \in \mathfrak{B}_2$ for all k. We write $\rho(x)$ in the form

$$\rho(x) = X_1 \sum_{m=0}^{n} A_0^m B_m$$

where the induction hypothesis has been used. Substitution of $X_1 = B_1 A_0$, and rearrangement of the summation index, yields

$$\rho(x) = \sum_{m=1}^{n+1} B_1 A_0^m B_{m-1}$$

$$= \sum_{m=1}^{n+1} A_0^m B_1 B_{m-1} + \sum_{m=1}^{n+1} \sum_{k=1}^{m} (-1)^k \begin{bmatrix} m \\ k \end{bmatrix} A_0^{m-k} (\mathrm{ad}\ A_0)^k (B_1) B_{m-1}.$$

Since the right hand side is a polynomial of degree $n+1$ in the Hermitian element A_0 with coefficients in the ring $\mathfrak{B}(\rho)$ of bounded operators, the induction proof is completed. Formula (5.5.24) does indeed hold for all n, and, moreover, we have an algorithm for calculating the coefficients B_m.

We are now ready to finish the proof of Theorem 5.5.1. We have shown that the restricted representation ρ_0 is essentially selfadjoint. By Proposition 5.2.2, this implies that the operator $\overline{A}_0$ is selfadjoint. Let

$$\overline{A}_0 = \int_{\mathbb{R}} \lambda \, dE(\lambda) \tag{5.5.25}$$

be the corresponding spectral resolution. We now consider vectors of the form

$$a = \int_J dE(\lambda) b \tag{5.5.26}$$

where $J \subset \mathbb{R}$ is a compact interval, and $b \in \mathfrak{D}(\rho)$. Vectors of this form are analytic for $\overline{A}_0$, i.e., the power series

$$\sum_{n=0}^{\infty} \frac{t^n}{n!} \|(\overline{A}_0)^n a\| \tag{5.5.27}$$

has positive radius of convergence. This is well known [Ne 3, Lemma 5.1], and it can easly be checked by direct substitution of (5.5.26) into (5.5.27).

In view of the estimates (5.5.9), it follows that a is

also analytic for the two operators X_1 and X_2. To be specific, let the interval J in (5.5.26) be $J = [-R,R]$. Then

$$\|A_0^n a\| \leq R^n \|b\|,$$

and when this is substituted into (5.5.9) we get

$$\|X_k^n a\| \leq c_1 \|a\| + c(c_2 R)^n \|b\|.$$

We get the stronger conclusion that vectors of the form (5.5.26) are entire analytic for the three operators $\{X_k : 0 \leq k \leq 2\}$, i.e., the power series

$$\sum_{n=0}^{\infty} \frac{t^n}{n!} \|X_k^n a\|$$

have infinite radius of convergence.

The integrability of the representation ρ, or equivalently, the real Lie algebra spanned by $\{X_k : 0 \leq k \leq 2\}$, now follows from [GJ 2, Theorem 3.1], or [JSi]. Alternatively, we may get the integrability from Nelson's theorem [Ne 3, Theorem 5]. Nelson's Laplace operator

$$\Delta = \sum_{k=0}^{2} X_k^2$$

simplifies in view of $\Omega = qI$ to

$$\Delta = -2A_0^2 - qI,$$

so Δ is essentially selfadjoint if A_0^2 is. But the essential selfadjointness of A_0^2 is implied by Proposition 5.2.2 and $\overline{\rho_0} = \rho_0^*$.

Additional References for Chapter 5

Sect. 5.1: [AG], [Fu 1], [GℓJa], [JM], [Jo 25], [JMu].

Sect. 5.2: [Gå 1], [Goo 1], [Jo 28], [Pow 1], [Rao].

Sect. 5.3: [Aa], [Goo 1], [Po 2-3].

Sect. 5.4: [Barg], [BB], [Jo 25], [Pow 2], [Rao].

CHAPTER 6. OPERATORS IN THE ENVELOPING ALGEBRA

Let G be a Lie group with Lie algebra $\mathfrak{g}$, and let $\mathfrak{U}_{\mathbb{C}}(\mathfrak{g})$ be the corresponding universal enveloping algebra. We saw in Section 5.3 that the derived representation dU (of a given unitary representation U of G) is a selfadjoint representation of the $*$-algebra $\mathfrak{U}_{\mathbb{C}}(\mathfrak{g})$, and we noted that not every selfadjoint representation of $\mathfrak{U}_{\mathbb{C}}(\mathfrak{g})$ is exact, i.e., of the form dU. (We have the Nelson type counterexamples.)

The representation dU is important because the operators $dU(x)$ for $x \in \mathfrak{U}_{\mathbb{C}}(\mathfrak{g})$ include important Quantum Mechanical operators, i.e., Hamiltonians with various kinds of potentials, interaction potentials, magnetic field potentials, etc. Some of these examples, associated with nilpotent Lie groups, are worked out in the recent paper [JK 1].

Quantum numbers are eigenvalues of operators of the form $dU(x)$ with $x = x^*$ in the center of $\mathfrak{U}_{\mathbb{C}}(\mathfrak{g})$. If ρ is just a selfadjoint representation of $\mathfrak{U}_{\mathbb{C}}(\mathfrak{g})$ we saw in Chapter 5 that the operator $\rho(x)$ need not be essentially selfadjoint on the domain $\mathfrak{D}(\rho)$. We shall prove in Section 6.1 below that $dU(x)$ is essentially selfadjoint. This is a pioneering result of I.E. Segal [Se 7]. (It follows from this that not every selfadjoint representation is exact.)

Operators of the form $dU(x)$ for more general elements x in $\mathfrak{U}_{\mathbb{C}}(\mathfrak{g})$ include many important partial differential operators. The distinction between various types of ellipticity and hypoellipticity comes out particularly well in this approach. If $x \in \mathfrak{U}_{\mathbb{C}}(\mathfrak{g})$ is of second degree, and $\mathfrak{g}$ is the Heisenberg Lie algebra, we get the famous Lewy operator [Le], and if x is of degree one, and $\mathfrak{g}$ is general, then we get important first order partial differential operators. We note that properties of the operator $dU(x)$ depend both on the choice of U and of x. For fixed x, the operator $dU(x)$ can have absolutely continuous (Lebesgue, in fact) spectrum for one choice of U, and purely discrete spectrum for a different choice of U. Frequently, but not always, the discreteness of the spectrum results from taking U to be irreducible. This happens in our spectral analysis (Section 7.1) of the curved magnetic field Hamiltonians. But if G is a symmetry group containing the Poincaré group, or a "similar" Lie group, then the elementary particle mass operator is of the form dU(x) for U irreducible. In this case, the operator dU(x) has no isolated components in its spectrum. This is the content of the O'Raifeartaigh theorem [O'R 3]. See also [Se 11], [Goo 3], [Ste 2], and [JM, Ch. 4]. Mathematical physicists refer to this as the O'Raifeartaigh problem because mass-splitting is observed for elementary particles. Since different values of the mass are experimentally observed, this denies the "conventional" interpretation of elementary particles as irreducible representations of the extended symmetry group. In resolving the paradox, I.E. Segal proposed a different group which includes the Lorentz group as a subgroup. This is the so-called conformal group which is 15-dimensional, and includes a candidate $x \in \mathfrak{u}_{\mathbb{C}}$(conformal Lie algebra) for the mass operator, i.e., dU(x) allows isolated components in the spectrum for irreducible unitary representations U.

We finally mention that model theory, and spectral theory, within "pure" operator theory have recently been proposed in this setting by the author, see, e.g., [Jo 29].

If $\mathfrak{g}$ is Abelian, then dU(x+iy), $x,y \in \mathfrak{g}$, is essentially normal on the Gårding space, and all essentially normal operators may be obtained this way. Essential normality may be expressed this way:

$$\overline{dU(x+iy)} = -dU(x-iy)^{*}. \tag{6.1}$$

It is understood that the domain of dU(x+iy) is the Gårding space.

We show in Section 6.3 that (6.1) also holds when $\mathfrak{g}$ is non-Abelian, and that, for many Lie algebras, (6.1) is characteristic for integrability of a representation. We also discuss other properties that the operator dU(x+iy) shares with the essentially normal operators.

Finally, we give in [Jo 29] a condition for an operator of the form $\rho(x+iy)$ to be *subnormal*, i.e., to be the restriction to $\mathcal{D}(\rho)$ of a normal operator, or a formally normal one, in a generally bigger Hilbert space containing the representation space of ρ.

6.1 Central Elements

Let U be a unitary representation of a Lie group G in a Hilbert space $\mathcal{H}$ and let dU be the corresponding representation of the universal enveloping algebra $\mathfrak{U}_{\mathbb{C}}(\mathfrak{g})$ where $\mathfrak{g}$ is the Lie algebra of G. The domain of this representation $\mathcal{D}(dU)$ is the Gårding space, and we have seen that the Gårding space agrees with the space of C^{∞}-vectors. The reader is referred to Section 5.3 for further details on dU.

Before starting, we recall a few definitions:

A unitary representation U of a group G is said to be *irreducible* if it has no nontrivial closed invariant subspaces, the trivial subspaces being the extreme ones, zero and the whole space. If ω is in the center of $\mathfrak{U}_{\mathbb{C}}(\mathfrak{g})$ and if U is irreducible, then it follows that $dU(\omega)$ is a scalar times the identity operator. (This is immediate from Schur's lemma.) Since $dU(\omega)$ is closable, $\Omega = \overline{dU(\omega)}$ is well defined. The two closed subspaces $N(\Omega)$ and $\overline{R(\Omega)}$ are invariant. Consider the decomposition

$$\omega = \omega_1 + i\omega_2$$

where $\omega_1 = \frac{1}{2}(\omega+\omega^*)$, $\omega_2 = (2i)^{-1}(\omega-\omega^*)$. Then the two operators $dU(\omega_1)$ and $dU(\omega_2)$ commute strongly by the result in Section 3.9. It follows that Ω is normal.

Let

$$\Omega = \int \lambda \, dP(\lambda) \tag{6.1.1}$$

be the spectral resolution. We may restrict to the case $N(\Omega) = \{0\}$ and $\overline{R(\Omega)} = \mathcal{H}$. If there is more than one point in the spectrum of Ω, then the spectral resolution produces nontrivial invariant subspaces. (The case $N(\Omega) = \{0\}$, $\overline{R(\Omega)} = \mathcal{H}$ occurs only when $\mathcal{H}$ is one-dimensional.)

We now turn to the main case when U is not assumed to be irreducible. Two typical examples are the *left-regular representation* L, and the *right-regular representation* R. They are defined as follows: Let $\mathcal{H}_L$, resp., $\mathcal{H}_R$, be the Hilbert space of square integrable functions on G relative to "the" left-invariant, resp., right-invariant, Haar measure on G. Then

$$(6.1.2) \qquad (L(g)f)(g') = f(g^{-1}g'), \qquad f \in \mathcal{H}_L,\ g,g' \in G,$$

$$(6.1.3) \qquad (R(g)f)(g') = f(g'g), \qquad f \in \mathcal{H}_R,\ g,g' \in G.$$

We are ready to state the result of the present section:

Theorem 6.1.1 (I.E.Segal [Se 7]). Let U be a unitary representation of a Lie group G. Let $\mathfrak{g}$ be the Lie algebra, $\mathfrak{U}_{\mathbb{C}}(\mathfrak{g})$ the enveloping algebra, and $\mathfrak{Z}$ the center of $\mathfrak{U}_{\mathbb{C}}(\mathfrak{g})$. Then the operator $dU(\omega)$ is essentially normal on the Gårding space for all $\omega \in \mathfrak{Z}$.

Proof. We wish to invoke Theorem 3.7, and we show that the closed operator $\Omega := \overline{dU(\omega)}$ is affiliated with the center of the commutant $dU(\mathfrak{U}_{\mathbb{C}}(\mathfrak{g}))'$. Recall we showed in Theorem 5.3.1 that dU is a selfadjoint representation, so the commmutant is a von Neumann algebra by Lemma 5.2.3.

We base the proof on the following lemma which is of independent interest.

Lemma 6.1.2. Let G_0 denote the connected component of e in G, and let B be a bounded operator in the representation Hilbert space $\mathcal{H}$.

Then the following five conditions are equivalent:

(i) B commutes with the operators $\{U(g) : g \in G_0\}$.

(ii) B commutes with the operators $\{dU(x) : x \in \mathfrak{g}\}$.

(iii) B commutes weakly with the operators in (ii).

(iv) B commutes with the operators $\{dU(x) : x \in \mathfrak{U}_{\mathbb{C}}(\mathfrak{g})\}$.

(v) B commutes weakly with the operators in (iv).

Proof. The only nontrivial part of the lemma which is not already contained in the two results, Theorem 5.3.1 and Lemma 5.2.3, is the bi-implication (i) $\Leftrightarrow$ (ii) which we proceed to prove.

The first step is the familiar fact that the subgroup generated by the exponential mapping $\exp : \mathfrak{g} \longrightarrow G$ is identical to G_0, or equivalently G_0 is the smallest subgroup of G which contains $\{\exp x : x \in \mathfrak{g}\}$. The reader is referred to [Hel 1, Ch. II, Thm. 2.1] for a proof, or to any book on Lie groups. (We note that this property is special to finite-dimensional groups. It is known to fail, for example, for the infinite-dimensional Lie group of orientation preserving diffeomorphisms of the circle. This latter Lie group is known to be simple [Her] and it has as a Lie algebra the Lie algebra of all real polynomial vector fields on the circle.)

Recall that the Gårding space $\mathfrak{D}$ is spanned by vectors a of the form

$$a = \int_{G_0} \varphi(g)U(g)b\, dg \quad \text{(left-invariant Haar measure)},$$

where $\varphi \in C_c^\infty(G_0)$, and $b \in \mathcal{H}$. If $B \in \mathcal{B}(\mathcal{H})$ satisfies (i), then

$$Ba = \int_{G_0} \varphi(g)U(g)Bb\, dg$$

is again in $\mathfrak{D}$, and, for $x \in \mathfrak{g}$, we have

$$\begin{aligned} dU(x)Ba &= dU(x)\int_{G_0} \varphi(g)U(g)Bb\, dg \\ &= \int_{G_0} (dL(x)\varphi)(g)U(g)Bb\, dg \end{aligned}$$

$$= B\int_{G_0} (dL(x)\varphi)(g)U(g)b\ dg$$

$$= B\ dU(x)\int_{G_0} \varphi(g)U(g)b\ dg$$

$$= B\ dU(x)a$$

where formula (5.3.4) was used twice in the computation. Property (ii) of the lemma follows.

Now assume (ii). To show (i), it is enough to check the identity

$$U(g)Ba = BU(g)a \tag{6.1.4}$$

for $g = \exp x$, $x \in \mathfrak{g}$, and $a \in \mathfrak{D}$. We need only verify (6.1.4) for vectors a in a dense subspace since the operators are bounded, and we may restrict to $g \in \exp \mathfrak{g}$ since $\exp \mathfrak{g}$ generates G_0.

Let

$$F(t) = U(\exp tx)BU(\exp((1-t)x))a.$$

Then $F(0) = BU(\exp x)a$, and $F(1) = U(\exp x)Ba$. The function F is differentiable with continuous derivative. In fact, an application of the product rule yields

$$F'(t) = U(\exp tx)[dU(x),B]U(\exp(1-t)x))a$$
$$= 0.$$

We have used the known invariance of $\mathfrak{D}$ together with assumption (ii). Since

$$F(1) - F(0) = \int_0^1 F'(t)dt = 0,$$

(i) follows.

Proof of Theorem 6.1.1, continued. We now have two commutants $dU(\mathfrak{U}_{\mathbb{C}}(\mathfrak{g}))'$ and $\{U(g) : g \in G_0\}'$, and they coincide by virtue of the lemma. We claim that $\Omega = \overline{dU(\omega)}$ is affiliated with the center of this von Neumann algebra. The only part of the claim which has not already been proved is this: Ω commutes with every unitary operator in the double commutant

$$\{U(g) : g \in G_0\}''.$$

But, by von Neumann's theorem (3.7.4), the double commutant is the weak closure of $\{U(g) : g \in G_0\}$ so it is enough to check that Ω commutes with $U(g)$ for $g \in G_0$, and that follows from the implication (i) $\Leftarrow$ (iv) in Lemma 6.1.2.

Remark 6.1.3. The special case when $\omega = \omega^* \in \mathfrak{Z}(\mathfrak{U}_{\mathbb{C}}(\mathfrak{g}))$ is included in the theorem. This is really the result from Segal's paper [Se 8]. The conclusion, in this special case, states that $dU(\omega)$ is essentially selfadjoint on $\mathfrak{D}$.

In an earlier paper [Se 7], Segal had proved that the operators $dU(x)$, $x \in \mathfrak{g}$, are essentially skew-adjoint on $\mathfrak{D}$.

6.2 Second Order Elements

Nelson and Stinespring [NS] considered the element $\sum_{i=1}^{d} x_i^2$ where $\{x_i : 1 \leq i \leq d\}$ is a given basis for $\mathfrak{g}$. They noted that the operator $dL(\sum_1^d x_i^2)$ is an elliptic partial differential operator on G, and used this to show that, for any unitary representation U of G, the operator $dU(\sum x_i^2)$ is essentially selfadjoint on the Gårding space $\mathfrak{D}$. Moreover, if $x, y \in \mathfrak{U}_{\mathbb{C}}(\mathfrak{g})$ are given with x elliptic, and $y^*y = yy^*$, $xy = yx$, then the operator $dU(y)$ is essentially normal on $\mathfrak{D}$.

We shall prove a stronger result below. As for the result on formally normal operators in the commutant of an elliptic element, we will just note that the proof is similar to our proof above of Theorem 6.1.1, and it is also based on Theorem 3.7.

Theorem 6.2.1 (Jorgensen [Jo 1]). Let U be a unitary representation of a Lie group with Lie algebra $\mathfrak{g}$, and let

$y_1, \cdots, y_n$ be arbitrary elements in g, and let z be in the real linear span of the elements y_k. Then the operator

$$(6.2.1) \qquad dU\left(\sum_{k=1}^{n} y_k^2 + iz\right)$$

is essentially selfadjoint on the Gårding space.

Example 6.2.2 (Jorgensen [Jo 1]). The assumptions in the theorem cannot be relaxed in an obvious way. In fact there is a unitary representation U with $g = s\ell_2(\mathbb{R})$ and $y, z \in g$ such that

$$(6.2.2) \qquad dU(y^2 + iz)$$

is not essentially selfadjoint on the Gårding space $\mathcal{D}$.

The proof of Theorem 6.2.1 is very detailed in the original paper [Jo 1] and we shall only sketch the distinct ideas involved in the successive steps of the proof.

It will first be assumed that $z = 0$, and that the elements $y_1, \cdots, y_n$ *generate* the whole Lie algebra g. By this we mean that g is the real linear span of the set

(6.2.3) $\{y_1, \cdots, y_n, [y_i, y_j], [[y_i, y_j], y_k]$, and all possible iterated commutators formed from the y_i's$\}$.

We begin by considering just the left-regular representation L, cf. (6.1.2), of G on the Hilbert space $\mathcal{H}_L$ of all square integrable functions on G with respect to the left-invariant Haar measure on G. Hörmander showed that the operator $S := dL(\sum_{k=1}^{n} y_k^2)$ is *hypoelliptic* when the y_i's are assumed to generate g. An operator S is said to be hypoelliptic if it satisfies the familiar smoothing property: If u, v are two distributions on G and if $Su = v$ (where this is taken in the weak sense of distribution), then the solution u is smooth on any open set where v is smooth.

Now let U be an arbitrary unitary representation of G on a Hilbert space $\mathcal{H}$, and let $\mathcal{D}$ denote the Gårding space. For vectors $a \in \mathcal{D}$, we have

$$\begin{aligned}\langle dU(\sum_1^n y_k^2)a,a\rangle &= \sum_k \langle dU(y_k)^2 a,a\rangle \\ &= -\sum_k \langle dU(y_k)a, dU(y_k)a\rangle \\ &= -\sum_k \|dU(y_k)a\|^2 \leq 0.\end{aligned}$$

It follows that $dU(\sum_k y_k^2)$ is semibounded. To verify essential selfadjointness it is enough to check a single deficiency space condition, cf. [DS, vol.II, p. 1232, Thm. 19]. We must check that if $b \in \mathcal{H}$ satisfies

$$\langle a - \sum_k dU(y_k)^2 a, b\rangle = 0 \tag{6.2.4}$$

for all $a \in \mathcal{D}$, then $b = 0$. But we may substitute

$$a = \int_G \varphi(g)U(g)b\, dg, \qquad \varphi \in C_c^\infty(G), \tag{6.2.5}$$

into (6.2.4), and the following equation results:

$$\int_G (\varphi(g)-(S\varphi)(g))\langle U(g)b,b\rangle dg = 0$$

where $\varphi \in C_c^\infty(G)$, $S = dL(\sum_k y_k^2)$, and dg is the left-invariant Haar measure. It follows that the positive-definite function

$$p(g) := \langle U(g)b,b\rangle, \qquad g \in G,$$

is a weak (distribution) solution on G to the partial differential equation

$$Sp = p. \tag{6.2.6}$$

Since the operator S is hypoelliptic and p is continuous, it follows that p is C^∞ everywhere on G, and that (6.2.6) is satisfied in the strong pointwise sense, i.e., the appropriate derivatives of p may be evaluated in the usual sense of "limit of differences."

We shall need now two facts about positive definite C^∞-functions on Lie groups. Both may be viewed as *maximum principles*, and both are worked out in detail in [Jo 1]. The first is that $|p|$ attains its maximum at the origin e in G, i.e.,

(i) $$|p(g)| \leq p(e), \qquad g \in G,$$

and the second one states that

(ii) $$(Sp)(e) \leq 0.$$

When the two are combined, we get

$$|p(g)| \leq p(e) = (Sp)(e) \leq 0.$$

It follows that $p \equiv 0$. In particular, $p(e) = \langle U(e)b,b\rangle = \|b\|^2 = 0$. We have the desired conclusion $b = 0$.

We now drop the assumption that the elements $y_1,\cdots,y_n$ generate $\mathfrak{g}$. Let $\mathfrak{h}$ be the real linear span of the elements from (6.2.3); in other words, $\mathfrak{h}$ is the smallest Lie subalgebra of $\mathfrak{g}$ which contains the y_i's. Let $H \subset G$ be the group generated by $\{\exp x : x \in \mathfrak{h}\}$. Then H is a Lie group, and its Lie algebra is $\mathfrak{h}$.

Let $\mathfrak{D}_G$ be the Gårding space spanned by the vectors (6.2.5), and let $\mathfrak{D}_H$ be spanned by vectors a of the form

$$(6.2.7) \qquad a = \int_H \varphi(h)U(h)b\,dh$$

with $\varphi \in C_c^\infty(H)$, $b \in \mathcal{H}$. It is understood that the integral in (6.2.7) is with respect to the left-invariant Haar measure dh on the subgroup H. Since $U|_H$ is also strongly continuous, $\mathfrak{D}_H$ is dense in $\mathcal{H}$ and the argument from the first part of the proof shows that $\sum_{k=1}^{n} dU(y_k)^2$ is essentially selfadjoint on $\mathcal{D}_H$. While the right-invariant operator $dL(\sum y_k^2)$ is not hypoelliptic on G, it is on H since Hörmander's theorem [Hö 1, Thm. 1.1] applies to H. We have set up the assumptions so that the Lie algebra of all right-invariant vector fields on H is generated by the vector fields $dL(y_k)$, $1 \leq k \leq n$.

Let

$$T = \overline{dU(\sum_1^n y_k^2)\Big|_{\mathcal{D}_H}}.$$

We have proved that T is selfadjoint, and we must check that $\mathcal{D}_G$ is a core for T, i.e., that

$$T = \overline{T\big|_{\mathcal{D}_G}}. \tag{6.2.8}$$

Consider the bounded operators

$$U_0(\varphi) = \int_H \varphi(h)U(h)dh, \qquad \varphi \in C_c^\infty(H). \tag{6.2.9}$$

We shall need the following three facts:

(i) $$U_0(\varphi)\mathcal{D}_G \subset \mathcal{D}_G,$$

(ii) $$TU_0(\varphi) = U_0(dL(\sum_k y_k^2)\varphi),$$

(iii) $$U_0(\varphi)U(\psi) = U(\varphi *_H \psi), \quad \varphi \in C_c^\infty(H),\ \psi \in C_c^\infty(G),$$

with

$$(\varphi *_H \psi)(g) = \int_H \varphi(h)\psi(h^{-1}g)dh \in C_c^\infty(G) \tag{6.2.10}$$

and

$$U(\psi) = \int_G \psi(g)U(g)dg.$$

The reader is referred to [Jo 1] for details of proof. Suffice it to say that (iii) $\Rightarrow$ (i), in view of (6.2.10).

We shall consider the domain $D(T)$ relative to the graph

norm

$$a \longrightarrow \|a\| + \|Ta\|, \tag{6.2.11}$$

and we shall show that every $a \in D(T)$ may be approximated with a net of vectors from $\mathcal{D}_G$ relative to the graph norm (6.2.11). If $a \in D(T)$ is given we may pick a Dirac delta approximation of functions $\varphi \in C_c^\infty(H)$ such that

$$U_0(\varphi)a \longrightarrow a \qquad \text{(convergence)}$$

in the graph-norm.

If $\psi \in C_c^\infty(G)$, then we also have the approximation

$$\lim_{\varphi} U_0(\varphi)U(\psi)a = U(\psi)a,$$

relative to the graph-norm.

Using (ii) and (iii), we have

$$\begin{aligned} TU_0(\varphi)U(\psi)a &= U((dL(\textstyle\sum y_k^2)\varphi) *_H \psi)a \\ &= U_0(dL(\textstyle\sum y_k^2)\varphi)U(\psi)a. \end{aligned}$$

Now fix φ, and let ψ run over a delta approximation in $C_c^\infty(G)$. Then

$$\lim_{\psi} U_0(\varphi)U(\psi)a = U_0(\varphi)a$$

with the limit being relative to the graph-norm approximation.

But the vectors $U_0(\varphi)U(\psi)a$ are in $\mathcal{D}_G$ by virtue of (i). This concludes the proof that $\mathcal{D}_G$ is a core, i.e., that (6.2.8) holds.

If the Lie algebra element z from (6.2.1) of the theorem is nonzero, we apply the following perturbation argument to get the essential selfadjointness.

We consider the two selfadjoint operators

$$T = \overline{dU(\textstyle\sum y_k^2)}$$

and

$$T_z = \overline{dU(iz)}$$

where $^-$ refers to graph closure of the respective operators on the Gårding space.

Using the assumption

$$z \in \operatorname{span}_{\mathbb{R}}\{y_k\} \tag{6.2.12}$$

we now show that T_z is infinitesimally small relative to T. Once this is established the conclusion follows from a familiar theorem of Kato [Kat, Ch. V, Thm. 4.3]. More specifically, for all $\epsilon > 0$ there is a constant C_ϵ such that the estimate

$$\|T_z a\| \leq \epsilon\|Ta\| + C_\epsilon\|a\|, \qquad a \in \mathfrak{D}, \tag{6.2.13}$$

holds.

Suppose $z = \sum \alpha_k y_k$ with $\alpha_k \in \mathbb{R}$. Then we have the following estimate valid for arbitrary pairs of vectors a,b in $\mathfrak{D}$:

$$
\begin{aligned}
|\langle T_z a, b\rangle| &= \Big|\sum_k \alpha_k \langle dU(y_k)a, b\rangle\Big| \leq \|b\|\sum_k |\alpha_k|\,\|dU(y_k)a\| \\
&\leq \|b\|\Big(\sum_k \alpha_k^2\Big)^{\frac{1}{2}}\Big(\sum_k \|dU(y_k)a\|^2\Big)^{\frac{1}{2}} \\
&= A\|b\|\langle -Ta, a\rangle^{\frac{1}{2}}
\end{aligned}
$$

where

$$A = \Big(\sum \alpha_k^2\Big)^{\frac{1}{2}}.$$

It follows that, if $\epsilon \in \mathbb{R}_+$ is given, then

$$
\begin{aligned}
\|T_z a\| &\leq A\langle -Ta, a\rangle^{\frac{1}{2}} \\
&\leq \epsilon\|Ta\| + A^2\epsilon^{-1}\|a\|
\end{aligned}
$$

which is the desired estimate, $C_\epsilon = \dfrac{A^2}{\epsilon}$.

Open Problem. Rather than assuming the condition (6.2.12) in Theorem 6.2.1, one might assume just the weaker condition that z is in the span of the larger set of elements (6.2.3) including also all commutators, and iterated commutators, formed from the given set of elements $\{y_i\}$ in g. It is not known whether or not $dU(\sum y_k^2 + iz)$ is still essentially self-adjoint on $\mathcal{D}$.

We now turn to the details of

Example 6.2.2. First recall the Schrödinger representation U_λ given by formula (4.2.5). The Hilbert space $\mathcal{H}$ is the space of all square integrable functions on $\mathbb{R}$. Let y_1, y_2, y_3 be a basis for the Heisenberg Lie algebra of upper triangluar real 3×3 matrices with zero in the diagonal (and with 1 and 0 in the appropriate entries) so that

$$[y_1, y_2] = y_3, \quad 0 = [y_1, y_3] = [y_2, y_3]. \tag{6.2.14}$$

It is known that the Schwartz-space of rapidly decreasing smooth functions on $\mathbb{R}$ is precisely equal to the space of C^∞-vectors, alias Gårding vectors $\mathcal{D}$, for U_λ. We have

$$\begin{cases} dU_\lambda(y_1)f(t) = \frac{d}{dt} f(t) \\ dU_\lambda(y_2)f(t) = \lambda i t f(t) \\ dU_\lambda(y_3)f(t) = \lambda i f(t), \qquad t \in \mathbb{R}, \quad f \in \mathcal{D}. \end{cases} \tag{6.2.15}$$

The two operators

$$P = -i \frac{d}{dt}, \quad Q = \text{multiplication by } t,$$

are called respectively the Quantum Mechanical momentum and position operators.

Example 6.2.2 will be constructed from the Shale-Weil-Segal [Sha, Ve 4] harmonic representation. We recall the construction of this representation in terms of the P-Q operators.

Let $\{x_k : k = 0,1,2\}$ be the basis (5.4.9) for $s\ell_2(\mathbb{R})$. It is easy to check that a Hermitian representation ρ is specified by:

$$(6.2.16) \qquad \begin{cases} \rho(x_0) := \frac{i}{4}(P^2+Q^2) \\ \rho(x_1) := \frac{i}{4}(PQ+QP). \\ \rho(x_2) := \frac{i}{4}(P^2-Q^2) \end{cases}$$

Since $\frac{1}{2}(P^2+Q^2-I)$ is the harmonic oscillator Hamiltonian, it is also called the oscillator representation. The Hermite functions on $\mathbb{R}$ diagonalize the oscillator Hamiltonian. Let $\{h_n(t) : n = 0,1,2,\cdots\}$ denote the normalized Hermite functions:

$$(6.2.17) \qquad h_n(t) = (\sqrt{\pi}\, 2^n n!)^{-\frac{1}{2}} e^{-\frac{t^2}{2}} H_n(t),$$

$$H_n(t) = (-1)^n e^{t^2} (\tfrac{d}{dt})^n e^{-t^2},$$

satisfying:

$$(6.2.18) \qquad \tfrac{1}{2}(P^2+Q^2-I)h_n = nh_n.$$

Using formulas (6.2.16) and (6.2.18), we get for the Casimir operator

$$\Omega = \rho(x_0^2-x_1^2-x_2^2)$$

$$\Omega h_n = \frac{3}{16} h_n.$$

Hence the Casimir operator is a scalar, in fact, the scalar q is $\frac{3}{16}$.

Theorem 5.4.1 now implies integrability of the representation ρ. In this case Nelson's theorem [Ne 3, Thm. 5] may also be applied directly since a direct verification yields:

$$\rho(\Delta)h_n = -(\tfrac{3}{16} + 2(\tfrac{1}{4} + \tfrac{n}{2})^2)h_n;$$

so the Hermite functions are analytic vectors.

The representation is *not* irreducible. The *two* closed subspaces spanned by the Hermite functions h_n with n even, respectively, n odd, are invariant.

The integrated unitary representation is not a representation of $SL_2(\mathbb{R})$, but rather of a two-sheeted covering, the so-called *metaplectic group*.

To see this, let G denote the simply connected covering group, and let U be the unitary representation of G satisfying $dU = \rho$. Finally let $\exp_G : s\ell_2(\mathbb{R}) \longrightarrow G$ be the corresponding exponential mapping, and let

$$\varphi : G \longrightarrow SL_2(\mathbb{R})$$

be the covering mapping (from "loops" to points). U is not a representation of $SL_2(\mathbb{R})$ because it is not trivial on the kernel of φ as follows from below:

Let $m \in \mathbb{Z}$. A calculation yields

$$U(\exp_G(4\pi m x_0)) = e^{i\pi m} I.$$

It follows that the discrete subgroup

$$\Gamma := \{\exp_G(4\pi m x_0) : m \in 2\mathbb{Z}\}$$

is normal in G, and U passes to a representation of the quotient group G/Γ. Since $\pi_1(G/\Gamma) \simeq \mathbb{Z}/2\mathbb{Z} = \mathbb{Z}_2$, the group G/Γ is a double-covering of $SL_2(\mathbb{R})$, and is a concrete realization of the metaplectic group.

We now consider the two elements y, z in $s\ell_2(\mathbb{R})$ given as follows:

$$y = 2(x_0 - x_2)$$

$$z = 2(x_0 + x_2)$$

Then, by virtue of formulas (6.2.16), we get

$$dU(y) = iQ^2$$

$$dU(z) = iP^2$$

and

$$dU(y^2 + iz) = -\left(t^4 + \frac{d^2}{dt^2}\right) \tag{6.2.19}$$

where U is the oscillator representation.

The operator (6.2.19) is again a Hamiltonian with a "very" repulsive potential. The "classical particle" shoots to $\pm\infty$ in finite time, since the integral $\int_{-\infty}^{\infty} v^{-1}\,dx = \int_{-\infty}^{\infty} (E + x^4)^{-\frac{1}{2}}\,dx$ is finite for $E > 0$. The singular Quantum Mechanical Hamiltonian is

$$-\frac{d^2}{dx^2} - x^4,$$

and the corresponding classical energy is

$$E = (\dot{x})^2 - x^4$$

for the trajectory $\{x(t) : t = \text{time}\}$.

The rigorous calculation of the deficiency indices of the operator (6.2.19) are sketched in [Jo 1, p. 107] with further details in [Win].

6.3 The Element $x+iy$

The subject of this section is the following:

Theorem 6.3.1. Let U be a unitary representation of a Lie group G, and let $\mathfrak{g}$ be the Lie algebra of G. Let dU be the derived representation of the universal enveloping algebra $\mathfrak{U}_{\mathbb{C}}(\mathfrak{g})$ with the Gårding space $\mathfrak{D}$ as domain.

Then we have

$$dU(x+iy)^* = -\overline{dU(x-iy)} \tag{6.3.1}$$

for all $x,y \in \mathfrak{g}$.

Before starting the proof we state a second result which is in fact equivalent to the first one. Both results have their origin in the study of operators associated with representations.

For certain representations U, and for certain elements $x,y \in \mathfrak{g}$, the operator $dU(x+iy)$ is a raising operator in the familiar mathematical-physics formalism. As an example, the operators $A_\pm$ stated in Section 5.5, cf. formulas (5.5.1), are of this form.

Our second result is about the following second degree element in $\mathfrak{U}_{\mathbb{C}}(\mathfrak{g})$:

$$x^2 + y^2 + i[x,y]. \tag{6.3.2}$$

While this element is similar to the one studied in Theorem 6.2.1, there is an important distinction: The first order part of (6.3.2) is $[x,y]$ and this commutator is generally not in the linear span of the two elements x,y.

We have the following:

Lemma 6.3.2. The following two conditions are equivalent:

(i) $dU(x+iy)^* = \overline{-dU(x-iy)}$

and

(ii) The operator $dU(x^2+y^2-i[x,y])$ is essentially self-adjoint on $\mathcal{D}$.

Proof of Lemma 6.3.2. We have $-(x+iy)(x+iy)^* = x^2+y^2-i[x,y]$ and it follows that the operator $dU(x^2+y^2-i[x,y])$ is semi-bounded. Specifically,

$$\langle dU(x^2+y^2-i[x,y])a,a\rangle \leq 0, \qquad a \in \mathcal{D}.$$

Hence (ii) is equivalent, by [DS; vol. II, p. 1232, Th. 19] or [RS; X.1], to the one-sided deficiency-index-condition: The equation

$$dU(x^2+y^2-i[x,y])^*b = b \tag{6.3.3}$$

has no nonzero solution b in $\mathcal{H}$.

The inclusion

$$-dU(x-iy) \subset dU(x+iy)^*$$

is trivially satisfied since

$$(x+iy)^* = -(x-iy) \qquad \text{for } x,y \in \mathfrak{g}.$$

Hence, condition (i) is just stating that $\mathcal{D}$ is a *core* for the closed operator $dU(x+iy)^*$, which is to say: $\mathcal{D}$ is dense in

$D(dU(x+iy)^*)$ relative to the graph norm defined by $dU(x+iy)^*$. We must show that if $b \in D(dU(x+iy)^*)$ is orthogonal to $\mathfrak{D}$ in the graph-inner product, then b is necessarily zero. But when the graph-inner product is written out, the orthogonality amounts to the identity

$$\langle a-dU(x^2+y^2-i[x,y])a,b\rangle = 0, \qquad a \in \mathfrak{D},$$

and this is the statement that b is a solution to (6.3.3). Hence, (i) and (ii) are equivalent. It follows that Theorem 6.3.1 may be regarded as an extension (in one direction) of Theorem 6.2.1.

But the methods of proof which were used in 6.2.1 break down manifestly when applied to Theorem 6.3.1. This is because there are many examples of Lie groups, and pairs x,y in the Lie algebra, where the right invariant partial differential operator

$$dL(x^2+y^2 \pm i[x,y]) \tag{6.3.4}$$

is *not* hypoelliptic. Such examples have been recorded recently in, e.g., [Mel 1] and [Jo 25].

The following corollary results when 6.3.1, 6.3.2, and 6.2.1 are combined.

Corollary 6.3.3. If $[x,y] = 0$, then the operator $dU(x+iy)$ is essentially normal on $\mathfrak{D}$.

Proof. By 6.3.2, the conclusion in 6.3.3 is equivalent to the essential selfadjointness of the operator $dU(x^2+y^2)$ which, in turn, is the conclusion of 6.2.1.

We refer to [Jo 25, Section 3] for the proof of Theorem 6.3.1. Instead we turn to the details of the converse problem. We present a class of Hermitian representations ρ of certain Lie algebras $\mathfrak{g}$ such that the assumption

$$\rho(x+iy)^* = -\overline{\rho(x-iy)}, \qquad x,y \in \mathfrak{g}, \tag{6.3.5}$$

implies integrability.

As a corollary we get a result on integrability of *Heisenberg's canonical commutation relations.*

Let $\mathcal{H}$ be a Hilbert space, and $\mathcal{D}$ a dense subspace. Let P,Q be a pair of Hermitian operators defined on $\mathcal{D}$ and leaving $\mathcal{D}$ invariant. If

$$PQa - QPa = -ia, \qquad a \in \mathcal{D}, \tag{6.3.6}$$

then we say that the pair satisfies the canonical commutation relations.

Let

$$A_{\pm} = \frac{1}{\sqrt{2}}(Q \pm iP). \tag{6.3.7}$$

Then the pair P,Q generate a Hermitian representation of the Heisenberg Lie algebra $\mathfrak{h}_3$ on generators y_1, y_2, y_3 satisfying:

(i) $$[y_1, y_2] = y_3,$$

(ii) y_3 is central.

If we define:

$$\begin{cases} \rho(y_1) = iP \\ \rho(y_2) = iQ , \\ \rho(y_3) = iI \end{cases} \tag{6.3.8}$$

then ρ extends to a Hermitian representation of $\mathfrak{U}_{\mathbb{C}}(\mathfrak{h}_3)$, and a special case of (6.3.5) amounts to the condition

$$A_-^* = \overline{A_+}, \tag{6.3.9}$$

or equivalently, to the statement that $\mathcal{D}$ is a core for the operator A_-^*.

The two operators $A_{\pm}$ are called raising, resp., lowering, operators and we have

$$[A_+, A_-] = I. \tag{6.3.10}$$

The precise meaning of (6.3.10) is this:

$$A_+A_-a - A_-A_+a = a, \qquad a \in \mathcal{D}.$$

Hence, (6.3.10) is a reformulation of the canonical commutation relations.

We shall prove that the representation ρ is exact if and only if (6.3.9) holds.

In view of Theorem 6.3.1 above, the following criterion for integrability of a Hermitian representation is both sufficient and necessary. We shall only consider the following two distinct cases for the Lie algebra $\mathfrak{g}$.

Case 1: $\mathfrak{g}$ is isomorphic to the 3-dimensional real Heisenberg Lie algebra $\mathfrak{h}_3$.

Case 2: $\mathfrak{g} \simeq s\ell_2(\mathbb{R})$.

In case 1, the center of $\mathfrak{g}$ is one-dimensional. We shall always denote by z a nonzero element in the center of $\mathfrak{h}_3$.

In case 2, $\mathfrak{g} \simeq s\ell_2(\mathbb{R})$, there is no nonzero element in the center of $\mathfrak{g}$, but we have the Casimir element in $\mathfrak{U}_{\mathbb{C}}(\mathfrak{g})$. The Casimir element may be constructed directly from the Killing form of $\mathfrak{g}$, and it is independent of the choice of basis in $\mathfrak{g}$. The Casimir element will be denoted ω.

Theorem 6.3.4. Let ρ be a Hermitian representation of $\mathfrak{g}$ where $\mathfrak{g} \simeq \mathfrak{h}_3$, or $\mathfrak{g} \simeq s\ell_2(\mathbb{R})$. If $\mathfrak{g} \simeq \mathfrak{h}_3$, we assume that $\rho(iz)$ is essentially selfadjoint, and if $\mathfrak{g} \simeq s\ell_2(\mathbb{R})$, we assume that $\rho(\omega)$ is essentially selfadjoint.

If the additional assumption

$$\rho(x+iy)^* = -\overline{\rho(x-iy)} \tag{6.3.11}$$

is satisifed for $x,y \in \mathfrak{g}$, then it follows that ρ is exact.

Remarks. $\mathfrak{h}_3$ is a central extension of a 2-dimensional Abelian Lie algebra V. If y_1,y_2,y_3 is a basis for $\mathfrak{h}_3$ such that $[y_1,y_2] = y_3$ is nonzero and central, then we may take $V = \text{span}(y_1,y_2)$. It is then enough to assume that condition

(6.3.5) in the theorem is satisfied for $x, y \in V$.

Similarly, consider the *Cartan decomposition* $\mathfrak{g} = \mathfrak{k} + \mathfrak{p}$ of $\mathfrak{g} \simeq s\ell_2(\mathbb{R})$. Then it is enough to have (6.3.5) satisfied for pairs of elements in $\mathfrak{p}$.

We recall the *Cartan decomposition* in the general case. Let $\mathfrak{g}$ be a Lie algebra over $\mathbb{R}$, and let B denote the Killing form. It is defined as:

$$B(x,y) = \text{trace}(\text{ad}x\ \text{ad}y), \qquad x, y \in \mathfrak{g},$$

It is known that $\mathfrak{g}$ is *semisimple* if and only if B is non-degenerate. In that case, we may choose a basis $s_1, \cdots, s_n$, $t_1, \cdots, t_m$ for $\mathfrak{g}$ such that

$$(6.3.12) \qquad B(s_i, s_j) = -\delta_{ij}, \qquad B(t_i, t_j) = \delta_{ij}.$$

Let $\mathfrak{k} = \text{span}_{\mathbb{R}}\{s_i\}$ and $\mathfrak{p} = \text{span}_{\mathbb{R}}\{t_j\}$. Then

$$[\mathfrak{k}, \mathfrak{k}] \subset \mathfrak{k},$$

$$[\mathfrak{k}, \mathfrak{p}] \subset \mathfrak{p},$$

and

$$[\mathfrak{p}, \mathfrak{p}] \subset \mathfrak{k}.$$

Moreover, $\mathfrak{k}$ and $\mathfrak{p}$ are orthogonal relative to B. This follows from invariance of B under inner automorphisms:

$$B((\text{ad } x)(y), z) + B(y, (\text{ad } x)z) = 0.$$

If a basis $t_1, \cdots, t_m$, $s_1, \cdots, s_n$ has been chosen satisfying the bi-orthogonal relations (6.3.12), then the element

$$(6.3.13) \qquad \omega = \sum_{k=1}^{m} t_j^2 - \sum_{k=1}^{n} s_k^2$$

is central in $\mathfrak{U}_{\mathbb{C}}(\mathfrak{g})$, as can be checked using basic facts on bilinear forms, see, e.g., [Ser; LA 6.5]. Moreover, ω is independent on the choice of basis. It is called the Casimir element.

With the basis x_0, x_1, x_2, given by (5.4.9), the Cartan decomposition of $s\ell_2(\mathbb{R})$ is:

$$\mathfrak{k} = \mathbb{R}x_0, \qquad \mathfrak{p} = \mathrm{span}_{\mathbb{R}}(x_1, x_2).$$

We will prove Theorem 6.3.4 by a reduction to the case where the respective operator $\rho(iz)$, resp., $\rho(\omega)$, is a scalar times the identity operator. This reduction can be carried out along the lines which are sketched in Section 5.4.

We will be using a direct integral based on the spectral resolution of the two respective central operators. In fact, the reduction is a bit simpler in the present argument, where the only property of $\rho(iz)$, resp., $\rho(\omega)$, which is used is boundedness.

The proof of Theorem 6.3.4 is based on the following basic lemma on normed linear spaces.

Before stating the lemma, some convenient notation is introduced:

Let $\mathfrak{D}$ be a normed linear space. We shall consider finite subsets of $\mathrm{End}(\mathfrak{D})$ = the ring of linear endomorphisms of $\mathfrak{D}$. We shall denote by $\mathcal{O}(\mathfrak{D})$ the free Abelian semigroup generated by symbols $|X|$, $X \in \mathrm{End}(\mathfrak{D})$. Let $\xi = |X_1|+\cdots+|X_\ell|$ and $\eta = |Y_1|+\cdots+|Y_m|$ be elements in $\mathcal{O}(\mathfrak{D})$. We make the definitions:

$$\xi\eta := \sum_{ij} |X_i Y_i|,$$

$$\mathrm{ad}\xi(\eta) := \sum_{ij} |X_i Y_j - Y_j X_i|,$$

$$\|\xi a\| = \sum_i \|X_i a\|, \qquad a \in \mathfrak{D},$$

and we define $\xi \leq \eta$ to mean

$$\|\xi a\| \leq \|\eta a\|, \qquad a \in \mathfrak{D}.$$

A vector a is said to be *analytic* for ξ if the power series

$$\sum_n t^n/n! \, \|\xi^n a\|$$

has a positive radius of convergence.

We shall also need the following simple fact from [Ne 3]:

$$\eta^n\xi \le \sum_{k=0}^{n} \begin{bmatrix} n \\ k \end{bmatrix} (\mathrm{ad}\eta)^k(\xi)\eta^{n-k}. \tag{6.3.14}$$

We note that a special case of (6.3.14) has already been used in the proof of Theorem 5.5.1 in Section 5.5.

Lemma 6.3.5 (F.M. Goodman and P.E.T. Jorgensen [GJ 2, Thm. 2.1]). Let $\mathcal{D}$ be a linear space, and $\xi,\eta \in \mathcal{O}(\mathcal{D})$. Suppose there are constants $c,p > 0$ such that

$$\xi \le c\eta \tag{6.3.15}$$

and

$$(\mathrm{ad}\eta)^n(\xi) \le cp^n\eta, \qquad n \in \mathbb{N}. \tag{6.3.16}$$

Then it follows that

$$\xi^n \le \pi_n(\eta), \qquad n \in \mathbb{N},$$

where

$$\pi_n(\eta) := c^n\eta(\eta+p)\cdots(\eta+(n-1)p).$$

In particular, we note that every analytic vector for η is also analytic for ξ. For, if there are constants M_0, M such that

$$\|\eta^n a\| \le M_0 M^n\, n!, \qquad n \in \mathbb{N},$$

then it follows that

$$\|\xi^n a\| \le \|\pi_n(\eta)a\| \le M_0(M+p)^n c^n n!, \qquad n \in \mathbb{N}.$$

We refer to [GJ 2] for the proof of the lemma.

Proof of Theorem 6.3.4. To prove exactness of the given representation ρ, we may pick the two elements x,y in the Lie algebra $\mathfrak{g}$ such that the commutator $z = [x,y]$ is nonzero and the triple x,y,z is a basis for $\mathfrak{g}$. For the $\mathfrak{h}_3$ case, any nonzero $z = [x,y]$ will work. For the case $\mathfrak{g} \simeq s\ell_2(\mathbb{R})$, we may pick the three elements such that $z = [x,y]$ and x,y,z is a bi-orthogonal system, i.e.,

$$B(x,x) = B(y,y) = 1, \qquad B(z,z) = -1,$$

and the different elements are mutually orthogonal relative to the Killing form B. Then

$$\omega = x^2+y^2-z^2 \tag{6.3.17}$$

will be central.

We define

$$v_+ = (x+iy)(x-iy) = x^2+y^2-iz$$

$$v_- = (x-iy)(x+iy) = x^2+y^2+iz$$

and note that

$$[v_+,v_-] = 2i[x^2+y^2,z]. \tag{6.3.18}$$

But if the element ω is central, then

$$[v_+,v_-] = 2i[\omega,z] = 0.$$

So, in both of the cases which are listed in the theorem, we have the commutativity relations

$$[v_+,v_-] = [v_\pm,z] = 0,$$

and furthermore,

$$v_+-v_- = -2iz.$$

We have additional commutation relations for the two elements $a_\pm$ defined by:

$$a_\pm = x \pm iy.$$

They are

$$[a_+,a_-] = -2iz,$$

$$[v_+,a_+] = 2ia_+z,$$

$$[v_-,a_-] = -2ia_-z$$

$$[v_+,a_-] = -2iza_-$$

and

$$[v_-,a_+] = 2iza_+.$$

In case $\mathfrak{g} \simeq \mathfrak{h}_3$, and $\rho(iz) = qI$, $q \in \mathbb{R}$, we get for the four operators:

$$A_{\pm} = \rho(a_{\pm}),$$

$$V_{\pm} = \rho(v_{\pm}),$$

the following commutator-formulas:

$$[V_{\pm}, A_{+}] = 2qA_{+},$$

$$[V_{\pm}, A_{-}] = -2qA_{-}.$$

Then the assumptions of Lemma 6.3.5 are satisfied for the following choice of ξ, η:

$$(6.3.19) \qquad \begin{cases} \xi = |A_{+}|+|A_{-}|+|V_{+}|+|V_{-}| \\ \eta = |V_{+}| + |I| \qquad \text{(case 1)} \\ \eta = |Z| + |I| \qquad \text{(case 2)}. \end{cases}$$

In the case $\mathfrak{g} \simeq s\ell_2(\mathbb{R})$, the η is the same, but ξ will have to include more terms: Since the operator

$$Z = \rho(iz)$$

is not a scalar in this case, it will be included in ξ:

$$\xi = |A_{+}|+|A_{-}|+|V_{+}|+|V_{-}|+|Z|.$$

The verification of the *a priori* estimates which are implicit in the two assumptions in Lemma 6.3.5 is a computation using only the commutator formulas above. It is contained in [Jo 25] and we shall restrict attention here to the analytic side of the proof.

The estimates which result involve only vectors in the domain $\mathcal{D}(\rho)$ of the representation, and the assumption (6.3.11) of Theorem 6.3.4 has not yet been used. Consider now the two operators $A_{\pm}$ with domain $\mathcal{D}(\rho)$. Since $\overline{A_{-}}$ is a closed operator, $A_{-}^{*}\overline{A_{-}}$ is selfadjoint by Theorem 3.4.1. But $A_{-}^{*} = -\overline{A_{+}}$ by (6.3.11), so it follows that $\overline{A_{+}}\,\overline{A_{-}}$ is a selfadjoint operator. It is an extension of the Hermitian operator V_{+} on $\mathcal{D}(\rho)$. Let $M = \overline{A_{+}}\,\overline{A_{-}}$ and $N = \overline{A_{-}}\,\overline{A_{+}}$. We will show that N is also selfadjoint, and

$$D(M) = D(N).$$

The second part of the proof is the verification that Lemma 6.3.5 applies also to the larger domain $D_\infty(M) := \bigcap_{n=1}^{\infty} D(M^n)$ where the sets ξ, η in $\mathcal{O}(D_\infty(M))$ will now be taken to be:

$$(6.3.20)\qquad \begin{cases} \xi = |\overline{A}_+|+|\overline{A}_-|+|M|+|N|+|\overline{Z}| \\ \eta = |M| + |I| \qquad\qquad \text{(case 1)}. \\ \eta = |\overline{Z}| + |I| \qquad\qquad \text{(case 2)} \end{cases}$$

Let $D_\omega(M)$ be the analytic vectors for M. We have $D_\omega(M) \subset D_\infty(M)$, and $D_\omega(M)$ is dense in $\mathcal{H}$ since M is selfadjoint. But Lemma 6.3.5, applied to $D_\infty(M)$, implies that vectors in $D_\omega(M)$ are also analytic for the individual operators $\overline{A}_+, \overline{A}_-$. The integrability follows readily from this. In case 2, we shall use the space $D_\infty(\overline{Z})$, and $D_\omega(\overline{Z})$ in a similar construction.

We now turn to the project of extending the operators from ξ to be defined on the larger domain $D^\infty(M)$, (case 1), resp., $D_\infty(\overline{Z})$, (case 2), with the extended operators leaving invariant $D_\infty(M)$, resp., $D_\infty(\overline{Z})$. This way we will get a new representation ρ_∞ of the same $\mathfrak{g}$, and $\rho \subset \rho_\infty$ in the sense of inclusion of representations as discussed in Section 5.1. The extension is "extension by closure", and when the construction is completed we will have $\mathfrak{D}(\rho_\infty) = D_\infty(M)$, as well as $\rho_\infty(x) = \rho^*(x)\big|_{D_\infty(M)}$ where $\rho^*(x) = \rho(x^*)^*$, $x \in \mathfrak{U}_{\mathbb{C}}(\mathfrak{g})$. Since $\overline{A}_+ = -A_-^*$ by assumption, we begin by proving that A_-^* is defined on $D(M^n)$, and maps $D(M^n)$ into $D(M^{n-1})$ for all $n = 1,2,\cdots$. (We adopt the definitions $M^0 = I$, and $D(M^0) = \mathcal{H}$.)

Before starting the technical points, we note that this "extension by closure" has been used in related algebraic contexts recently, see, e.g., [JM, Chapter 7] and [BGJ, Theorem 5.10].

Note first that, if $\overline{A}_+ = -A_-^*$ is assumed, then the twin condition $\overline{A}_- = -A_+^*$ is a consequence, since

$$\overline{A}_- = A_-^{**} = -(\overline{A}_+)^* = -A_+^*.$$

Hence, both of the operators

$$M = \overline{A}_+\overline{A}_-$$

and

$$N = \overline{A}_-\overline{A}_+$$

are selfadjoint by the above argument.

We have the following stronger conclusion:

Lemma 6.3.6. If $\overline{A}_+ = -A_-^*$, then the operator

$$\mathcal{A} = \begin{bmatrix} 0 & \overline{A}_- \\ \overline{A}_+ & 0 \end{bmatrix}$$

is skew-adjoint in $\mathcal{H} \times \mathcal{H}$ with domain $D(\overline{A}_+) \times D(\overline{A}_-)$.

Proof. Since $\mathcal{A}$ is skew-Hermitian on $\mathcal{D}(\rho) \times \mathcal{D}(\rho)$, it is enough to show that the two equations, $\mathcal{A}^*b = \pm b$, do not have any nonzero solutions b in $\mathcal{H} \times \mathcal{H}$. We do the details only for the first equation, since the seond one is similar.

The first equation is equivalent to the system

$$\begin{cases} A_-^* b_1 = b_2 \\ A_+^* b_2 = b_1 \end{cases}.$$

But $A_-^* = -\overline{A}_+$, and $A_+^* = -\overline{A}_-$, so we get

$$\begin{cases} \overline{A}_+ b_1 = -b_2 \\ \overline{A}_- b_2 = -b_1 \end{cases}.$$

It follows that $b_1 \in D(N)$, and $Nb_1 = b_1$. Similarly, we conclude that $b_2 \in D(M)$, and $Mb_2 = b_2$.

But

$$\langle Nb_1, b_1\rangle = -\|\bar{A}_+ b_1\|^2,$$

and $b_1 = 0$. Similarly,

$$\langle Mb_2, b_2\rangle = -\|\bar{A}_- b_2\|^2,$$

and $b_2 = 0$.

If $\mathfrak{g} \simeq \mathfrak{h}_3$, and $Z = \lambda I$, $\lambda \in \mathbb{R}$, then

$$\begin{aligned} \text{(6.3.21)} \qquad \|A_- a\|^2 - \|A_+ a\|^2 &= \langle V_- a, a\rangle - \langle V_+ a, a\rangle \\ &= 2\langle Za, a\rangle \\ &= 2\lambda\|a\|^2, \qquad a \in \mathcal{D}(\rho), \end{aligned}$$

and it follows from this that

$$\text{(6.3.22)} \qquad D(\bar{A}_-) = D(\bar{A}_+).$$

For, if $a \in D(\bar{A}_-)$, and $\{a_n\} \subset \mathcal{D}(\rho)$ satisfies $a_n \longrightarrow a$, and $A_- a_n \longrightarrow \bar{A}_- a$, then (6.3.21) applies to the sequence $\{a_n - a_m\}$, and it follows that the limit, $\lim_n A_+ a_n$ exists. Hence, $a \in D(\bar{A}_+)$. The reverse inclusion $D(A_+) \subset D(A_-)$ follows by symmetry. We have proved (6.3.22).

We introduce the following shorthand notation

$$D_A := D(\bar{A}_+) = D(\bar{A}_-).$$

We now prove

$$\text{(6.3.23)} \qquad D(M) = D(N).$$

Assume $a \in D(M)$. Then the linear form

$$b \longrightarrow \langle A_- b, \bar{A}_- a\rangle$$

is continuous on $\mathcal{D}(\rho)$. But

$$\begin{aligned} \langle A_- b, \bar{A}_- a\rangle &= -\langle A_+ A_- \, b, a\rangle \\ &= -\langle A_- A_+ \, b, a\rangle + 2\lambda\langle b, a\rangle \end{aligned}$$

for all $b \in \mathscr{D}(\rho)$. It follows that the linear form

$$b \longrightarrow \langle A_+b,\ A_+a\rangle$$

is also continuous on $\mathscr{D}(\rho)$. Since

$$D(N) = \{c \in D_A : \bar{A}_+c \in D_A\},$$

and $D_A = D(A_+^*)$, we conclude that $a \in D(N)$. This proves the inclusion $D(M) \subset D(N)$. The other inclusion follows by symmetry, and we have (6.3.23).

An easy induction argument yields the related formula

$$D_\infty(M) = D_\infty(N),$$

and

$$\text{(6.3.24)} \qquad MNa = NMa \qquad a \in D_\infty(M).$$

If $a \in D(M^2) = D(\bar{A}_+\bar{A}_-\bar{A}_+\bar{A}_-)$, then

$$\bar{A}_-a \in D(\bar{A}_+\bar{A}_-\bar{A}_+) \subset D(\bar{A}_-\bar{A}_+) = D(N) = D(M)$$

which proves that $\bar{A}_-$ leaves invariant the domain $D_\infty(M)$. Naturally $D_\infty(M)$ is also invariant under $\bar{A}_+$ by the same argument.

We now show that the analytic vectors $D_\omega(M)$ are also invariant under the two operators $\bar{A}_\pm$.

The commutativity (6.3.24) implies

$$D_\omega(M) = D_\omega(N).$$

If $a \in D_\omega(M)$, there are constants c_0, c_1 such that

$$\|M^n a\| \le c_0\, c_1^n\, n!.$$

But $N^n\bar{A}_-a = \bar{A}_-M^na$, and

$$\begin{aligned}\|N^n\bar{A}_-a\|^2 &= \langle \bar{A}_-M^na, \bar{A}_-M^na\rangle \\ &= \langle A_-^*\bar{A}_-M^na, M^na\rangle \\ &= -\langle M^{n+1}a, M^na\rangle\end{aligned}$$

$$\leq \|M^{n+1}a\| \|M^n a\|$$
$$\leq c_0^2 \; c_1^{2n+1} (n!)^2 (n+1).$$

It follows that

$$\|N^n \overline{A}_- a\| \leq c_0 \; c_1^n \; n! \; \sqrt{(n+1)c_1}.$$

If $c_1 < c_2$, then

$$c_1^n \sqrt{(n+1)c_1} < c_2^n$$

for n sufficiently large. Hence, we may choose $c_3 \geq c_0$ such that

$$\|N^n \overline{A}_- a\| \leq c_3 \; c_2^n \; n! \qquad \text{for all } n.$$

It follows that $\overline{A}_- a \in D_\omega(N) = D_\omega(M)$ and the invariance of $D_\omega(M)$ is proved. The same argument, *mutatis mutandis*, yields invariance of $D_\omega(M)$ under $\overline{A}_+$.

We have proved that the original representation on $\mathcal{D}(\rho)$ extends to a representation on $D_\infty(M)$, and the extended representation, in turn, restricts to a representation on $D_\omega(M)$.

We now choose ξ and η as specified by formulas (6.3.20), and Lemma 6.3.5 applies. It follows that $D_\omega(M)$ is a space of analytic vectors for the operators $\rho(x)$ and $\rho(y)$. Integrability then follows from the analytic vector theorem [GJ, Thm. 3.1].

We now turn to the second case, $\mathfrak{g} \simeq s\ell_2(\mathbb{R})$, $\Omega = qI$, $q \in \mathbb{R}$, which is similar but technically more complicated.

We shall need the preliminary identity:

$$2(\|\rho(x)a\|^2 + \|\rho(y)a\|^2) = -\langle V_+ a + V_- a, a\rangle$$
$$= \|A_+ a\|^2 + \|A_- a\|^2, \qquad a \in \mathcal{D}(\rho),$$

which shows that the following two graph norms on $\mathcal{D}(\rho)$ are equivalent:

(i) $$a \longrightarrow \|a\| + \|\rho(x)a\| + \|\rho(y)a\|$$

and

(ii) $$a \longrightarrow \|a\| + \|A_+a\| + \|A_-a\|.$$

We shall show that both $D(M)$, and $D(N)$, are the completion of $\mathfrak{D}(\rho)$ with respect to any one of the two equivalent norms. The completion will be denoted $\tilde{\mathfrak{D}}$.

Since $\Omega = qI$, we have

$$\|Za\|^2 = \|\rho(x)a\|^2 + \|\rho(y)a\|^2 + q\|a\|^2, \qquad a \in \mathfrak{D}(\rho),$$

which immediately yields the identity

(6.3.25) $$D(\overline{Z}) = \tilde{\mathfrak{D}}$$

But we also have

$$\|A_-a\|^2 - \|A_+a\|^2 = 2\langle Za, a\rangle, \qquad a \in \mathfrak{D}(\rho).$$

We shall need the commutator formulas

$$[A_+, A_-] = -2Z$$

$$[Z, A_\pm] = \pm A_\pm$$

and $$V_+ - V_- = -2Z$$

which are all valid on the domain $\mathfrak{D}(\rho)$.

The methods from case 1 carry over to the present case, but we still need to show that Z is essentially selfadjoint, i.e., that

$$\overline{Z} = Z^*.$$

The easiest way to do this is to introduce the *bounded* operator $B_\pm$, from the proof of Theorem 5.5.1, satisfying

$$A_\pm = B_\pm Z$$

and

$$[Z, B_\pm] = \pm B_\pm.$$

Using the assumption (6.3.11), the commutator formulas, and induction we show that ρ is essentially selfadjoint. It follows from this that Z is essentially selfadjoint, in fact, all powers of Z are essentially selfadjoint. The result follows from this, as indicated above.

6.4 Higher Order Elements

Let U be a unitary representation, and let dU be the corresponding derived representation of $\mathfrak{U}_{\mathbb{C}}(\mathfrak{g})$. Then $\mathfrak{A} := dU(\mathfrak{U}_{\mathbb{C}}(\mathfrak{g}))$ is an algebra of (unbounded) operators. The Gårding space $\mathfrak{D}$ is dense in the representation space $\mathcal{H}$, and invariant under $\mathfrak{A}$.

We mentioned some examples of algebras $\mathfrak{A}$ of this type in Section 6.2. and we shall give three more below.

Example 6.4.1. Let U be the Schrödinger representation, given by formula (6.2.15), $\lambda = 1$. Recall that $\mathfrak{g} \simeq \mathfrak{h}_3$ then, and

$$(6.4.1) \qquad \begin{cases} dU(y_1) = iP \\ dU(y_2) = iQ \\ dU(y_3) = iI \end{cases}$$

where y_1, y_2, y_3 is a basis for $\mathfrak{h}_3$ with y_3 central, and P,Q are the Quantum Mechanical momentum and position operators. The Hilbert space $\mathcal{H}$ is $L^2(\mathbb{R})$, and the Gårding space $\mathfrak{D}$ is precisely the Schwartz space of rapidly decreasing smooth functions on $\mathbb{R}$. This latter characterization of the Gårding space is a corollary of Lemma 5.3.2. It follows from (6.4.1) that $\mathfrak{A}$ is the algebra of all ordinary differential operators on $\mathbb{R}$ with poynomial coefficients. This algebra is also called the *Weyl algebra*, and it is known to be simple, see [Dix 24; §4.6.6].

Example 6.4.2. Let G be a Lie group with Lie algebra $\mathfrak{g}$, and let R be the right regular representation of G. Then apply the construction to the unitary representation R of G on the Hilbert space $L^2(G)$ where L^2 is defined relative to the right invariant Haar measure. The resulting algebra $\mathfrak{A}_R := dR(\mathfrak{U}_{\mathbb{C}}(\mathfrak{g}))$ is the algebra of all left-invariant analytic vector fields on G.

Proof. Since $\mathfrak{A}$ is generated by $dR(g)$ it is enough to show that $\{dR(x) : x \in g\}$ is the full Lie algebra of all left invariant real analytic vector fields on G.

A vector field is said to be *analytic* if it is a derivation of the ring of analytic functions on G, i.e., a linear mapping V satisfying the familiar Leibniz rule,

$$V(f_1 f_2) = (Vf_1)f_2 + f_1(Vf_2). \tag{6.4.2}$$

We say that V is left-invariant if

$$L_g V = VL_g, \qquad g \in G. \tag{6.4.3}$$

where L denotes the left-regular representation. (The assertion in (6.4.2) implies in particular that a left-invariant vector field is automatically analytic.)

Let V be a given left-invariant vector field, i.e., a derivation of $C^\infty(G)$ satisfying (6.4.3). Then V_e defines a tangent vector at the identity element e in G. So V_e may be identified with a Lie algebra element, $V_e \sim x \in g$. By virtue of (6.4.3) we have

$$\begin{aligned}(Vf)(g) &= (L(g^{-1})Vf)(e) \\ &= VL(g^{-1})f(e) \\ &= \langle x, L(g^{-1})f\rangle\end{aligned} \tag{6.4.4}$$

where the notation $\langle x, \cdot\rangle$ means the tangent vector applied to $\cdot$. Now

$$\begin{aligned}(dR(x)f)(g) &= \frac{d}{dt}\Big|_{t=0} f(ge^{tx}) \\ &= \frac{d}{dt}\Big|_{t=0} (L(g^{-1})f)(e^{tx}) \\ &= \langle x, L(g^{-1})f\rangle,\end{aligned}$$

and it follows that

$$Vf = dR(x)f, \qquad f \in C^\infty(G). \tag{6.4.5}$$

But if, for given $x \in g$, we define V by (6.4.4), then V is a derivation since

$$V(f_1f_2)(g)$$

$$= \frac{d}{dt}\Big|_{t=0} L(g^{-1})f_1(e^{tx})L(g^{-1})f_2(e^{tx})$$

$$= (\frac{d}{dt}\Big|_{t=0} L(g^{-1})f_1(e^{tx}))f_2(g) + f_1(g)(\frac{d}{dt}\Big|_{t=0} L(g^{-1})f_2(e^{tx}))$$

$$= V(f_1)(g)f_2(g)+f_1(g)V(f_2)(g), \quad f_1 \in C^\infty(G), \quad f_2 \in C^\infty(G), \quad g \in G.$$

We now check that this derivation (defined by (6.4.5)) maps the ring of analytic functions to itself. We show that, if f is analytic on G, then so is Vf.

Let $g \in G$ be given and consider two coordinate charts $(v_1,\cdots,v_n)$ for a neighborhood of g, and $(u_1,\cdots,u_n)$ for a neighborhood of e. Let φ, resp., ψ, be the coordinate maps from Euclidean space to the respective neighborhoods in G. We shall assume that $\psi(0,\cdots,0) = e$. Then

$$x = \sum_{k=1}^{n} c_k \frac{\partial}{\partial u_k}$$

where the coefficients c_k are real numbers, and it follows that

$$\text{(6.4.6)} \qquad (Vf) \circ \varphi(v) = \sum_{k=1}^{n} c_k \frac{\partial}{\partial u_k}\Big|_{u=0} f(\varphi(v)\psi(u)).$$

Since G is a Lie group, the G-multiplication is analytic. We picked f to be analytic, and it follows that the mapping

$$v,u \longrightarrow f(\varphi(v)\psi(u))$$

is analytic in a suitable neighborhood in $\mathbb{R}^{2n}$. The $\frac{\partial}{\partial u_k}$ derivations, restricted to $u = 0$, are therefore analytic in v above. This proves that Vf is analytic. We have found a coordinate system, where it is given by an analytic expression (6.4.6).

Example 6.4.3. Let G be the Lie group of upper triangular $4{\times}4$ real matrices with ones in the diagonal

(6.4.7)
$$g = \begin{bmatrix} 1 & x_1 & y_2 & z \\ 0 & 1 & x_3 & y_1 \\ 0 & 0 & 1 & x_2 \\ 0 & 0 & 0 & 1 \end{bmatrix}$$

and let U be the unitary representation of G on $L^2(\mathbb{R}^3)$ which is parameterized by $(\beta_1, \beta_2, \gamma) \in \mathbb{R}^3$, and given as follows:

(6.4.8a)
$$(U_g f)(v_1, v_2, v_3) = e^{iE} f(v_1+x_1, v_2+x_2, v_3+x_3)$$

where

(6.4.8b)
$$E = \beta_1(y_1+v_3x_2) + \beta_2(y_2+v_1x_3) - \gamma(z+v_1y_1-y_2x_2-y_2v_2-v_1x_2x_3-v_1v_2x_3).$$

The Lie algebra of G is labeled by the matrix entries in (6.4.7), and we have

$$[x_3, x_2] = y_1 \qquad [x_1, y_1] = z$$

$$[x_1, x_3] = y_2 \qquad [x_2, y_2] = -z$$

with the understanding that all other commutators (not listed) are zero. It follows that

(6.4.8c)
$$\begin{cases} dU(x_1) = \dfrac{\partial}{\partial v_1} \\ dU(x_2) = \dfrac{\partial}{\partial v_2} + i\beta_1 v_3 \\ dU(x_3) = \dfrac{\partial}{\partial v_3} + i(\beta_2 v_1 + \gamma v_1 v_2) \end{cases}$$

Moreover,

(6.4.9)
$$dU(x_1^2+x_2^2+x_3^2) = -H$$

where H is the following Quantum Mechanical Hamiltonian:

$$H = P_1^2 + (P_2+\beta_1 Q_3)^2 + (P_3+\beta_2 Q_1 + \gamma Q_1 Q_2)^2.$$

This is the Hamiltonian for a single particle in a curved magnetic field. The P's and Q's are the momentum and position operators, respectively, in the 3 coordinate directions of the physical space.

Note that (6.4.8b) specifies the unitary representation for the Hamiltonian H with vector potential

$$\mathbb{A} = (0,0,\gamma v_1 v_2) + (0,\beta_1 v_3,\beta_2 v_1)$$

with magnetic field

$$\mathbb{B} = \nabla \times \mathbb{A} = (\gamma v_1,-\gamma v_2,0) + (-\beta_1,\beta_2,0).$$

That is, the perturbation on $\mathbb{A}$ which results by taking non-zero parameters β_1,β_2, just amounts to a constant translation of the magnetic field $\mathbb{B}$ by the vector $(-\beta_1,\beta_2,0)$. In the analysis which follows we shall take $\beta_i = 0$, and the modifications needed for the general case are relatively minor. They will be omitted.

Theorem 6.2.1 applies to the operator (6.4.9), and we have a group theoretic proof of the essential selfadjointness of the Hamiltonian.

More specifically, the magnetic field $\mathbb{B}$ is given by the following affine expression:

$$\mathbb{B} = (\gamma Q_1-\beta_1,\beta_2-\gamma Q_2,0),$$

while the vector potential is

$$\mathbb{A} = (0,\beta_1 Q_3,\beta_2 Q_1+\gamma Q_1 Q_2)$$

polynomial of degree 2. We have $\mathbb{B} = \nabla \times \mathbb{A}$.

This is an example of a larger class of higher order elements in the enveloping algebra of a unitary representation, and we will show how to use representation theory and Section 6.2 to spectral analyze the Hamiltonian, and the operators in the class under consideration.

Before proceeding, we finally note that the representation U given by (6.4.8) is induced from the 3-dimensional normal subgroup N of G described by y_1,y_2,z, i.e., elements g in G determined by $x_1 = x_2 = x_3 = 0$. It is induced from the one-dimensional representation of N which is

$$\chi : (y_1, y_2, z) \longrightarrow e^{i(\beta_1 y_1 + \beta_2 y_2 + \gamma z)} .$$

Recall from Section 4.1 that this induced representation is obtained by a completion of functions φ on G satisfying

$$\varphi(hg) = \chi(h)\varphi(g), \qquad h \in N, \quad g \in G. \tag{6.4.10}$$

(Recall that G, being nilpotent, is unimodular, and, in fact, the Haar measure is identical to Lebesgue measure $dx_1 dx_2 dx_3 dy_1 dy_2 dz$. Similarly, the invariant measure μ on the quotient $N\backslash G$ is just $dx_1 dx_2 dx_3$.) The functions φ from (6.4.10) are completed in the norm

$$\left(\int_{N\backslash G} |\varphi|^2 d\mu\right)^{\frac{1}{2}}$$

and the resulting Hilbert space is denoted $\mathcal{H}(\chi)$. But the coordinates of the particle are given by Q_1, Q_2, Q_3 where Q_i is multiplication by v_i, $i = 1,2,3$, in $L^2(\mathbb{R}^3)$ with respect to $dv_1 dv_2 dv_3$.

Therefore we define a unitary equivalence W as follows:

$$(W\varphi)(v_1, v_2, v_3) = \varphi(v_1, v_2, v_3, 0, 0, 0), \qquad \varphi \in \mathcal{H}(\chi),$$

where we have chosen to write the group element (6.4.7) in a row vector $(x_1, x_2, x_3, y_1, y_2, z)$. The three physical coordinates are embedded into G on the first three coordinate slots, and the remaining coordinates (y_1, y_2, z) are set equal to zero.

Let U^χ be the induced representation of G on $\mathcal{H}(\chi)$ which is defined in Section 4.1. Then a computation shows that $WU^\chi W^*$ is the representation which was given by formula (6.4.8a) at the very outset.

In the recent paper [JK 1], we show that this is a special case of a very general method of analyzing Quantum Mechanical Hamiltonians involving polynomial potentials of magnetic, or scalar, form. We obtain detailed results for specific physical problems of interest.

We conclude with a note on terminology. We studied two representations $(U, L^2(\mathbb{R}^3))$ and $(U^{\chi}, \mathscr{H}(\chi))$, along with a unitary isomorphism

$$W : \mathscr{H}(\chi) \longrightarrow L^2(\mathbb{R}^3)$$

satisfying

$$U_g = WU_g^{\chi}W^*, \qquad g \in G. \tag{6.4.11}$$

We say that W *intertwines* the two representations. When a unitary isomorphism W can be found, satisfying (6.4.11), then we say that the two representations are unitarily equivalent.

From the beginning, the study of operators in the enveloping algebra of a unitary representation was a tool for harmonic analysis and spectral theory of Quantum Mechanical observables. The reader is referred to [Se 7] where many of the pioneering ideas (now frequently taken for granted by workers in representation theory) are developed.

Hermitian elements in the envelopiong algebra are studied, and the two primary questions are: (i) "when is a given Hermitian element essentially selfadjoint on the Gårding space?" (ii) In the affirmative, "what is the spectral representation, and the spectral theory, of the resulting selfadjoint operator?"

Already, Example 6.2.2 shows a negative answer to (i) is more typical. When the answer is however affirmative, a significant physical observable is identified. The corresponding spectrum yields the quantum numbers.

Naturally, questions (i) and (ii) must be asked also for formally normal elements in the enveloping algebra. The question is again, "when does the closure of the operator (defined on the Gårding space) have a spectral resolution?" Or equivalently, "when is the formally normal operator essentially normal?" If yes, "what is the nature of the spectrum and the spectral resolution?"

Theorem 3.9 shows that the two dual questions are closely related.

In the remainder of this section we shall restrict attention to a special class of nilpotent Lie groups G, and a distinguished family of elements x in the universal enveloping algebra, with the property that the answer to question (i) above is affirmative for all unitary representations U of G.

We shall apply this result, in particular, to the representations discussed in Examples 6.4.2 and 6.4.3 above. We shall study in detail the case when U is irreducible, and show that a certain heat semigroup associated with $dU(x)$ is trace class.

The assumptions on G are as follows:

(i) G is simply connected nilpotent.

(ii) The Lie algebra $\mathfrak{g}$ of G is graded with a *scaling semigroup* of automorphisms. Specifically, $\{\delta_s : s \in \mathbb{R}_+\}$ is a subsemigroup of $\mathrm{Aut}(\mathfrak{g})$, $\delta_{st} = \delta_s \circ \delta_t$, $s,t \in \mathbb{R}_+$, and there are spectral subspaces $\{\mathfrak{g}_\nu\}$, indexed by a subset $\{\nu\} \subset \mathbb{R}$, satisfying

$$\mathfrak{g}_\nu = \{x \in \mathfrak{g} : \delta_s(x) = s^\nu x,\ s \in \mathbb{R}_+\}.$$

If $\{\delta_s : s \in \mathbb{R}_+\} \subset \mathrm{Aut}(\mathfrak{g})$ is given as in (ii), then the subset of indices ν, such that $\mathfrak{g}_\nu \neq (0)$, is called the *spectrum*.

It follows that

$$\mathfrak{g} = \mathrm{span}\{\mathfrak{g}_\nu : \nu \text{ in the spectrum}\}. \tag{6.4.12}$$

Indeed, the spectrum is a finite subset of $\mathbb{R}$ since $\mathfrak{g}$ is finite-dimensional, and $s \longrightarrow \delta_s(x)$ is almost periodic [Katn; Ch. 5] on $\mathbb{R}_+$ for all $x \in \mathfrak{g}$. There is a compactification $\overline{\mathbb{R}}_+$ of $\mathbb{R}_+$ with Haar measure $d\mu$ such that

$$\pi_\nu(x) = \int_{\overline{\mathbb{R}}_+} s^{-\nu}\delta_s(x)\,d\mu(s) \tag{6.4.13}$$

and (6.4.12) follows from elementary Fourier analysis on $\overline{\mathbb{R}}_+$. This is because

$$\pi_\nu(x) \in \mathfrak{g}_\nu, \qquad \nu \text{ in the spectrum.}$$

Indeed, using that the Mellin integral transform (6.4.13) is the Fourier transform of the group $\overline{\mathbb{R}}_+$, we have

$$\begin{aligned}
\delta_t(\pi_\nu(x)) &= \int_{\overline{\mathbb{R}}_+} s^{-\nu}\ \delta_t\delta_s(x)d\mu(s) \\
&= \int_{\overline{\mathbb{R}}_+} s^{-\nu}\ \delta_{ts}(x)d\mu(s) \\
&= \int_{\overline{\mathbb{R}}_+} (t^{-1}s)^{-\nu}\ \delta_s(x)d\mu(s) \\
&= t^\nu \int_{\overline{\mathbb{R}}_+} s^{-\nu}\ \delta_s(x)d\mu(s) \\
&= t^\nu\ \pi_\nu(x), \qquad t \in \mathbb{R}_+, \quad x \in \mathfrak{g}.
\end{aligned}$$

Let $\{\delta_s : s \in \mathbb{R}_+\} \subset \operatorname{Aut}(\mathfrak{g})$ be a given scaling semigroup, and let ν_1 be the smallest point in the spectrum. Let $\mathfrak{h}$ be the smallest Lie subalgebra $\subset \mathfrak{g}$ which contains $\mathfrak{g}_{\nu_1}$. Then $\{\delta_s\}$ restricts to a semigroup of automorphisms on $\mathfrak{h}$, i.e.,

$$\delta_s(x) \in \mathfrak{h} \qquad \text{for } s \in \mathbb{R}_+, \quad x \in \mathfrak{h}.$$

Note that $\mathfrak{h}$ is spanned by iterated commutators of the form

$$y = [\cdots[[x_1,[x_2,x_3]]\cdots,x_m]$$

with the x_i's in $\mathfrak{g}_{\nu_1}$. Evaluation of δ_s on such elements yields

$$\delta_s(y) = [[\cdots[\delta_s(x_1),[\delta_s(x_2),\delta_s(x_3)]]\cdots,\delta_s(x_m)] = s^{\nu_1 m} y.$$

It follows that the spectrum of δ_s restricted to $\mathfrak{h}$ is contained in an arithmetic progression $\{m\nu_1 : m = 1,2,\cdots\}$.

In the sequel, we shall frequently start with a given scaling $(\mathfrak{g},\{\delta_s\})$. But the problems of interest can be decided by passing to the restriction

$$(\mathfrak{h}, \{\delta_s|_{\mathfrak{h}}\}).$$

We shall also use the scaling semigroup on G. Since G is simply connected, the semigroup $\{\sigma_s : s \in \mathbb{R}_+\} \subset \mathrm{Aut}(G)$ which is determined uniquely in terms of δ_s by

$$\sigma_s(\exp x) = \exp \delta_s(x), \qquad s \in \mathbb{R}_+,\ x \in \mathfrak{g}. \tag{6.4.14}$$

Let $\mathfrak{h} \subset \mathfrak{g}$ be a Lie subalgebra and let $H \subset G$ be the closed subgroup generated by $\{\exp x : x \in \mathfrak{h}\}$. Then it follows from Lie theory [Hel 3, Ch.II, Thm. 2.3] that H is a Lie group with $\mathfrak{h}$ as its Lie algebra. Moreover,

$$\mathfrak{h} = \{x \in \mathfrak{g} : \exp(tx) \in H,\ t \in \mathbb{R}\}.$$

It follows from this, and (6.4.14), that the two conditions below are equivalent:

(a) $$\delta_s(\mathfrak{h}) = \mathfrak{h}, \qquad s \in \mathbb{R}_+,$$

(b) $$\sigma_s(H) = H, \qquad s \in \mathbb{R}_+.$$

Before stating the next result, we need a remark on positive elements in $\mathfrak{U}_{\mathbb{C}}(\mathfrak{g})$. An element $y \in \mathfrak{U}_{\mathbb{C}}(\mathfrak{g})$ is said to be *strongly positive* if there is a finite subset $\{z_j\}$ of $\mathfrak{U}_{\mathbb{C}}(\mathfrak{g})$ such that

$$y = \sum_j z_j^* z_j. \tag{6.4.15}$$

We say that y is *positive* if

$$\int_G dR(y)f(g)\overline{f(g)}dg \geq 0, \qquad f \in C_c(G), \tag{6.4.16}$$

where R is the right-regular representation, and dg is "the" right-invariant Haar measure on G. If G has properties (i) and (ii), i.e., is simply connected, nilpotent, with scaling semigroup, then it follows from a familiar application of the Plancherel theorem [Pu 2; II, III, §6], that y is positive if and only if, for all unitary irreducible representations U of G,

(6.4.17) $\langle dU(y)a,a\rangle \geq 0, \qquad a \in \mathfrak{D}(U),$

where $\mathfrak{D}(U)$ is the Gårding space of U.

It is immediate that every strongly positive element is positive, and if $\mathfrak{g}$ is Abelian, the converse implication is also known [Pow 2].

In the theorem we shall consider G simply connected nilpotent. We shall assume that G has a scaling semigroup, and consider the dual pair (G,σ), $(\mathfrak{g},\delta)$ where $\mathfrak{g}$ is the Lie algebra, and σ and δ are related through (6.4.14). We shall consider the subspace $\mathfrak{g}_\nu$ for some ν in the spectrum of $\{\delta_s\}$.

Let $\mathfrak{X}_\nu$ be the subalgebra of $\mathfrak{U}_{\mathbb{C}}(\mathfrak{g})$ generated by $\mathfrak{g}_\nu$. Then an element $y \in \mathfrak{U}_{\mathbb{C}}(\mathfrak{g})$ is said to be *admissible* if

(i) $Y \in \mathfrak{X}_\nu$ for some ν in the spectrum of $\{\delta_s\}$.

(ii) y is positive.

(iii) For some $d \in \mathbb{R}$, we have $\delta_s(y) = s^d y$, $s \in \mathbb{R}_+$.

(iv) Let ν be as in (i) and let $T_\nu \subset \mathfrak{g}_\nu$ be a minimal subset such that y is in the subalgebra of $\mathfrak{U}_{\mathbb{C}}(\mathfrak{g})$ generated by T_ν, and let $H \subset G$ be the closed subgroup generated by $\{\exp x : x \in T_\nu\}$. It is assumed that, for every irreducible representation U of H, different from the trivial one-dimensional representation, the operator $dU(y)$ is invertible on $\mathfrak{D}(U)$.

We have:

Theorem 6.4.4. Let G be a simply connected nilpotent Lie group with a scaling semigroup, and let U be a unitary representation of G. Let y be an admissible element in the universal enveloping algebra.

Then the operator $dU(y)$ is essentially selfadjoint and positive on $\mathfrak{D}(U)$.

Before starting the proof, we note that the essential self-adjointness conclusions in Example 6.4.3 and Theorem 6.2.2 follow from Theorem 6.4.4 above. Also note that Theorem 6.4.4 implies special cases of earlier results, namely, Theorem 6.3.1.

In some applications, the object of primary interest is the operator $dU(y)$ for some given element y and some given unitary representation. Example 6.4.3 above is typical in this regard. It is then possible to choose the Lie group G to coincide with the subgroup H which is used in the definition of *admissibility*.

We now give an example (the Hamiltonian of the Henon-Hailes potential) where the theory applies with suitable modifications to the operator $H = dU(y)$ even though y is not admissible.

Example 6.4.5. *The Henon-Hailes potential* (Jorgensen-Klink [JK 1].) This is a potential V defined on $\mathbb{R}^2$, two space coordinates $v = (v_1, v_2)$,

$$V = \frac{1}{2}(v_1^2+v_2^2) + v_1 v_2^2 - \frac{1}{3} v_2^3.$$

The corresponding Hamiltonian

$$H = (2m)^{-1}(P_1^2+P_2^2) + \frac{1}{2}(Q_1^2+Q_2^2) + Q_1 Q_2^2 - \frac{1}{3} Q_2^3 \tag{6.4.18}$$

may be resolved as

$$H = dU(y)$$

for some element y in the enveloping algebra, and a unitary representation U. In this particular case, the representation U may be taken to be irreducible. (This is in contrast to the curved magnetic field Hamiltonian from Example 6.4.3 above where the corresponding unitary representation U is *not* irreducible. For further details, see [JK 1].)

We may take $\mathfrak{g}$ (for the Henon-Hailes potential) to be the semidirect product of the following two Abelian Lie algebras; $\mathfrak{g}_1$, resp., $\mathfrak{g}_2$. Take $\mathfrak{g}_1$ to be spanned by $\frac{\partial}{\partial v_1}, \frac{\partial}{\partial v_2}$, and take $\mathfrak{g}_2$ to be

(6.4.19) $\mathfrak{g}_2 = i\ \text{span}_{\mathbb{R}}\{v_1, v_2, v_2^2, v_1 v_2, 1, V\}.$

It is possible to construct a scaling semigroup of automorphisms on the semidirect product Lie algebra as follows, as noted by S. Pedersen (private communication):

TABLE 6.4

Elements	Degree d
$\frac{1}{2}(v_1^2+v_2^2) + v_1 v_2^2 - \frac{1}{3} v_2^3$, $\frac{1}{3}\frac{\partial}{\partial v_1} + \frac{\partial}{\partial v_2}$	1
$\frac{1}{3} v_1 + v_2 + 2v_1 v_2 - \frac{2}{3} v_2^2$, $\frac{\partial}{\partial v_1}$	2
$v_1 + v_2^2$, $\frac{5}{9} + v_1 - \frac{1}{3} v_2$	3
$\frac{1}{3} + 2v_2$	4
1	5

(When listing the elements in $\mathfrak{g}_2$, we have neglected the factor $i = \sqrt{-1}$. An element x is said to have degree d if $\delta_s(x) = s^d x$, $s \in \mathbb{R}_+$.)

It follows from general theory [Goo 9, Appendix A2] that $\mathfrak{g}$ carries a scaling semigroup $\{\delta_s\}_{s\in\mathbb{R}_+}$ such that the respective elements in $\mathfrak{g}$ have the degrees which are assigned in (6.4.19) and the table, for the basis elements for $\mathfrak{g}_2$.

Let G be the simply connected Lie group with Lie algebra $\mathfrak{g}$. It is the semidirect product of two Abelian groups G_1, resp., G_2, where G_2 is normal. It is easy to represent G concretely as a matrix group but we shall not need this.

It follows from (6.4.1) that the Schrödinger representation U of the 5-dimensional Heisenberg group extends to a unitary representation (also denoted by U) of the larger 8-dimensional group G.

Since the Schrödinger representation is irreducible (Lemma 4.2.1, §4.2), it follows that the extended representation is also irreducible. Recall that the Hilbert space is now $L^2(\mathbb{R}^2)$, and recall that the 5-dimensional Heisenberg group is the group of 4×4 real triangular matrices:

$$\begin{bmatrix} 1 & u_1 & u_2 & w \\ & 1 & 0 & v_2 \\ & & 1 & v_1 \\ & 0 & & 1 \end{bmatrix}.$$

It follows from (6.4.18) that $H = dU(y)$. But an inspection of Table 6.4 reveals that y is not admissible. It is, in fact, not homogeneous, and, moreover, the operator $dU(y)$ is not semibounded.

To see that $H = dU(y)$ is essentially selfadjoint we may invoke a result of Nelson [Ne 3, Thm. 1].

Let $\mathcal{L}_2$ be the Lie algebra of the Shale-Weil harmonic representation of the symplectic group $Sp(2,\mathbb{R})$, Recall from Section 6.2 that the operator Lie algebra $\mathcal{L}_2$ is spanned by all second order noncommutative monomials in the Quantum Mechanical operators P_j, Q_k, $1 \leq j,k \leq 2$. It follows from the Canonical Commutation Relations (6.3.6) that the derivation ad H normalizes $\mathcal{L}_0$, i.e., $[H,\mathcal{L}_0] \subset \mathcal{L}_0$. It follows that the square of the Harmonic oscillator Hamiltonian

$$K = P_1^2+P_2^2+Q_1^2+Q_2^2 \quad \text{(in two space dimensions)}$$

analytically dominates H, i.e., every analytic vector for K^2 is also analytic for H. Indeed, an inspection shows that

$$\|Hf\| \leq \|K^2 f\| + \text{const.}\|f\|$$

for Schwartz functions f on $\mathbb{R}^2$. It can be checked that the second assumption in [Ne, Theorem 1] follows from the commutator formula $[H,\mathcal{L}_0] \subset \mathcal{L}_0$. Recall that K is an element in $\mathcal{L}_0$, and that K analytically dominates $\mathcal{L}_0$ by virtue of Lemma 6.3.5.

We include a sketch of the details: Since $[H,\mathcal{L}_0] \subset \mathcal{L}_0$, and $K \in \mathcal{L}_0$, there is a constant $b > 0$ such that

$$(\mathrm{ad}\ H)^m(K) \leq b^m K, \qquad m \in \mathbb{N}.$$

We have used the symbolic notation from Section 6.3 for commutators of families of operators. To verify Nelson's second assumption, we prove that

$$(\mathrm{ad}\ H)^n(K^2) \leq (2b)^n K^2, \qquad n \in \mathbb{N}.$$

We have:

$$(\mathrm{ad}\ H)^n(K^2) \leq \sum_{m=0}^{n} \binom{n}{m} (\mathrm{ad}\ H)^m(K)(\mathrm{ad}\ H)^{n-m}(K)$$

$$\leq \sum_{0}^{n} \binom{n}{m} b^m K b^{n-m} K = (2b)^n K^2,$$

as claimed.

Proof of Theorem 6.4.4. We have an arbitrary unitary representation U and an admissible element y, and we wish to show that $dU(y)$ is essentially selfadjoint on the Gårding space $\mathcal{D}(U)$. A result of Helffer-Nourrigat [HN, Théorème 2.3] implies that the operator $dL(y)$ is hypoelliptic on the subgroup H described in the paragraph preceding the statement of Theorem 6.4.4.

We must check that the range of the operator $I + dU(y)$ is dense in $\mathcal{H}$. But if $a \in \mathcal{H}$, and $\langle (I+dU(y))U(\varphi)a,a\rangle = 0$ for all $\varphi \in C_c^\infty(H)$, then as noted in the proof of Theorem 6.2.2, the positive definite function

$$p(h) = \langle U(h)a,a\rangle \qquad h \in H,$$

is a weak solution to the partial differential equation

$$(I+dL(y))p = 0.$$

By hypoellipticity, it follows that p is a C^∞-function on H, and the differential equation

$$dL(y)p(h) = -p(h), \qquad h \in H, \tag{6.4.20}$$

is valid in the pointwise sense. A standard result on positive definite functions [Hel 1, p. 414, Thm. X4.4] yields the existence of a new unitary representation V of H and a cyclic vector Ω in the Gårding space $\mathfrak{D}(V)$ such that

$$\langle V(h)\Omega,\Omega\rangle = p(h), \qquad h \in H. \tag{6.4.21}$$

Since y is positive, the operator $dV(y)$ is positive on $\mathfrak{D}(V)$. It follows from (6.4.21) that

$$dL(y)p(e) = \langle dV(y)\Omega,\Omega\rangle \geq 0.$$

Equation (6.4.20) now implies that $p(e) = 0$, and $p \equiv 0$ since p is positive definite. But $p(e) = \|a\|^2$. So $a = 0$, and we have proved that $dU(y)$ is essentially selfadjoint on the Gårding space of $U|_H$. But the argument at the end of the proof of Theorem 6.2.2 then implies that $dU(y)$ is then also essentially selfadjoint on the Gårding space of U, i.e., the Gårding space obtained by integration over the whole group G, rather than just the subgroup H.

This completes the proof of Theorem 6.4.4.

When the theorem is applied to the right regular representation R, we get essential selfadjointness of the operator $dR(y)$ for all admissible elements y. In the next result we shall obtain a spectral representation for $dR(y)$ which implies in particular that $dR(y)$ has (absolutely continuous) Lebesgue spectrum, infinite multiplicity, on the Hilbert space $L^2(H)$ of square-integrable functions on the subgroup H.

Theorem 6.4.6. Let G be a simply connected nilpotent Lie group with a scaling semigroup $\{\sigma_s\}_{s\in\mathbb{R}_+}$, and let U be a unitary representation of G. Let y be an admissible element in $\mathfrak{U}_{\mathbb{C}}(\mathfrak{g})$ with associated subgroup $H \subset G$.

Then there is a separable Hilbert space $\mathcal{V}$, a conjugation, $v \longrightarrow v^*$, of $\mathcal{V}$, and a unitary isomorphism W of $L^2(H)$ onto $L^2(0,\infty;\mathcal{V})$ such that

$$WdR(y)W^*f(\lambda) = \lambda f(\lambda), \quad f \in L^2(0,\infty,\mathcal{V}),\ 0 < \lambda < \infty, \tag{6.4.22}$$

(6.4.23) $$\int_H |\psi(g)|^2 dg = \int_{\mathbb{R}_+} \|W\psi(\lambda)\|_{\mathscr{V}}^2 \frac{d\lambda}{\lambda}, \qquad \psi \in L^2(H),$$

and there is a family of generalized eigenfunctions $\varphi(\lambda,g)$, $\lambda \in \mathbb{R}_+$, $g \in H$, which is determined by

(6.4.24) $$W\psi(\lambda) = \int_H \psi(g)\varphi^*(\lambda,g)dg, \qquad y \in L^2(H).$$

Proof. Let dg denote the Haar measure of H. Using uniqueness of dg, up to a scalar factor, we get the formula

$$d(\sigma_s(g)) = s^m dg, \qquad s \in \mathbb{R}_+,$$

where the exponent m $(\in \mathbb{R})$ is called the homogeneous degree of H.

Now define a representation U of $\mathbb{R}_+$ as follows:

(6.4.25) $$U_s\psi(g) = \psi(\sigma_s(g))s^{\frac{m}{2}}, \qquad \psi \in L^2(H),\ g \in H,\ s \in \mathbb{R}_+.$$

Since $\delta_s(y) = s^d y$, $s \in \mathbb{R}_+$, it follows that

$$U_s^* \, dR(y)U_s = s^d dR(y), \qquad s \in \mathbb{R}_+.$$

Indeed,

$$\begin{aligned} U_s^* \, dR(y)U_s\psi(g) &= dR(y)U_s\psi(\sigma_s^{-1}(g))s^{-\frac{m}{2}} \\ &= s^d \, dR(y)\psi(\sigma_s\sigma_s^{-1}(g)) \\ &= s^d \, dR(y)\psi(g), \end{aligned}$$

for functions ψ in the Gårding space of $R|_H$, $s \in \mathbb{R}_+$, and $g \in H$.

We proved in Theorem 6.4.4 that $\overline{dR(y)}$ is selfadjoint with spectrum contained in $[0,\infty)$. Let E be the corresponding spectral resolution, i.e., a measure on $\mathbb{R}$, supported on $[0,\infty)$, and with values in projections of the Hilbert space $L^2(H)$. Consider the transformed measure

$$\lambda \longrightarrow E(s^{-d}\lambda).$$

Then

(6.4.26) $$dE(s^{-d}\lambda) = U_s^* \, dE(\lambda)U_s$$

since

$$\int \lambda \, dE(s^{-d}\lambda) = \int s^d\lambda \, dE(\lambda)$$
$$= s^d \, \overline{dR(y)}$$
$$= U_s^* \, \overline{dR(y)}U_s$$
$$= \int \lambda \, U_s^* \, dE(\lambda)U_s, \qquad s \in \mathbb{R}_+.$$

We now define the integrated form $E(\varphi)$, for $\varphi \in C_c(\mathbb{R}_+)$, by

$$E(\varphi) = \int \varphi(\lambda)dE(\lambda).$$

It follows from (6.4.26) that

(6.4.27) $$U_s^* \, E(\varphi)U_s = E(\tau_s\varphi), \qquad s \in \mathbb{R}_+,$$

where

$$\tau_s\varphi(\lambda) = \varphi(s^d\lambda).$$

We have proved that the pair (U,E) is a system of imprimitivity over the group $\mathbb{R}_+$, and the Hilbert space $L^2(H)$.

We now show that it is nondegenerate, i.e., that

(6.4.28) $$E(C_c(\mathbb{R}_+))L^2(H) \text{ is dense in } L^2(H).$$

But if a given $\psi \in L^2(H)$ is orthogonal to the space (6.4.28), then

$$\langle E(\varphi)\psi,\psi\rangle_{L^2(H)} = 0, \qquad \varphi \in C_c(\mathbb{R}_+),$$

i.e., $\int\varphi(\lambda)\|dE(\lambda)\psi\|^2 = 0$ for all $\varphi \in C_c(\mathbb{R}_+)$. It follows from this that the support of $\|E(\cdot)\psi\|^2$ must be contained in $(-\infty,0]$. Since we already know that it is contained in $[0,\infty)$, the only possible point in the support is $\lambda = 0$. It follows that $dR(y)\psi = 0$ since

$$\|dR(y)\psi\|^2 = \int|\lambda|^2\|dE(\lambda)\psi\|^2.$$

Let $\{U^\xi\}$ be a parametrization of the unitary dual of H, i.e., the unitary irreducible representations of H, and let $d\mu(\xi)$ denote the Plancherel measure. It follows from [Pu 2; II, Ch. III] and Kirillov-Kostant theory that μ may be regarded as a measure on the space of co-adjoint orbits, i.e., the orbits of the adjoint representation of H on the dual of the Lie algebra $\mathfrak{h}$.

The Plancherel formula [Pu 2, p. 171] yields a direct integral decomposition of $L^2(H)$:

$$\|\psi\|^2 = \int_\xi \|\psi_\xi\|^2 d\mu(\xi), \tag{6.4.29}$$

$$\|dR(y)\psi\|^2 = \int_\xi \|dU^\xi(y)\psi_\xi\|^2 d\mu(\xi). \tag{6.4.30}$$

It follows that $\|dU^\xi(y)\psi_\xi\|^2 = 0$, μ almost all ξ, i.e., $dU^\xi(y)\psi_\xi = 0$ except on a set of points ξ of measure zero relative to μ. But $dU^\xi(y)$ is invertible whenever U^ξ is not the trivial one-dimensional representation. It follows from (6.4.29) that $\psi = 0$. This is because the constant function on H is not in $L^2(H)$ since the Haar measure on H is equivalent to Lebesgue measure.

We have verified the assumptions in the Imprimitivity Theorem, Theorem 4.1.1, or its variant, Theorem 4.2.1. As a conclusion we get a representation of the system $\{U_s, E(\cdot), L^2(H)\}$ onto a canonical system based on $\mathbb{R}_+$, $\hat{\mathbb{R}}_+ \simeq \mathbb{R}$, and a separable Hilbert space $\mathscr{V}$. With the terminology from the proof of Theorem 4.1.1, we may define

$$(W\psi)(\lambda) = [U_{\lambda^{-1/d}}(\psi)],$$

for $\psi \in L^2(H)$, and $\lambda \in \mathbb{R}_+$. It then follows that

$$\int_{\mathbb{R}_+} \varphi(\lambda) \|W\psi(\lambda)\|_{\mathscr{V}}^2 \frac{d\lambda}{\lambda} = \langle E(\varphi)\psi, \psi\rangle_{L^2(H)}$$

for $\varphi \in C_c(\mathbb{R}_+)$, $\psi \in L^2(H)$, and moreover that (6.4.23) holds.

We may apply the L. Schwartz kernel theorem [Schw 2] to the unitary isomorphism

(6.4.31) $$W : L^2(H) \longrightarrow L^2(\mathbb{R}_+, \mathcal{V}),$$

and we get a distribution K on $\mathbb{R}_+ \times H$ with values in $\mathcal{V}$ satisfying the equation

$$dR(y)K(\lambda,\cdot) = \lambda K(\lambda,\cdot)$$

on H. Since $dR(y)$ is hypoelliptic by [HN 10, Théorème 2.3], it follows that $K(\lambda,\cdot)$ is in $C^\infty(H)$ for each λ. It follows, by going to an orthonormal basis for $\mathcal{V}$, that a conjugation, $v \longrightarrow v^*$, may be chosen such that the kernel K of (6.4.31) takes the form

$$K(\lambda,g) = \varphi^*(\lambda,g), \qquad \lambda \in \mathbb{R}_+,\ g \in H.$$

The conjugation, $v \longrightarrow v^*$ in $\mathcal{V}$, is determined by the conjugation on $L^2(H)$,

$$\psi \longrightarrow \overline{\psi},$$

which is just pointwise complex conjugation. The equation is:

(6.4.32) $$W(\overline{\psi})(\lambda) = W(\psi)(\lambda)^*, \qquad \psi \in L^2(H),\ \lambda \in \mathbb{R}_+.$$

Additional References for Chapter 6

Sect. 6.1: [AK], [DoMe], [DS], [Se 14].

Sect. 6.2: [BB], [BRac], [BO], [GJ 2], [JM], [Kat 2], [NS], [Rao], [RS], [Wm], [Win].

Sect. 6.3: [MS], [Sj].

Sect. 6.4: [AR], [Ar 1], [BG], [He 6], [Hul 1], [Hö 1], [Pan], [Stö].

CHAPTER 7. SPECTRAL THEORY

In this chapter, we shall continue the study of operators associated with unitary representations of Lie groups, and, in the first half of the chapter, we shall be concerned with a class of Hamiltonians which includes the examples from Chapter 6.

In Chapter 6, we specialized to unitary representations of simply connected nilpotent Lie groups G. We assumed the existence of a *scaling semigroup* $\{\sigma_s : s \in \mathbb{R}_+\} \subset \mathrm{Aut}(G)$. We studied the corresponding semigroup $\{\delta_s : s \in \mathbb{R}_+\} \subset \mathrm{Aut}(\mathfrak{g})$, given by

$$\exp(\delta_s(x)) = \sigma_s(\exp x), \qquad s \in \mathbb{R}_+,\ x \in \mathfrak{g}.$$

We shall add here an assumption on the spectral subspace

$$\mathfrak{g}_\upsilon = \{x \in \mathfrak{g} : \delta_s(x) = s^\upsilon x,\ s \in \mathbb{R}_+\}$$

corresponding the the smallest point, υ_1 say, in the spectrum of $\{\delta_s\}_{s\in\mathbb{R}_+}$. We shall assume that $\mathfrak{g}_{\upsilon_1}$ generates $\mathfrak{g}$ as a Lie algebra, i.e., that the smallest Lie algebra of $\mathfrak{g}$, which contains $\mathfrak{g}_{\upsilon_1}$, is $\mathfrak{g}$ itself. A consequence of this is that the spectrum of $\{\delta_s\}_{s\in\mathbb{R}_+}$ is contained in an arithmetic progression

$$\{m\upsilon_1 : m \in \mathbb{N}\}.$$

An additional consequence is that for any choice of basis $x_1,\cdots,x_r$ for $\mathfrak{g}_{\upsilon_1}$ the two operators, $dL(\sum_1^r x_i^2)$ and $dR(\sum_1^r x_i^2)$, are hypoelliptic on G. This is because the subgroup H which was introduced in the statement of Theorems 6.4.4 and 6.4.6 concides with G itself. Hence, Hörmanders' theorem [Hö 1, Thm. 1.1], or the Helffer-Nourrigat theorem [HN 10, Théorème 2.3], applies to G. Our Theorem 6.4.6 yields additional information concerning the spectrum of the two operators: Lebesgue spectrum with uniform infinite multiplicity. In Section 7.4, we shall give an explicit formula for the spectral transform in case G is the Heisenberg group.

We begin this chapter by noting that the added restriction on G, and the Lie algebra $\mathfrak{g}$, is *not* a constraint as far as the quantum mechanical applications are concerned. In those applications, we will simply construct the pair $(G,\mathfrak{g})$ from the problem at hand, and it will be evident that the new added assumption holds. (The construction of the pair $(G,\mathfrak{g})$, with scaling semigroup, was carried out, in a very special case, already in Example 6.4.5 in the previous chapter.)

Consider a spinless particle of mass m in an external magnetic field $\mathbb{B} = \mathbb{B}(v)$, and let $\mathbb{A}$ be a vector potential. Then the Hamiltonian is

$$H = (2m)^{-1}(\mathbb{P} - \tfrac{e}{c}\mathbb{A})^2$$

where

$$\mathbb{P} = h/i\ \nabla_{\mathbf{v}}.$$

We have $\mathbb{B} = \nabla \times \mathbb{A}$ and the commutator formulas.

$$\left[P_i - \tfrac{e}{c}A_i,\ P_j - \tfrac{e}{c}A_j\right] = -\frac{h}{i}\frac{e}{c}\epsilon_{ijk}B_k \tag{7.0.1}$$

$$\left[P_i - \tfrac{e}{c}A_i,\ B_j\right] = \frac{h}{i}\frac{\partial B_j}{\partial v_i} := \frac{h}{i}B_{ij}$$

$$\left[P_i - \tfrac{e}{c}A_i,\ B_{jk}\right] = \frac{h}{i}\frac{\partial B_{jk}}{\partial v_i} := \frac{h}{i}B_{ijk}$$

$$\vdots \qquad\qquad \vdots$$

where we have used the familiar summation convention on the right-hand side of the first equation, and the symbol ϵ_{ijk} is one when the triple index $1 \leq i,j,k \leq 3$ is an even permutation of 1,2,3; it is minus one for odd permutations, and zero otherwise. If we take the components A_i of the vector potential $\mathbb{A}$ to be polynomial functions of v_1, v_2, v_3, the chain of commutators must terminate eventually since the degree of the respective polynomials is decreased by one at each step.

It is also transparent from the chain of commutator formulas how to choose the nilpotent Lie group G and the unitary representation U such that

$$H = -dU(x_1^2 + x_2^2 + x_3^2) \tag{7.0.2}$$

with the x_i's in the Lie algebra $\mathfrak{g}$.

For $\mathfrak{g}$ we take the semi-direct product (or rather semi-direct sum at the Lie algebra level) of two Abelian Lie algebras $\mathfrak{g}_1$ and $\mathfrak{g}_2$. We take $\mathfrak{g}_1$ to be spanned by symbols x_1, x_2, x_3 corresponding to $P_i - \frac{e}{c} A_i$, $i = 1,2,3$, and we take $\mathfrak{g}_2$ to be spanned by symbols $y_k, y_{ij}, y_{ijk}, y_{ijk\ell}$, etc., corresponding to the terms (functions) on the right hand side of the string of equations (7.0.1) above:

$$\begin{array}{lcl} y_k & \longleftrightarrow & \frac{h}{i}\frac{e}{c} B_k \\ y_{ij} & \longleftrightarrow & \frac{h}{i} B_{ij} \\ y_{ijk} & \longleftrightarrow & \frac{h}{i} B_{ijk} \\ \vdots & & \vdots \end{array}$$

with indices i,j,k, etc. satisfying $1 \leq i,j,k \leq 3$, and the i in the denominator meaning $\sqrt{-1}$. We have $[\mathfrak{g}_1, \mathfrak{g}_2] \subset \mathfrak{g}_2$, so $\mathfrak{g}_2$ is an ideal whose dimension depends on the combined degree of the polynomial entries in the vector potential A. The relations for the basis elements $x_1, x_2, x_3, y_k, y_{ij}, y_{ijk}$, etc., are

$$\begin{array}{rcl} [x_i, x_j] & = & \epsilon_{ijk} y_k \\ [x_i, y_j] & = & y_{ij} \\ [x_i, y_{jk}] & = & y_{ijk} \\ \vdots & & \vdots \end{array} \tag{7.0.3}$$

It is assumed that all unlisted commutator brackets are zero.

We now construct a unitary representation U of the corresponding simply connected Lie group with Lie algebra g which is specified by the generator conditions

(7.0.4)
$$dU(x_i) = \sqrt{-1}\,(P_i - \frac{e}{c} A_i)$$
$$dU(y_k) = \frac{h}{\sqrt{-1}} \frac{e}{c} B_k$$
$$\vdots \qquad\qquad \vdots$$

so the commutator relations,

$$[dU(x_i), dU(x_j)] = \epsilon_{ijk} dU(y_k)$$

are preserved.

To see that such a unitary representation U exists, satisfying (7.0.4), we may refer to one of the integrability theorems for Lie algebras of operators, such as [JM, Thm. 9.2], or [GJ 2, Thm. 3.1], see (the present) Appendix for details.

Alternatively, U may be explicitly constructed as an induced representation: Corresponding to the semidirect sum decomposition

(7.0.5)
$$g = g_1 + g_2$$

we have a semi-direct product decomposition of the corresponding Lie group G, and we shall write

(7.0.6)
$$G = G_1 G_2$$

since every element g in G decomposes uniquely as

(7.0.7)
$$g = (\exp x)(\exp y), \qquad x \in g_1, \quad y \in g_2.$$

Let $v = (v_1, v_2, v_3)$ be the space coordinates (position) of the particle. Choose coordinate labels for elements g in G consistent with the labeling (7.0.3) of the Lie algebra elements:

$$g \sim (u_1, u_2, u_3, w_k, w_{ij}, w_{ijk}, \text{etc.} \cdots).$$

The embedding $u = (u_1, u_2, u_3) \longrightarrow (u, 0, 0, \cdots)$ is the embedding

$G_1 \longrightarrow G = G_1G_2$ making the short exact sequence

$$(e) \longrightarrow G_2 \longrightarrow G \longrightarrow G_1 \longrightarrow (e)$$

split, cf. formulas (7.0.1) and (7.0.3). We shall induce from one-dimensional representations χ of the normal subgroup G_2 as in Example 6.4.3,

$$\chi : (w_k, w_{ij}, w_{ijk}, \cdots) \longrightarrow \exp\sqrt{-1}(\sum \beta_k w_k + \beta_{ij} w_{ij} + \beta_{ijk} w_{ijk} + \cdots),$$

and the Hilbert space $\mathscr{H}(\chi)$ of the induced representation U_g^χ is the completion of the space of functions φ on G satisfying

$$\varphi(hg) = \chi(h)\varphi(g), \qquad h \in G_2, \quad g \in G, \tag{7.0.8}$$

relative to the integral norm

$$(\int_{G_1} |\varphi|^2 d\mu)^{1/2}.$$

The measure μ on G_1 is merely $du = du_1 du_2 du_3$, i.e., Lebesgue measure on $\mathbb{R}^3$. We define a unitary isomorphism W as follows:

$$W : \mathscr{H}(\chi) \longrightarrow L^2(\mathbb{R}^3),$$

$$(W\varphi)(\mathbf{v}) = \varphi(\mathbf{v}, 0, 0, \cdots), \qquad \varphi \in \mathscr{H}(X),$$

and calculate the representation

$$U_g := WU_g^\chi W^*, \qquad g \in G,$$

on $L^2(\mathbb{R}^3)$. For the derived representations, we have

$$dU(x) = WdU^\chi(x)W^*, \qquad x \in \mathfrak{g}, \tag{7.0.9}$$

and the parameters

$$\beta_k, \beta_{ij}, \beta_{ijk}, \cdots$$

for the character χ may be determined by the system of differential equations which is implicit in (7.0.9).

By virtue of (7.0.3), the three basis elements x_1, x_2, x_3

for $\mathfrak{g}_1$ generate the whole Lie algebra $\mathfrak{g}$, and the representation U is determined by (7.0.4) for x in this three element basis.

Using (7.0.4), we get the three equations:

$$\frac{\partial}{\partial v_i} - \sqrt{-1}\,\frac{e}{c}\,A_i(v) = WdU^{\chi}(x_i)W^*, \quad i = 1,2,3, \tag{7.0.10}$$

from which χ may be determined by comparison of coefficients. Although there are only three equations, all the parameters $\beta_k, \beta_{ij}, \beta_{ijk}, \cdots$ of χ can be determined from this system because the coefficients of the three polynomials $A_i(v)$ must be matched with the same number of parameters on the right hand side of the system of equations (7.0.10). When the parameters are found, it follows, as in Example 6.4.3 that U is of the form

$$(U_g f)(v) = \exp(\sqrt{-1}\,E(g,v))f(v+u) \tag{7.0.11}$$

where $g = (u,\cdots)$ and $u = (u_1,u_2,u_3)$.

Finally note that G, resp. $\mathfrak{g}$, carries scaling semigroups:

$$\{\sigma_s\}_{s\in\mathbb{R}_+} \subset \mathrm{Aut}(G),$$

and

$$\{\delta_s\}_{s\in\mathbb{R}_+} \subset \mathrm{Aut}(\mathfrak{g})$$

which are given as follows:

$$\sigma_s(u,w_k,w_{ij},w_{ijk},\cdots) = (su,s^2w_k,s^3w_{ij},s^4w_{ijk},\cdots)$$

and

$$\begin{aligned} \delta_s(x_i) &= sx_i \\ \delta_s(y_k) &= s^2y_k \\ \delta_s(y_{ij}) &= s^3y_{ij} \\ \delta_s(y_{ijk}) &= s^4y_{ijk} \\ &\vdots \end{aligned}$$

We mention, without going into details, that the representation theoretic methods from above also work on multiparticle Hamiltonians with polynomial interactions. Consider, for example, the Hamiltonian

$$H = \frac{1}{2} \sum_{i=1}^{n} P_i^2 + V(x_1, \cdots, x_n)$$

where the (scalar) potential V is a polynomial in $x_1, \cdots, x_n$. Then, as discussed in [JK 1], the commutators

$$\left[\frac{\partial}{\partial x_i}, V\right] = \frac{\partial V}{\partial x_i} := V_i$$

$$\left[\frac{\partial}{\partial x_i}, V_j\right] = \frac{\partial V_j}{\partial x_i} := V_{ij}$$

$$\left[\frac{\partial}{\partial x_i}, V_{jk}\right] = \frac{\partial V_{jk}}{\partial x_i} := V_{ijk}$$

$$\vdots \qquad \vdots$$

generate Lie algebra elements $\frac{\partial}{\partial x_i}, V_i, V_{ij}, V_{ijk}, \cdots$ that finally close to give a finite-dimensional nilpotent Lie algebra since the iterated partial derivatives of V must eventually be zero.

In the special case where V is the Henon-Hailes potential, we discussed in Example 6.4.5 how to construct the Lie group, the scaling semigroup $\{\sigma_s\}_{s\in\mathbb{R}_+} \subset \mathrm{Aut}(G)$, and the unitary representation U.

When the process is completed, in the general case, we get

$$H = -dU(x_1^2 + \cdots + x_n^2 + iz) \tag{7.0.12}$$

where $x_1, \cdots, x_n$, and z, are elements in the Lie algebra $\mathfrak{g}$ of G. Note that (7.0.12) is of the form (6.2.1) discussed in Section 6.2, but z is *not* in the span of the Lie algebra elements $x_1, \cdots, x_n$.

This adds a great deal of complication to the spectral

analysis of $H = P^2 + V$. *Not* all polynomials $V(\cdot)$ yield a selfadjoint operator H. As an example, $n = 1$, take

$$(7.0.13) \qquad H = P^2 - Q^4, \quad \text{i.e.,} \quad V(v) = -v^4.$$

We saw in Example 6.2.2 that H is *not* selfadjoint, in fact, it has deficiency indices (2,2), cf. [Jo 1, p. 107].

We now return to the Hamiltonians (7.0.2) and work out the spectral analysis.

7.1 Discrete Spectrum

We shall continue the study of the Hamiltonian for a nonrelativistic particle in a magnetic field. When the factor $\frac{e}{c}$ is absorbed into the vector potential $\mathbb{A} = (A_1, A_2, A_3)$, we have

$$(7.1.1) \qquad H = \frac{1}{2m}(\mathbb{P} - \mathbb{A})^2$$

with $\mathbb{P} = \frac{1}{\sqrt{-1}} \nabla$. In a coordinate system $v = (v_1, v_2, v_3) \in \mathbb{R}^3$, we have

$$(7.1.2) \qquad H = -\frac{1}{2m} \sum_{i=1}^{3} \left(\frac{\partial}{\partial v_i} - \sqrt{-1}\, A_i\right)^2.$$

Let $\mathfrak{g}_1$ be a 3-dimensional Abelian Lie algebra. Let $\mathfrak{g}_2$ be the Abelian Lie algebra which was constructed in the previous section. Recall that $\mathfrak{g}_2$ is spanned by the components B_i, $i = 1,2,3$ of the magnetic field $\mathbb{B} = \nabla \times \mathbb{A}$ and all the (non-zero) partial derivatives of the entries B_i. It is assumed that $\mathbb{A}$ is polynomial, so it follows that the dimension of $\mathfrak{g}_2$ is the "combined" degree of the vector polynomial $\mathbb{A}$, see formulas (7.0.1) for a precise statement.

In the previous section, we constructed an extension $\mathfrak{g}$ of $\mathfrak{g}_1$ by $\mathfrak{g}_2$, i.e., a short exact sequence of Lie algebras:

$$0 \longrightarrow \mathfrak{g}_2 \longrightarrow \mathfrak{g} \longrightarrow \mathfrak{g}_1 \longrightarrow 0$$

associated with the simply connected Lie group with Lie algebra

$\mathfrak{g}$, and a canonical unitary representation U of G satisfying

$$dU(x_i) = \frac{\partial}{\partial v_i} - \sqrt{-1}\, A_i, \qquad i = 1,2,3, \tag{7.1.3}$$

where $\{x_i : i = 1,2,3\}$ is a basis for $\mathfrak{g}_1$. The scaling semigroup $\{\delta_s : s \in \mathbb{R}_+\} \subset \operatorname{Aut} \mathfrak{g}$ was constructed to satisfy:

$$\delta_s(x_i) = sx_i; \qquad s \in \mathbb{R}_+, \quad i = 1,2,3. \tag{7.1.4}$$

We are now ready to state our first result:

Theorem 7.1.1 [JK 1]. If the representation U is irreducible, then the spectrum of H is purely discrete, and, in fact, the operator

$$e^{-tH} \tag{7.1.5}$$

is of trace class whenever $t > 0$.

Proof. By virtue of (7.1.3) we have

$$-2mH = dU(x_1^2+x_2^2+x_3^2), \tag{7.1.6}$$

and the three Lie algebra elements x_i, $i = 1,2,3$, generate the whole Lie algebra $\mathfrak{g}$. (Even when the dimension of $\mathfrak{g}$ is very large, there is no proper Lie subalgebra $\mathfrak{g}' \subset \mathfrak{g}$ containing the x_i's.)

We proved in [Jo 1, Section 3] that there is a function p on $\mathbb{R} \times G$ which solves the following Cauchy problem for the heat equation:

$$dL\left(\sum_{i=1}^{3} x_i^2\right)p(t,g) = 2m \frac{\partial}{\partial t} p(t,g) \tag{7.1.7}$$

and

$$p(t,g) = \delta(g), \qquad t = 0,$$

where $\delta(g)$ denotes the Dirac delta function on G. The following properties were established in [Jo 1] for the heat kernel $p(t,g)$:

(i) $\quad p(\cdot,\cdot) \in C^{\infty}(\mathbb{R}\times G\setminus\{(0,e)\}.$

(ii) $\quad p(t,g) = 0, \qquad t < 0, \quad g \in G.$

(iii) $\quad p(t,g) > 0, \qquad t > 0, \quad g \in G.$

(iv) $\quad g \longrightarrow p(t,g) \in \mathcal{L}^2(G) \cap \mathcal{L}^1(G) \qquad$ for all $t > 0,$

and

$$\int_G dg\; p(t,g) = 1.$$

We should note that L in (7.1.7) refers to the left-regular representation of G, and $\mathcal{L}^p(G)$ is the usual $\mathcal{L}^p$ space relative to the Haar measure on G. In fact, the Haar measure on G was calculated in the previous section. When coordinates (u,w) in G are chosen relative to the decomposition

(7.1.8)
$$\mathfrak{g} = \mathfrak{g}_1 + \mathfrak{g}_2.$$
$$u \longleftrightarrow \mathfrak{g}_1, \qquad w \longleftrightarrow \mathfrak{g}_2,$$

then

(7.1.9)
$$dg = du\, dw$$

where du is Lebesgue measure on $\mathbb{R}^3$ and dw is Lebesgue measure on $\mathbb{R}^d$, $d = \dim \mathfrak{g}_2$.

The group G carries a homogeneous norm defined as follows:

(7.1.10)
$$g = (u,w)$$
$$u = (u_1,u_2,u_3)$$
$$w = (w_i,w_{ij},w_{ijk},\text{etc.}\cdots),$$

$$|g| = \Big(\sum u_i^{2d} + \sum w_i^{\frac{2d}{2}} + \sum w_{ij}^{\frac{2d}{3}} + \sum w_{ijk}^{\frac{2d}{4}} + \cdots\Big)^{\frac{1}{2d}}$$

and satisfying

(7.1.11)
$$|\delta_s g| = s|g|, \qquad s \in \mathbb{R}_+, \quad g \in G.$$

The fifth property of the heat kernel which was established in [Jo 1] is:

(v) For all $t,c > 0$ there is a constant K such that

$$\int_{|g|>r} dg\ p(s,g) \le Ke^{-cr}, \qquad 0 < s < t.$$

It follows, in particular, from (v) that the the integral

$$\int_G dg\ e^{c|g|}\ p(t,g) \tag{7.1.12}$$

is finite for all $c,t > 0$.

The purpose of introducing the heat kernel is that the semigroup generated by $dU(\sum_1^3 x_i^2)$ may be given as an integral expression involving $p(t,g)$. When applied to the unitary representation U, given by (7.1.3), we get

$$e^{-tH} = \int_G dg\ p(t,g)U_g. \tag{7.1.13}$$

Let $S(t)$ be the operator defined by the integral on the right hand side of (7.1.13). Then, using (i)-(v) above, we get, for $f \in \mathcal{L}^2(\mathbb{R}^3)$ and $t \in \mathbb{R}_+$,

$$\begin{aligned} HS(t)f &= -\frac{1}{2m}\ dU(\sum_1^3 x_i^2)S(t)f \\ &= -\frac{1}{2m}\int_G dg\ dL(\sum_1^3 x_i^2)p(t,g)U_g f \\ &= -\int_G dg\ \frac{\partial}{\partial t}\ p(t,g)U_g f \\ &= -\frac{d}{dt}\ S(t)f. \end{aligned}$$

Standard semigroup theory [Yo, Chapter IX, §3] now implies the identity

$$e^{-tH}f = S(t)f, \qquad t \in \mathbb{R}_+, \quad f \in \mathcal{L}^2(\mathbb{R}^3). \tag{7.1.14}$$

Indeed,

$$\frac{d}{dt}\ e^{-(t-\tau)H}S(\tau)f = 0, \qquad 0 \le \tau \le t, \tag{7.1.15}$$

so (7.1.14) follows because the function, $\tau \longrightarrow e^{-(t-\tau)H}S(\tau)f$, must be constant by (7.1.15).

We are in a position to state and use the following result

of J. Dixmier.

Theorem 7.1.2 [Dix 2]. Let G be a simply connected nilpotent Lie group and let U be a unitary irreducible representation of G on a Hilbert space $\mathcal{H}$. Let $\varphi \in \mathcal{L}^1(G)$. Then the operator

$$U(\varphi) = \int_G dg\ \varphi(g)U_g \tag{7.1.16}$$

is trace class on $\mathcal{H}$.

We shall omit the proof, but only remark that it applies to the operator e^{-tH} since we just proved that

$$e^{-tH} = U(p(t,\cdot)) \tag{7.1.17}$$

for a particular unitary representation U, and we noted that

$$p(t,\cdot) \in \mathcal{L}^1(G), \qquad t \in \mathbb{R}_+.$$

If U is assumed irreducible, it follows that e^{-tH} is trace class.

As an application of Theorem 7.1.1, we get that the operator H, given by

$$-H = \left(\frac{\partial}{\partial v_1}\right)^2 + \left(\frac{\partial}{\partial v_2}\right)^2 + \left(\frac{\partial}{\partial v_3} - \sqrt{-1}\ v_1v_2v_3\right)^2 \tag{7.1.18}$$

on $\mathcal{L}^2(\mathbb{R}^3)$, has purely discrete spectrum. In fact e^{-tH} is trace class for $t \in \mathbb{R}_+$.

This is because the unitary representation U on $\mathcal{L}^2(\mathbb{R}^3)$, determined by

$$\begin{cases} dU(x_i) = \dfrac{\partial}{\partial v_i}, & i = 1,2 \\ dU(x_3) = \dfrac{\partial}{\partial v_3} - \sqrt{-1}\ v_1v_2v_3 \end{cases}, \tag{7.1.19}$$

is irreducible. The irreducibility assertion can be checked directly, and it also follows from the next result.

As a second application we get that the operator H on $\mathcal{L}^2(\mathbb{R}^2)$, give, for $\xi \in \mathbb{R}$, by

$$(7.1.20) \qquad -H = \left(\frac{\partial}{\partial v_1}\right)^2 + \left(\frac{\partial}{\partial v_2}\right)^2 - (\xi + v_1 v_2)^2,$$

has purely discrete spectrum, and in fact e^{-tH} is trace class, $t \in \mathbb{R}_+$.

The irreducible representation U on $\mathcal{L}^2(\mathbb{R}^2)$ which is used here is determined by

$$(7.1.21) \qquad \begin{cases} dU(x_j) = \frac{\partial}{\partial v_j}, & j = 1,2 \\ dU(x_3) = i(\xi + v_1 v_2), & i = \sqrt{-1}. \end{cases}$$

The irreducibility can be verified directly, or obtained from the sequel.

Verifications: Irreducibility of (7.1.19). This is the magnetic field problem with $\mathbb{A} = (0,0,v_1v_2v_3)$ and $\mathbb{B} = \nabla \times \mathbb{A} = (v_1v_3, -v_2v_3, 0)$. By taking commutators, we show that the operator Lie algebra $\mathcal{L} = dU(\mathfrak{g})$ contains the three multiplication operators, $f(v_1,v_2,v_3) \longrightarrow \sqrt{-1}\, v_j f(v_1,v_2,v_3)$, $j = 1,2,3$. The operators $\frac{\partial}{\partial v_i}$, $i = 1,2$, are already included in the primary list (7.1.19). So, if a projection E in $\mathcal{L}^2(\mathbb{R}^3)$ commutes with the operators in $\mathcal{L} = dU(\mathfrak{g})$, it commutes with multiplication by $\sqrt{-1}\, v_1v_2v_3$. Since E commutes with $dU(x_3)$, it must commute with $\frac{\partial}{\partial v_3} = dU(x_3) + \sqrt{-1}\, v_1v_2v_3$. We have proved that E commutes with the generators of the Schrödinger representation, for 3 degrees of freedom, and it follows from the familiar irreducibility of the latter representation on $\mathcal{L}^2(\mathbb{R}^3)$ that E must be 0 or I, cf. Lemma 4.2.1.

Irreducibility of (7.1.21).

Consider a projection E in $\mathcal{L}^2(\mathbb{R}^2)$ which commutes with the operators listed in (7.1.21). Since

$$[dU(x_1), dU(x_3)] = \sqrt{-1}\ v_2$$

and

$$[dU(x_2), dU(x_3)] = \sqrt{-1}\ v_1,$$

it follows that E commutes with the generators for the Schödinger representation, for 2 degrees of freedom. Lemma 4.2.1 again implies that E must be one of the trivial operators 0 or I on $\mathcal{L}^2(\mathbb{R}^2)$. This concludes the proof of irreducibility.

Proposition 7.1.3. Consider a polynomial vector field $\mathbb{A}$ on $\mathbb{R}^3$, and the corresponding nilpotent Lie group G with Lie algebra $\mathfrak{g}$ as in (7.1.8), and basis elements x_1, x_2, x_3 for $\mathfrak{g}_1$. Let U be the unitary representation of G on $\mathbb{R}^3$ determined by

$$dU(x_i) = \frac{\partial}{\partial v_i} - \sqrt{-1}\ A_i, \qquad i = 1,2,3,$$

and finally let $\mathbb{B} = \nabla \times \mathbb{A}$.

Then if $(\mathbf{w}\cdot\nabla)\mathbb{B} \neq 0$ for all $\mathbf{w} \in \mathbb{R}^3\backslash(0)$, then U is irreducible.

Proof. Pick $i = 1, 2,$ or 3 such that $(\mathbf{w}\cdot\nabla)B_i \neq 0$ for $\mathbf{w} \in \mathbb{R}^3\backslash(0)$. Then B_i must contain a monomial term $v_1^{m_1} v_2^{m_2} v_3^{m_3}$ with $m_1 > 0$, $m_2 > 0$, $m_3 > 0$. Then, by continuing the chain of commutator formulas (7.0.1), we can recover the three monomials $c_i v_i$, $i = 1,2,3$, with constants $c_i \neq 0$. If a projection E commutes with the Lie algebra of operators $dU(\mathfrak{g})$, then E must itself be a multiplication operator, i.e.,

$$(Ef)(v) = \psi(v)f(v), \qquad v \in \mathbb{R}^3,\ f \in \mathcal{L}^2(\mathbb{R}^3)$$

for some function ψ. This is a familiar argument on multiplication operators, and it is recalled in Lemma 4.2.1. But E also commutes with $dU(x_i)$, $i = 1,2,3$, and

$$[dU(x_i),E] = \frac{\partial}{\partial v_i}\psi$$

where the right hand side indicates the multiplication operator determined by the function $\frac{\partial}{\partial v_i}\psi$. It follows that ψ is a constant function. Since E is a projection, $\psi = 0$, or 1, and the irreducibility follows.

Remark 7.1.4. We do not know whether the converse of the proposition is true, but the following *fact* serves as a partial converse.

Suppose there is a nonzero smooth vector field $\mathbb{F}$ which commutes with the operator Lie algebra $dU(\mathfrak{g})$ given by (7.1.3), then $\mathbb{F}$ is constant, and, after a rotation of the coordinate system, we have

$$\mathbb{A} = \mathbb{A}(v_1,v_2). \tag{7.1.22}$$

In this case, the Fourier transform in v_3 yields a direct integral decomposition of the representation U.

We also note that the direct integral decomposition is utilized in [JK 1] in the analysis of the curved magnetic field

$$\mathbb{B} = (0,0,\ \gamma\ v_1v_2), \qquad \gamma \text{ constant.}$$

The results carry over to the general case (7.1.1)-(7.1.3), with (7.1.22), above. The partial Fourier transform

$$\int_{\mathbb{R}} e^{-i\xi v_3} f(v_1,v_2,v_3)dv_3 \tag{7.1.23}$$

yields a decomposition of the Hamiltonians

$$H = \frac{1}{2m}(\mathbb{P}-\mathbb{A})^2$$

into a direct integral over Hamiltonian

$$H(\xi) = \frac{1}{2m}\{(P_1-A_1)^2 + (P_2-A_2)^2 + (\xi-A_3)^2\}, \qquad \xi \in \mathbb{R}, \tag{7.1.24}$$

on the Hilbert space $\mathcal{L}^2(\mathbb{R}^2)$. Our irreducibility test may then be applied again to the unitary representation U^ξ determined by:

$$dU^\xi(x_j) = \sqrt{-1}(P_j - A_j), \qquad j = 1,2,$$

$$dU^\xi(x_3) = \sqrt{-1}(\xi - A_3), \qquad \xi \in \mathbb{R}.$$

If U^ξ is irreducible, then $e^{-tH(\xi)}$, $t \in \mathbb{R}_+$, is trace class by Theorem 7.1.1 above.

We now prove the assertion concerning the given vector field $\mathbb{F}$ on $\mathbb{R}^3$. By assumption, we have

$$(7.1.25) \qquad [\frac{\partial}{\partial v_i} - \sqrt{-1}\, A_i, \mathbb{F}] = 0, \quad i = 1,2,3.$$

Suppose $\mathbb{F} = \sum_{k=1}^{3} F_k \frac{\partial}{\partial v_k}$ with $F_k \in C^\infty(\mathbb{R}^3, \mathbb{R})$. Substitution into (7.1.25) shows that the Jacobian matrix $\left[\frac{\partial F_k}{\partial v_i}\right]$ vanishes identically, and that $\sum_k F_k \frac{\partial A_i}{\partial v_k} = 0$, $i = 1,2,3$. It follows that the three functions F_i are constant, i.e., that $\mathbb{F}$ is a constant vector field. Let $\mathbf{w}$ be the vector given by

$$w_i = F_i(\sum F_i^2)^{-\frac{1}{2}}, \qquad i = 1,2,3.$$

Then

$$(\mathbf{w}\cdot\nabla)A_i \equiv 0, \qquad i = 1,2,3.$$

Let $\mathcal{O}$ be the rotation which turns the basis vector $\mathbf{k} = (0,0,1)$ into $\mathbf{w}$, and let the new coordinates also be denoted (v_1, v_2, v_3). Then the Hamiltonian (7.1.1) transforms into

$$H = \frac{1}{2m}(\mathcal{O}\mathbb{P} - \mathbb{A}(v_1, v_2))^2.$$

as claimed.

7.2 Trace Formula: High Energy Behavior

We continue the study of the Hamiltonian operator

$$H = \frac{1}{2m}(\mathbb{P}-\mathbb{A})^2$$

on $\mathcal{L}^2(\mathbb{R}^3)$ with $\mathbb{P} = \frac{1}{\sqrt{-1}}\nabla_{\mathbf{v}}$, and $\mathbb{A} = (A_1, A_2, A_3)$ a polynomial vector potential. Recall the parametrization $g = (\mathbf{u}, \mathbf{w})$ of the nilpotent Lie group G, and the unitary representation U of G on $\mathcal{L}^2(\mathbb{R}^3)$ determined by

$$dU(x_i) = \frac{\partial}{\partial v_i} - \sqrt{-1}\, A_i, \qquad i = 1,2,3. \tag{7.2.1}$$

The Lie algebra elements x_i form a basis for $\mathfrak{g}$ in the short exact sequence of Lie algebras

$$0 \longrightarrow \mathfrak{g}_2 \longrightarrow \mathfrak{g} \longrightarrow \mathfrak{g}_1 \longrightarrow 0$$

and $d = \dim \mathfrak{g}_2$ is the combined degree of $\mathbb{A}$, as calculated in formulas (7.0.3).

Examples.

7.2.1 If $\mathbb{A} = (0,0,v_1v_2)$, $\mathbb{B} = \nabla \times \mathbb{A} = (v_1, -v_2, 0)$, and $d = 3$. As a result, $\dim G = 6$.

7.2.2 If $\mathbb{A} = (0,0,v_1v_2v_3)$, $\mathbb{B} = \nabla \times \mathbb{A} = (v_1v_3, -v_2v_3, 0)$, and $d = 6$. As a result, $\dim G = 9$.

In the first example, U is not irreducible, but it is a direct integral of a one-parameter family of irreducible representations U^{ξ}, where the direct integral is merely the Fourier transform in the third variable v_3. We get a direct integral decomposition

$$H = \int_{\mathbb{R}}^{\oplus} H(\xi)d\xi \tag{7.2.2}$$

and we proceed to show that: (i) H has continuous spectrum, while the component operators $H(\xi)$ have purely discrete spectrum. We also proved in Remark 7.1.4 that the decomposition (7.2.1) is typical for those curved magnetic field

Hamiltonians where the corresponding operators $P_i - \sqrt{-1}\, A_i$, $i = 1,2,3$, commute with some nonzero smooth vector field on $\mathbb{R}^3$.

In the second example, U is irreducible and H has purely discrete spectrum.

We now turn to the problem of finding $\operatorname{trace}(e^{-tH(\xi)})$ and $\operatorname{trace}(e^{-tH})$ in a general setting which includes the two types of examples.

In general, the representation U from (7.2.2) is induced from the subgroup G_2, and it may be expanded as follows:

$$(U_g f)(v) = e^{iE(g,v)} f(v+u), \qquad g \in G,\ g = (u,w),\ f \in \mathcal{L}^2(\mathbb{R}^3) \tag{7.2.3}$$

where the function $E : G \times \mathbb{R}^3 \longrightarrow \mathbb{R}$ may be determined from (7.2.1) and a system of polynomial equations, as described in Section 7.1.

Proposition 7.2.1. Let the Hamiltonian H be given by $H = \frac{1}{2}(\mathbb{P}-\mathbb{A})^2$, and the unitary representation U by (7.2.1). Let $p(t,g) = p_t(u,w)$ be the heat kernel on G satisfying (7.1.7), and let $d = \dim G_2$.

Then

$$\operatorname{trace}(e^{-tH}) = \int_{\mathbb{R}^d} dw\, p_t(0,w) \int_{\mathbb{R}^d} dv\, e^{iE(0,w,v)} \tag{7.2.4}$$

Proof. We established the formula

$$e^{-tH} = \int_G dg\, p(t,g) U_g$$

where dg is the Haar measure on G, and $p(t,g)$ is the heat kernel. We introduce the parametrization $g = (u,w)$, formula (7.1.9) for dg, and (7.2.3) for U. Then

$$e^{-tH} f(v) = \int_{\mathbb{R}^{3+d}} du dw\, p_t(u,w) e^{iE(u,w,v)} f(v+u).$$

It follows that e^{-tH} is an integral operator with kernel,

(7.2.5) $$K_t(\mathbf{v},\mathbf{u}) = \int_{\mathbb{R}^d} d\mathbf{w} \;\; p_t(\mathbf{u}-\mathbf{v},\mathbf{w})e^{iE(\mathbf{u}-\mathbf{v},\mathbf{w},\mathbf{v})}.$$

Since e^{-tH} is trace class by Theorem 7.1.1, and the kernel is continuous, in fact smooth, in both variables, it follows from Mercer's theorem [DS, p. 1088, Exercises 56-58] that

$$\begin{aligned}\mathrm{trace}(e^{-tH}) &= \int_{\mathbb{R}^3} d\mathbf{v}\; K_t(\mathbf{v},\mathbf{v}) \\ &= \int_{\mathbb{R}^d} d\mathbf{w}\; p_t(0,\mathbf{w})\int_{\mathbb{R}^3} d\mathbf{v}\;\; e^{iE(0,\mathbf{w},\mathbf{v})},\end{aligned}$$

which is the conclusion of the proposition.

We now turn to the trace formula for $H(\xi)$. We recall that the group G associated with Example 7.2.1 is 6-dimensional. In (6.4.7), we chose parameters $g = (x_1,x_2,x_3,y_1,y_2,z)$ corresponding to the matrix entries in the 4×4 triangular matrix. The calculations from Proposition 7.2.1 (application of Mercer's theorem and Fubini's theorem) yield

(7.2.6) $$\mathrm{trace}(e^{-tH(\xi)}) = \int_{\mathbb{R}^2} dv\;\; \hat{p}_t(0,0,\xi-v_1v_2,v_1,-v_2,1)$$

where $\hat{p}_t$ denotes the Fourier transform of the heat kernel in the last four variables, i.e.,

$$\begin{aligned}&\hat{p}_t(x_1,x_2,\varphi,\eta_1,\eta_2,\zeta) \\ &\quad = \int_{\mathbb{R}^3} dx_3dy_1dy_2dz\;\; e^{i(x_3\varphi+y_1\eta_1+y_2\eta_2+z\zeta)} \times p_t(x_1,x_2,x_3,y_1,y_2,z).\end{aligned}$$

We sketch the proof. First recall the formula for U^ξ on $\mathcal{L}^2(\mathbb{R}^2)$. With $g = (x_1,x_2,x_3,y_1,y_2,z)$, and $(v_1,v_2) \in \mathbb{R}^2$, we have

(7.2.7) $$U_g^\xi f(v_1,v_2) = e^{iE}\;\; f(v_1+x_1,v_2+x_2)$$

where

$$(7.2.8) \qquad E = \xi x_3 + z + y_1 v_1 - (y_2 + x_3 v_1)(v_2 + x_2).$$

Substitution into

$$e^{-tH(\xi)} f(v_1, v_2) = \int_G dg\ p_t(g) U_g^{\xi} f(v_1, v_2)$$
$$= \int_{\mathbb{R}^2} dx_1 dx_2\ K_t(v_1, v_2; x_1, x_2) f(x_1, x_2)$$

yields

$$K_t(v_1, v_2; x_1, x_2) = \int dx_3 dy_1 dy_2 dz\ \ p_t(x_1 - v_1, x_2 - v_2, x_3, y_1, y_2, z)$$
$$\times\ e^{iE(x_1 - v_1, x_2 - v_2, x_3, y_1, y_2, z, v_1, v_2)}$$

and

$$\operatorname{trace}(e^{-tH(\xi)})$$
$$= \int_{\mathbb{R}^2} dv\ K_t(v, v)$$
$$= \int_{\mathbb{R}^2} dv \int_{\mathbb{R}^4} dx_3 dy_1 dy_2 dz$$
$$p_t(0, 0, x_3, y_2, y_1, z) e^{i(\xi x_3 + z + y_1 v_1 - (y_2 + x_3 v_1) v_2)}$$
$$= \int_{\mathbb{R}^2} dv\ \hat{p}_t(0, 0, \xi - v_1 v_2, v_1, -v_2, 1).$$

When the formulas for the fields $\mathbb{A}$ and $\mathbb{B}$ are substituted, we get

$$(7.2.9) \qquad \operatorname{trace}(e^{-tH(\xi)}) = \int_{\mathbb{R}^2} dv\ \hat{p}_t(0, 0, \xi - A_3(v), \mathbb{B}(v) + k)$$

where $k = (0,0,1)$ is the conserved quantity direction which decomposes the Hamiltonian, cf. Example 6.4.3.

Note that (7.2.9) implies, in particular, continuity of $\xi \longrightarrow \operatorname{trace}(e^{-tH(\xi)})$ on $\mathbb{R}$; we shall prove analyticity in the next section.

One of the applications of trace formulas like (7.2.4) and (7.2.6) is the determination of eigenvalue asymptotics via

Tauberian methods. If the list of eigenvalues is denoted $\lambda_1, \lambda_2, \cdots$, with corresponding multiplicities $m_1, m_2, \cdots$, then the trace expression is also equal to

$$m_1 e^{-t\lambda_1} + m_2 e^{-t\lambda_2} + \cdots$$

which means that the spectrum, at least in principle, may be obtained by applying an inverse Laplace transform to the expression for $\operatorname{trace}(e^{-tH})$.

Explicit results along these lines are still modest in nature, because we have only been able to find formulas for the heat kernel for the case of Lie groups of rather small dimension. However, even in the general case, interesting asymptotics result when the scaling semigroup is substituted into the trace formulas.

By virtue of the scaling formulas, $\delta_s(x_i) = sx_i$, $\delta_s(y_i) = s^2 y_i$, and $\delta_s(z) = s^3 z$, $s \in \mathbb{R}_+$, we have, for the heat kernel,

$$\begin{aligned} &p_t(x_1, x_2, x_3, y_1, y_2, z) \\ &\qquad = t^{-5}\ p_1(t^{-\frac{1}{2}}x_1, t^{-\frac{1}{2}}x_2, t^{-\frac{1}{2}}x_3, t^{-1}y_1, t^{-1}y_2, t^{-\frac{3}{2}}z). \end{aligned}$$

For the Fourier transform in the last four variables, we therefore have

$$\hat{p}_t(0,0,\varphi,\eta_1,\eta_2,\zeta) = t^{-1}\ \hat{p}_1(0,0,t^{\frac{1}{2}}\varphi, t\eta_1, t\eta_2, t^{\frac{3}{2}}\zeta),$$

and substitution into (7.2.6) yields,

$$\begin{aligned} \operatorname{trace}(e^{-tH(\xi)}) &= t^{-1} \int_{\mathbb{R}^2} dv\ \hat{p}_1(0,0,t^{\frac{1}{2}}(\xi - v_1 v_2), tv_1, -tv_2, t^{\frac{3}{2}}) \\ &= t^{-1} \int_{\mathbb{R}^2} dv\ \hat{p}_1(0,0,t^{\frac{1}{2}}(\xi - A_3(v)), t\mathbb{B}(v) + t^{\frac{3}{2}}\mathbf{k}). \end{aligned}$$

As shown in [JK 1], this formula is useful in obtaining explicit information on $\lim_{t\to 0_+} \operatorname{trace}(e^{-tH(\xi)})$. Estimates on this quantity are called *high energy estimates*, because they give information about the upper part of the spectrum of $H(\xi)$.

Using methods from partial differential equations, it is possible to get direct estimates on the eigenvalues $\{\lambda_n(\xi) : n \in \mathbb{N}\}$ of the Hamiltonian

(7.2.10) $$H(\xi) = P_1^2 + P_2^2 + (\xi - v_1 v_2)^2$$

on $\mathcal{L}^2(\mathbb{R}^2)$: Let $V(v_1, v_2) = (\xi - v_1 v_2)^2$. For every $j = (j_1, j_2) \in \mathbb{Z}^2$, let C_j^λ be the square of side length $\lambda^{-\frac{1}{2}}$, centered at $j\lambda^{-\frac{1}{2}} = (j_1 \lambda^{-\frac{1}{2}}, j_2 \lambda^{-\frac{1}{2}})$. Let $N(\lambda)$ be the number of squares C_j^λ with

$$\max_{v \in C_j^\lambda} V(v) \leq \lambda.$$

An elementary counting argument yields the asymptotic estimate

$$N(\lambda) \leq 2\lambda^{\frac{3}{2}} \ln(\lambda|\xi| + \lambda^{\frac{3}{2}}).$$

The Fefferman-Phong theorem [FP] states that $N(\lambda)$ is equivalent to the spectral function for the operator. More specifically, let $M(\lambda)$ denote the number of eigenvalues of $H(\xi)$, counted with multiplicity, less than or equal to λ. Then there is an intrinsic constant c such that

$$M(\lambda) \leq cN(\lambda).$$

Combining the two results, we get

$$M(\lambda) \leq 2c\lambda^{\frac{3}{2}} \ell n(\lambda|\xi| + \lambda^{\frac{3}{2}})$$

for the spectral function $M(\lambda)$.

When this is substituted into the formula

$$\operatorname{trace}(e^{-tH(\xi)}) = \sum_{n=1}^{\infty} m_n e^{-t\lambda_n(\xi)}$$

we get

$$\mathrm{trace}(e^{-tH(\xi)}) \leq 2c \sum_n e^{-nt}(n+1)^{\frac{3}{2}} \ell n(|\xi|(n+1)+(n+1)^{\frac{3}{2}}),$$

from which we obtain the high-energy behavior $(t \longrightarrow 0_+)$:

$$\mathrm{trace}(e^{-tH(\xi)}) \leq \mathrm{const.}\ t^{-(\frac{5}{2}+\epsilon)}, \tag{7.2.11}$$

or

$$\mathrm{trace}(e^{-tH(\xi)}) \leq \mathrm{const.}\ t^{-\frac{5}{2}}|\ell n\ t|. \tag{7.2.12}$$

7.3 Continuous Spectrum

We saw in Section 7.1 that the spectral theory of

$$H = \frac{1}{2m}(\mathbb{P}-\mathbb{A})^2 \tag{7.3.1a}$$

may be decided by an analysis of the unitary representation U given by

$$dU(x_i) = \sqrt{-1}(P_i - A_i). \qquad i = 1,2,3. \tag{7.3.1b}$$

If the magnetic field $\mathbb{B} = \nabla \times \mathbb{A}$ is a function of all three space variables, there is no *conserved quantity*, the representation U is irreducible, and H has purely discrete spectrum. In fact, we obtained a formula for

$$\mathrm{trace}(e^{-tH}).$$

On the other hand, we also showed that the case of a conserved quantity reduces to the Hamiltonian

$$H = \frac{1}{2m}(\mathcal{O}\mathbb{P}-\mathbb{A}(v_1,v_2))^2 \tag{7.3.2}$$

where $\mathcal{O}$ is an orthogonal 3×3 real matrix. We showed that the partial Fourier transform

$$\int_{\mathbb{R}} dv_3\ e^{-iv_3\xi}\ f(v_1,v_2,v_3) \tag{7.3.3}$$

gives rise to a direct integral decomposition of H:

$$H = \int_{\mathbb{R}}^{\oplus} d\xi H(\xi)$$

where $H(\xi)$ is obtained from H by replacing $\mathbb{P} = (P_1, P_2, P_3)$ by $\mathbb{P}(\xi) = (P_1, P_2, \xi I)$, i.e.,

$$H(\xi) = \frac{1}{2m}(\mathcal{O}\mathbb{P}(\xi) - \mathbb{A}(v_1, v_2))^2. \tag{7.3.4}$$

Note that $H(\xi)$ is only acting in the remaining two space variables v_1, v_2, and we shall regard it as a selfadjoint operator in $\mathcal{L}^2(\mathbb{R}^2)$, rather than in $\mathcal{L}^2(\mathbb{R}^3)$ which is the Hilbert space for the original physical Hamiltonian H, given in (7.3.1).

Finally, recall that $H(\xi)$ is determined by a unitary representation U^ξ of G on $\mathcal{L}^2(\mathbb{R}^2)$ satisfying

$$dU^\xi(x_i) = \sqrt{-1}((\mathcal{O}\mathbb{P}(\xi))_i - A_i(v_1, v_2)), \quad i = 1,2,3. \tag{7.3.5}$$

If the representation U^ξ is irreducible, then the spectrum of $H(\xi)$ is discrete by Theorem 7.1.1; in fact, $e^{-tH(\xi)}$ is trace class on $\mathcal{L}^2(\mathbb{R}^2)$ for $t \in \mathbb{R}_+$, and in particular, the operator $z-H(\xi)$ has a compact inverse when z is not in the spectrum of $H(\xi)$.

The compactness of $(z-H(\xi))^{-1}$ for $\mathrm{Re}\, z > 0$ follows from [Jo 1] and Theorem 7.1.2. (It is not hard to extend to the general case.) Indeed,

$$(z-H(\xi))^{-1} = \int_G dg\; r_z(g) U_g^\xi \tag{7.3.6}$$

where the integral kernel $r_z(\cdot)$ is the Laplace transform of the heat kernel:

$$r_z(g) = \int_0^\infty dt\; e^{-tz}\; p_t(g) \in \mathcal{L}^1(G). \tag{7.3.7}$$

In this section, we shall give an irreducibility criterion for U^ξ, and show that the spectrum of H is continuous when it is satisfied for all $\xi \in \mathbb{R}$.

Proposition 7.3.1. The representation U^ξ is irreducible if

$$\mathbf{w} \cdot \nabla \mathbb{B} \neq 0 \quad \text{for all } \mathbf{w} \in \mathbb{R}^2 \setminus \{0\}. \tag{7.3.8}$$

Remark 7.3.2. Note that

$$\mathbb{B} = \left[\frac{\partial A_3}{\partial v_2}, -\frac{\partial A_3}{\partial v_1}, \frac{\partial A_2}{\partial v_1} - \frac{\partial A_1}{\partial v_2}\right]$$

in the present situation where $\mathbb{A} = \mathbb{A}(v_1, v_2)$, and it follows that the condition is satisfied whenever

$$(7.3.9) \qquad \mathbf{w}\cdot\nabla A_3 \neq 0 \quad \text{for all} \quad \mathbf{w} \in \mathbb{R}^2\backslash\{0\},$$

i.e., A_3 is not a function only on a line in $\mathbb{R}^2$.

We shall omit the proof of Proposition 7.3.1 since it is only a little different from the proof in Proposition 7.1.3.

We shall restrict attention to the case when (7.3.9) is satisfied, and turn to the spectral theory of H.

We begin by the

Lemma 7.3.3. Let the vector potential $\mathbb{A}$ satisfy the condition in Remark 7.1.4. (We say that $\mathbb{A}$ has one conserved quantity.) Assume U^{ξ} is irreducible for $\xi \in \mathbb{R}$, i.e., there is no more than one conserved quantity. Let $z \in \mathbb{C}$, $\operatorname{Re} z > 0$. Then the resolvent $(z-H(\xi))^{-1}$ depends analytically on ξ.

Remark 7.3.4. Lemma 7.3.3 was conjectured by W.H. Klink and the author [JK 1] in the special case of the setting of Example 7.2.1, where $\mathbb{A} = (0, 0, v_1 v_2)$, and it was proved in this special case, by Helffer [He 6]. Helffer's proof uses p.d.e. (a priori estimates) techniques and does not generalize to the general case. The present representation theoretic proof (based on Theorem 7.1.1) is new, as is Lemma 7.3.3 in the general case, and has not appeared in print (earlier).

Proof of Lemma 7.3.3. Let $g = (\mathbf{u}, \mathbf{w})$ be the familiar parametrization of G, $\mathbf{u} = (u_1, u_2, u_3)$, $\mathbf{w} = (w_i, w_{ij}, w_{ijk}, \cdots)$, and let $\mathbf{v} = (v_1, v_2, v_3) \in \mathbb{R}^3$. After Fourier transform in v_3, we arrive at the space $\mathcal{L}^2(\mathbb{R}^2)$ in the remaining variables v_1, v_2. The irreducible representation U^ξ is induced from the normal subgroup of G parametrized by (u_3, w), i.e., the dimension of this subgroup is one more than $d = \dim G_2$. This subgroup is still normal in G as well as Abelian, as can be checked by direct inspection, and U^ξ is induced from a one-dimensional representation of the larger subgroup $\simeq \mathbb{R}^{d+1}$. The representation U^ξ is determined by the three generators

$$dU^\xi(x_i) = \sqrt{-1}\{(\mathcal{O}\mathbb{P}(\xi))_i - A_i(v_1, v_2)\}, \qquad i = 1,2,3. \tag{7.3.10}$$

It follows that there is a polynomial function

$$E : \mathbb{R} \times G \times \mathbb{R}^2 \longrightarrow \mathbb{R}$$

such that

$$(U_g^\xi f)(v_1, v_2) = e^{iE(\xi, g, v_1, v_2)} f(v_1+u_1, v_2+u_2), \tag{7.3.11}$$

$$\xi \in \mathbb{R}, \quad g \in G, \quad f \in \mathcal{L}^2(\mathbb{R}^2).$$

Indeed, let H be the $d+1$ dimensional subgroup, parametrized by $(u_3, w_i, w_{ij}, w_{ijk}, \cdots)$. Then we can use (7.3.5) to determine a unitary character (depending on ξ) χ_ξ on H such that U^ξ is unitarily equivalent to the representation $U(\chi_\xi)$ induced from $\chi_\xi \in \hat{H}$, and acting on functions

$$\varphi : G \longrightarrow \mathbb{C}$$

satisfying

$$\varphi(hg) = \chi_\xi(h)\varphi(g), \qquad h \in H, \quad g \in G, \tag{7.3.12}$$

as completed in the norm

$$\left[\int_{H\backslash G} |\varphi|^2 \, d\mu\right]^{\frac{1}{2}}.$$

The mapping W defined by

$$(W\varphi)(u_1, u_2) = \varphi(u_1, u_2, 0, 0) \tag{7.3.13}$$

is then a unitary isomorphism

$$W : \mathcal{H}(\chi_\xi) \longrightarrow \mathcal{L}^2(\mathbb{R}^2)$$

and

$$U_g^\xi = WU_g(\chi_\xi)W^*, \qquad g \in G. \tag{7.3.14}$$

We now derive the formula for the unitary cocycle

$$e^{iE(\chi,g,v_1,v_2)}$$

from (7.3.11) in terms of χ_ξ. Let v denote the element in G with coordinates $(v_1,v_2,0,0)$. Pick a 2-dimensional subgroup K of G such that every g in G factors uniquely, $g = hu$, $h \in H$, $u \in K$. Then, according to (7.3.14), we have

$$(U_g^\xi f)(v) = WU_g(\chi_\xi)W^* f(v)$$

where $v = (v_1,v_2)$ is now treated (dually) as an element in the subgroup K as well as a space "position" vector in $\mathbb{R}^2$. We get

$$\begin{aligned}
U_g^\xi f(v) &= U_g(\chi_\xi)W^* f(v) \\
&= W^* f(vg) \\
&= W^* f(vhu) \\
&= W^* f((vhv^{-1})vu) \\
&= \chi_\xi(vhv^{-1})W^* f(vu) \\
&= \chi_\xi(vhv^{-1})f(v+u) \\
&= e^{iE(\xi,g,v_1,v_2)} f(v_1+u_1,v_2+u_2).
\end{aligned}$$

We conclude that the mapping $\xi \longrightarrow U^\xi$ is analytic on $\mathbb{R}$, i.e., that it is analytic into the space of unitary representations with the "compact"—"open" topology. "Compact" refers to compact subsets of G, and "open" refers to the strong operator topology on the unitary operators on the fixed Hilbert space $\mathcal{L}^2(\mathbb{R}^2)$.

But we saw in (7.3.6) that

$$(z-H(\xi))^{-1} = \int_G dg\ r_z(g)U_g^\xi \tag{7.3.15}$$

with a smooth $\mathcal{L}^1(G)$-kernel $r_z(\cdot)$ given by (7.3.7). Since the right hand side of (7.3.11) depends analytically on ξ (for ξ on the real line), it follows that $\xi \longrightarrow (z-H(\xi))^{-1}$ is analytic. Since it also takes values in the compact operators on $\mathcal{L}^2(\mathbb{R}^2)$, it follows from the Rellich-Kato theorem [Kat 2, VII.5, Thm. 3.9, p. 392] that the spectrum of $H(\xi)$ depends analytically on ξ. More specifically, there is a sequence of functions

$$\lambda_n : \mathbb{R} \longrightarrow \mathbb{R}, \qquad n \in \mathbb{N}, \tag{7.3.16}$$

and

$$f_n : \mathbb{R} \longrightarrow \mathcal{L}^2(\mathbb{R}^2), \qquad n \in \mathbb{N}, \tag{7.3.17}$$

which may be taken both to be *analytic* on $\mathbb{R}$ and satisfying (i)-(iv) below:

(i) $$\lambda_1(\xi) \le \lambda_2(\xi) \le \cdots,$$

(ii) $$\lim_{n\to\infty} \lambda_n(\xi) = \infty,$$

(iii) $\{f_n(\xi)\}_{n\in\mathbb{N}}$ is an *orthonormal* basis for $\mathcal{L}^2(\mathbb{R}^2)$,

and

(iv)

$$H(\xi)f_n(\xi) = \lambda_n(\xi)f_n(\xi), \qquad \xi \in \mathbb{R},\quad n \in \mathbb{N}. \tag{7.3.18}$$

We are ready to prove the

Theorem 7.3.5. Let

$$H = \frac{1}{2m}(\mathbb{P}-\mathbb{A})^2$$

be a Hamiltonian with precisely one conserved quantity for the field. Then the spectrum of H is continuous.

Proof. Consider a solution to the eigenvalue problem

(7.3.19) $$Hf = \lambda f$$

with $f \in D(H) \subset \mathcal{L}^2(\mathbb{R}^3)$, and $\lambda \in \mathbb{R}$. Let U be the unitary representation (reducible) of G on $\mathcal{L}^2(\mathbb{R}^3)$ associated with H. We proved in [Jo 1, Theorem 3.1] that $D(H) = C^\infty(U)$. The proof of this result is similar to the proof of Lemma 5.3.2 in Section 5.3, and we shall not include the details here. It follows that f is in the domain of all operator monomials formed from elements in $dU(g)$. An inspection of the operators in $dU(g)$ reveals that f must then be in the domain of all monomials formed from the generators of the Schrödinger representation of the 7-dimensional Heisenberg group on $\mathcal{L}^2(\mathbb{R}^3)$, i.e., in the space of C^∞-vectors for this Schrödinger representation. We noted in Section 4.2 that the latter C^∞-space coincides with the Schwartz-space of rapidly decreasing smooth functions on $\mathbb{R}^3$, i.e., $f \in \mathcal{S}(\mathbb{R}^3)$.

We shall use the validity of the Fourier-inversion for such functions:

(7.3.20) $$f(v_1,v_2,v_3) = \int_{\mathbb{R}} dv_3 e^{iv_3\xi} \hat{f}(v_1,v_2,\xi)$$

where $\hat{f}(v_1,v_2,\xi)$ denotes the partial Fourier transform of f in the third variable.

Substitution of (7.3.19) into (7.3.20) yields

$$\begin{aligned}\int_{\mathbb{R}} dv_3 e^{iv_3\xi} H(\xi)\hat{f}(v_1,v_2,\xi) &= \int_{\mathbb{R}} dv_3 e^{iv_3\xi} (Hf)^\wedge(v_1,v_2,\xi) \\ &= \int_{\mathbb{R}} dv_3 e^{iv_3\xi} \lambda\hat{f}(v_1,v_2,\xi) \\ &= \lambda\int_{\mathbb{R}} dv_3 e^{iv_3\xi} \hat{f}(v_1,v_2,\xi),\end{aligned}$$

and uniqueness of the Fourier-transform yields further

(7.3.21) $$H(\xi)\hat{f}(v_1,v_2,\xi) = \lambda\hat{f}(v_1,v_2,\xi)$$

where the operator $H(\xi)$ is acting in the two real variables v_1,v_2. Since we are working with Schwartz-functions, (7.3.21) holds pointwise everywhere. (The appropriate functions may not be of Schwartz type from the beginning, but they may be "adjusted" on sets of measure zero such that the resulting "modified" functions are of Schwartz type.

We shall now examine equation (7.3.21) more closely, and we shall rewrite (7.3.21) in the following more compact, but equivalent form:

$$H(\xi)\hat{f}(\xi) = \lambda\hat{f}(\xi), \qquad \xi \in \mathbb{R}. \tag{7.3.22}$$

We shall avoid "extraneous" technicalities, and assume that H is given by

$$H = \frac{1}{2}(\mathbb{P}-\mathbb{A}(v_1,v_2))^2,$$

and, therefore,

$$H(\xi) = \frac{1}{2}(\mathbb{P}(\xi)-\mathbb{A}(v_1,v_2))^2$$

where

$$\mathbb{P}(\xi) = (P_1,P_2,\xi) \qquad \xi \in \mathbb{R}.$$

It follows that

$$\frac{d}{d\xi} H(\xi) = \xi - A_3(v_1,v_2). \tag{7.3.23}$$

Now substitute this into the differentiated version of equation (7.3.22), i.e.,

$$\left(\frac{d}{d\xi} H(\xi)\right) \hat{f}(\xi) + H(\xi) \frac{d}{d\xi} \hat{f}(\xi) = \lambda \frac{d}{d\xi} \hat{f}(\xi).$$

Then take inner products with the basis functions $f_n(\xi)$ from (7.3.18) corresponding to values of the pair $(n,\xi) \in \mathbb{N} \times \mathbb{R}$ such that

$$\lambda = \lambda_n(\xi) \tag{7.3.24}$$

holds. After a simple algebraic reduction, using (7.3.18), we get

$$\langle A_3\hat{f}(\xi),f_n(\xi)\rangle_{\mathcal{L}^2(\mathbb{R}^2)} = (\lambda_n'(\xi)-\xi)\langle\hat{f}(\xi),f_n(\xi)\rangle_{\mathcal{L}^2(\mathbb{R}^2)}.$$

Let $E_n(\xi)$ be the projection onto the $\lambda_n(\xi)$-eigenspace. It follows that $E_n(\xi)\hat{f}(\xi) \in D(A_3)$, and that

(7.3.25) $$A_3E_n(\xi)\hat{f}(\xi) = (\lambda_n'(\xi)-\xi)E_n(\xi)\hat{f}(\xi).$$

We have used the symbol A_3 for the multiplication operator in $\mathscr{L}^2(\mathbb{R}^2)$ given by:

$$\psi(v_1,v_2) \longrightarrow A_3(v_1,v_2)\psi(v_1,v_2);$$

and the domain of this operator consists of $\psi \in \mathscr{L}^2(\mathbb{R}^2)$ such that

$$\int_{\mathbb{R}^2} dv_1dv_2|A_3(v_1,v_2)\psi(v_1,v_2)|^2 < \infty.$$

But $A_3(v_1,v_2)$ is a real-valued polynomial in two variables. If this polynomial is nonzero, there can be no solutions $\psi \in \mathscr{L}^2(\mathbb{R}^2)\backslash\{0\}$ to the equation

(7.3.26) $$A_3\psi = (\lambda_n'(\xi)-\xi)\psi.$$

In that case, (7.3.25) implies that $E_n(\xi)\hat{f}(\xi) = 0$. Since this holds for all values n,ξ such that (7.3.24) is satisfied, we conclude that $\hat{f} = 0$, and therefore $f = 0$.

It remains to consider the case where $A_3 = A_3(v_1,v_2) \equiv 0$. (In this case, we have $B_1(v_1,v_2) = B_2(v_1,v_2) \equiv 0$, and $B_3 = \frac{\partial A_2}{\partial v_1} - \frac{\partial A_1}{\partial v_2}$.)

The equation (7.3.25) reduces to

(7.3.27) $$0 = (\lambda_n'(\xi)-\xi)E_n(\xi)\hat{f}(\xi).$$

If $E_n(\xi)\hat{f}(\xi) \neq 0$ for some $\xi \in \mathbb{R}$, then it must be nonzero in an interval. In this interval, we would then have $0 = \lambda_n'(\xi)-\xi$ and, by analyticity, this would hold on $\mathbb{R}$. This means that $\lambda_n(\xi)$ is a polynomial of no more than second degree, and that there is only one value of n where (7.3.24) is satisfied. But then there can be no more than two values of ξ where

(7.3.24) holds (the roots of the corresponding second degree polynomial). This means that $\hat{f}(\xi)$ is *not* different from zero on a set of positive measure in the ξ-variable. Therefore $\hat{f} \equiv 0$ also in this case.

This concludes the proof of Theorem 7.3.4.

7.4 Lebesgue Spectrum

Let G be a simply connected nilpotent Lie group with Lie algebra $\mathfrak{g}$, and let L be the left-regular representation on the Hilbert space $\mathcal{L}^2(G)$ of square integrable functions on G. Suppose there are elements $\{x_i : i = 1,\cdots,r\}$ in $\mathfrak{g}$, and a scaling semigroup $\{\delta_s : s \in \mathbb{R}_+\} \subset \mathrm{Aut}(\mathfrak{g})$, such that

(a) $$\delta_s(x_i) = sx_i, \qquad 1 \leq i \leq r,$$

(b) the x_i's generate $\mathfrak{g}$ as a Lie algebra.

Then it follows, as a special case of Theorem 6.2.1, that the selfadjoint operator $dL\left(\sum_{i=1}^{r} x_i^2\right)$ has Lebesgue spectrum, i.e., it has a *spectral transform*, as specified in the conclusion of Theorem 6.4.6. We say that it is unitarily equivalent to a multiplication operator.

Definition 7.4.1. Let G be a Lie group with Lie algebra $\mathfrak{g}$. Let elements $x_1,\cdots,x_r$ in $\mathfrak{g}$ be given, and assume that conditions (a) and (b) hold, then we say that the second order right-invariant partial differential operator $-dL\left(\sum_1^r x_i^2\right)$ is a *sub-Laplacian* on G. We shall reserve the letter Δ for this operator. Note that we have chosen Δ to have positive spectrum.

In the special case where $\mathfrak{g}$, G, and U are associated with the curved magnetic field Hamiltonian, we saw that the spectral theory of

(7.4.1) $$H = (\mathbb{P}-\mathbb{A})^2 = -dU(x_1^2+x_2^2+x_3^2)$$

can be decided in terms of the operator $dL\left[\sum_1^3 x_i^2\right]$ on $\mathcal{L}^2(G)$.

The important first step in this analysis was the observation that the semigroup e^{-tH} on $\mathcal{L}^2(\mathbb{R}^3)$ can be calculated in terms of the Gaussian $p(t,g)$ on G, given by

(7.4.2) $$\frac{\partial}{\partial t} p(t,g) = dL\left[\sum_1^3 x_i^2\right]p(t,g)$$

$$p(0,g) = \delta(g).$$

We have

(7.4.3) $$e^{-tH} = \int_G dg\ p(t,g)U_g.$$

The further Fourier analysis of H is carried out in terms of U. A decomposition of U as a superposition of irreducibles U^ξ leads to a decomposition of H, and the corresponding Fourier components H_ξ also satisfy

(7.4.4) $$e^{-tH_\xi} = \int_G dg\ p(t,g)U_g^\xi.$$

The spectral transform, applied to $p(t,g)$ is just multiplication by $e^{-t\lambda}$, and finding the *spectral transform* is an important step in the solution of the Schrödinger equation for H. The spectral transform for the sub-Laplacian $\Delta = dL\left[\sum_1^3 x_i^2\right]$ also leads to a formula for the Gaussian convolution semigroup $p(t,g)$, $(t,g) \in \mathbb{R}_+ \times G$.

However, Theorem 6.4.6 is not particularly explicit as it stands. It does not, for instance, give much information on the separable multiplicity space $\mathscr{V}$.

Following [JK 2], we shall work out an explicit formula for the spectral transform in case G is the Heisenberg group.

Rather than using the coordinates on the Heisenberg group

from Section 4.2, or Example 6.4.1, the following alternative coordinates will facilitiate computations.

Let $(x_1,x_2,x_3) \in \mathbb{R}^3$, $z = x_1 + ix_2 \in \mathbb{C}$, and define a multiplication on $\mathbb{C} \times \mathbb{R}$ by

$$(7.4.5) \qquad (z,x_3)\cdot(z',x_3') = (z+z',x_3+x_3'+ 2\ \mathrm{Im}(z\bar{z}')).$$

It is well known that the resulting Lie group G is isomorphic to the Heisenberg group.

We shall consider the following two bases for the real Lie algebra of all right-invariant vector fields on G:

$$(7.4.6a) \qquad X_1 = \frac{\partial}{\partial x_1} - 2x_2\frac{\partial}{\partial x_3},$$

$$X_2 = \frac{\partial}{\partial x_2} + 2x_1\frac{\partial}{\partial x_3},$$

$$X_3 = \frac{\partial}{\partial x_3},$$

and

$$(7.4.6b) \qquad Z = \frac{1}{2}(X_1-iX_2)$$

$$\bar{Z} = \frac{1}{2}(X_1+iX_2)$$

$$D = X_3.$$

In terms of the Cauchy-Riemann operators, we have

$$\partial = \frac{\partial}{\partial z} = \frac{1}{2}(\frac{\partial}{\partial x_1} - i\frac{\partial}{\partial x_2})$$

$$\bar{\partial} = \frac{\partial}{\partial \bar{z}} = \frac{1}{2}(\frac{\partial}{\partial x_1} + i\frac{\partial}{\partial x_2})$$

and

$$Z = \frac{\partial}{\partial z} - i\bar{z}\frac{\partial}{\partial x_3} = \partial - i\bar{z}D,$$

$$\bar{Z} = \frac{\partial}{\partial \bar{z}} + iz\frac{\partial}{\partial x_3} = \bar{\partial} + izD.$$

The sub-Laplacian Δ is then

$$\Delta = -\frac{1}{4}(X_1^2+X_2^2) = -\frac{1}{2}(Z\bar{Z}+\bar{Z}Z)$$

$$= -\partial\bar{\partial} + iD(\bar{z}\,\bar{\partial} - z\partial) - |z|^2D^2$$

$$= -\frac{1}{4}\left(\left(\frac{\partial}{\partial x_1}\right)^2+\left(\frac{\partial}{\partial x_2}\right)^2\right) + D\left(x_2\frac{\partial}{\partial x_1} - x_1\frac{\partial}{\partial x_2}\right) - (x_1^2+x_2^2)D^2$$

$$= -\frac{1}{4}\left(\frac{1}{r}\frac{\partial}{\partial r}\left(r\frac{\partial}{\partial r}\right) + \frac{1}{r^2}\left(\frac{\partial}{\partial\varphi}\right)^2\right) - \frac{\partial}{\partial\varphi}D - r^2D^2$$

where, in the last step, polar coordinates $z = re^{i\varphi}$ have been introduced. Recall the transformation formulas:

$$\partial = \frac{1}{2}e^{-i\varphi}\left(\frac{\partial}{\partial r} + \frac{1}{r}\frac{\partial}{\partial\varphi}\right)$$

$$\bar{\partial} = \frac{1}{2}e^{i\varphi}\left(\frac{\partial}{\partial r} - \frac{1}{r}\frac{\partial}{\partial\varphi}\right).$$

Since the operator $\frac{1}{i}D$ transforms into multiplication by k under the partial Fourier transform

$$\int_{\mathbb{R}} dx_3\, e^{-ikx_3} f(r,\varphi,x_3), \tag{7.4.7}$$

it follows that the sub-Laplacian Δ transforms into

$$\Delta = -\frac{1}{4}\nabla^2 - ik\frac{\partial}{\partial\varphi} + r^2k^2 \tag{7.4.8}$$

in the new variables (r,φ,k). We have denoted the usual Laplacian on $\mathbb{R}^2 \sim \mathbb{C}$ by ∇^2, i.e.,

$$\nabla^2 = \left(\frac{\partial}{\partial x_1}\right)^2 + \left(\frac{\partial}{\partial x_2}\right)^2 = 4\partial\bar{\partial}.$$

We note that (7.4.8) involves the harmonic oscillator,

$$-\frac{1}{4}\nabla^2 + k^2r^2,$$

and the angular momentum operator,

$$\frac{\partial}{\partial\varphi} = x_1\frac{\partial}{\partial x_2} - x_2\frac{\partial}{\partial x_1}.$$

Formula (7.4.8) also reveals that the sub-Laplacian is itself a magnetic field Hamiltonian, this time with a constant magnetic field. When Theorem 7.3.5, from the previous section, is applied, we get an independent proof of the continuity of the spectrum. But we shall proceed to show that the spectrum is in fact Lebesgue, i.e., in particular, absolutely continuous.

For the vector potential, take

$$A = A(x_1, x_2) = (kx_2, -kx_1), \qquad (x_1, x_2) \in \mathbb{R}^2.$$

Then

$$\begin{aligned} H &= \frac{1}{4}\left(\frac{1}{i}\nabla - A\right)^2 \\ &= -\frac{1}{4}\nabla^2 - ik\left(x_1 \frac{\partial}{\partial x_2} - x_2 \frac{\partial}{\partial x_1}\right) + k^2(x_1^2 + x_2^2) \\ &= \Delta \qquad \text{(i.e., the sub-Laplacian).} \end{aligned}$$

We now show that there is a natural representation of $SU(2)$ such that Δ is also an element in the enveloping algebra of this representation. The $SU(2)$-representation will be constructed from a representation of the canonical commutation relations as follows: Define

$$A_i = \frac{1}{2\sqrt{|k|}} \frac{\partial}{\partial x_i} + \sqrt{|k|}\, x_i,$$

$$A_i^* = \frac{-1}{2\sqrt{|k|}} \frac{\partial}{\partial x_i} + \sqrt{|k|}\, x_i, \qquad i = 1,2.$$

We view A_i as an operator on $\mathcal{L}^2(G)$ but the coordinates (x_1, x_2, x_3) are used, and x_3 is dualized into the k-variable by virtue of the transform (7.4.7).

It is immediate that the CCR's are satisfied, i.e.,

$$[A_i, A_j^*] = \delta_{ij} I, \qquad 1 \le i,j \le 2, \tag{7.4.9}$$

and it follows that the three operators $J_\pm, J_3$, defined below, yield a representation of $SU(2)$:

$$J_+ = A_1^* A_2$$

$$J_- = J_+^* = A_2^* A_1$$

and

$$J_3 = \frac{1}{2}(A_1^*A_1 - A_2^*A_2).$$

Indeed, a direct computation yields the familiar commutation relations for the SU(2)-Lie algebra:

$$[J_3, J_\pm] = \pm J_\pm$$

and

$$[J_+, J_-] = 2J_3.$$

The Casimir operator Ω is therefore given by the formula

$$\Omega = \frac{1}{2}\{J_+, J_-\} + J_3^2$$

where $\{\cdot,\cdot\}$ is the anticommutator.

For the harmonic oscillator we get:

$$\begin{aligned} -\frac{1}{4}\nabla^2 + k^2r^2 &= \frac{|k|}{2}\sum_{i=1}^{2}\{A_i, A_i^*\} \\ &= |k|(I + \sum_{i=1}^{2} A_i^*A_i) \\ &= |k|(I + 4\Omega)^{\frac{1}{2}}. \end{aligned}$$

Similarly, for the angular momentum operator, we have:

$$\begin{aligned} \frac{\partial}{\partial\varphi} &= x_1\frac{\partial}{\partial x_2} - x_2\frac{\partial}{\partial x_1} \\ &= A_1^*A_2 - A_2^*A_1 \\ &= J_+ - J_-. \end{aligned}$$

Substitution into (7.4.8) yields:

(7.4.10) $$\Delta = |k|(I + 4\Omega)^{\frac{1}{2}} - ik(J_+ - J_-).$$

The second term on the right hand side is in the SU(2)-Lie algebra as can be seen from an application of the following base-change: Define J_i, $i = 1,2$, by $J_\pm = J_1 \pm \sqrt{-1}\,J_2$, and

$$K_{\pm} := J_3 \pm \sqrt{-1}\, J_1,$$

$$K_0 := J_2.$$

Then

$$[K_0, K_{\pm}] = \pm K_{\pm},$$

$$[K_+, K_-] = 2K_0.$$

This means that we have raising, resp., lowering, operators $K_{\pm}$ which can be used to simultaneously diagonalize the two operators on the right in the formula

$$\Delta = |k|(I + 4\Omega)^{\frac{1}{2}} + 2kK_0. \tag{7.4.10}$$

In terms of the coordinates on G, we have

$$K_+ = |k|z^2 - \frac{1}{|k|}(\bar{\partial})^2,$$

$$K_- = |k|(\bar{z})^2 - \frac{1}{|k|}\partial^2,$$

and

$$K_0 = \frac{1}{2\sqrt{-1}}\frac{\partial}{\partial\varphi} = J_2.$$

The familiar raising/lowering technique yields a basis $\{f_{j,m}\}$ satisfying

$$K_+ f_{j,m} = (j(j+1) - m(m+1))^{\frac{1}{2}} f_{j,m+1}, \tag{7.4.11}$$

$$K_- f_{j,m} = (j(j+1) - m(m-1))^{\frac{1}{2}} f_{j,m-1} \tag{7.4.12}$$

where $\Omega = j(j+1)I$, and

$$K_+ f_{j,j} = 0 = K_- f_{j,-j} \tag{7.4.13}$$

are the endpoint (highest/lowest weight) condition on the orthonormal basis vectors

$$\{f_{j,m} : -j \leq m \leq j\}, \qquad j = 0, \tfrac{1}{2}, 1, \cdots.$$

We solve this system of differential equations in [JK 2]. The solution is summarized in the following:

Theorem 7.4.2. The system (7.4.11)-(7.4.13) is solved by the following orthonormal family of functions

$$(7.4.14)\quad f^{k}_{j,m}(r,\varphi,x_3)$$

$$= \sqrt{\frac{(j-|m|)!\ 2^{2|m|+1}}{\Gamma(j+|m|+1)}}\ |k|^{|m|}\ r^{2|m|} \times e^{-|k|r^2}\ L^{2|m|}_{j-|m|}(2|k|r^2) \times \exp(\sqrt{-1}(m\varphi+kx_3)),$$

where $L^{2|m|}_{j-|m|}(\cdot)$ is the Laguerre polynomial with indices as indicated. For the eigenvalues $\lambda_{j,m}(k)$ of

$$(7.4.15)\qquad \Delta f_{j,m} = \lambda_{j,m}(k) f_{j,m}$$

we have

$$\lambda_{j,m}(k) = (2j+1)|k|+2mk, \qquad j = 0,\tfrac{1}{2},1,\cdots, \quad m = -j,\cdots,j.$$

It follows from the theorem that the spectral transform W which diagonalizes the sub-Laplacian Δ is determined by

$$(WF)(j,m,k) = \int_{\mathbb{R}_+} r\,dr \int_{-\pi}^{\pi} d\varphi \int_{\mathbb{R}} dx_3\ F(r,\varphi,x_3)\overline{f^{k}_{j,m}(r,\varphi,x_3)}.$$

We need to check the two properties listed in Theorem 6.4.6 for the spectral transform. One is immediate: Formula (7.4.15) shows that W does indeed diagonalize Δ. The other property amounts to the following Parseval type identity:

$$(7.4.16)\quad \sum_j \sum_m \int_{\mathbb{R}} dk\,|(WF)(j,m,k)|^2 = \|F\|^2_{\mathcal{L}^2(G)}$$

$$= \int_{\mathbb{R}_+} r\,dr \int_{-\pi}^{\pi} d\varphi \int_{\mathbb{R}} dx_3\,|F(r,\varphi,x_3)|^2$$

By Theorem 7.4.2, the functions $f_{j,m}$ are constructed to form an orthonormal basis for the representation space of the $(K_\pm, K_0)$ representation of $SU(2)$. This implies the inequality $\leq$ of (7.4.16) by virtue of the Bessel inequality. If the inequality were sharp, the projection onto $N(I-W^*W)$ would commute with the $SU(2)$-action. This projection, E say, also commutes with Δ. Let $\mathcal{H}(j,k)$ be the space spanned by

$$\{f^k_{j,m} : -j \leq m \leq j\}.$$

Then $\mathcal{H}(j,k)$ is invariant under E. But the $SU(2)$-action on $\mathcal{H}(j,k)$ is irreducible for all j,k by construction. It follows that E is the identity which says that W is isometric, and the Bessel inequality must be an equality. In other words, (7.4.16) holds.

As an application, we get the following formula for the Gaussian on the Heisenberg group, relative to the sub-Laplacian. Recall that $p(t,g)$ is the solution to

$$\frac{\partial}{\partial t} p(t,g) = -\Delta p(t,g)$$

$$p(0,g) = \delta(g), \qquad g \in G, \quad t \in \mathbb{R}_+.$$

We choose the coordinates (r,φ,x_3) for g in G:

$$p(t,r,\varphi,x_3) = \sum_j \frac{1}{\Gamma(2j+1)} \int_{\mathbb{R}} dk\, e^{-t(2j+1)|k|} f^k_{j,0}(r,\varphi,x_3)$$

where

$$f^k_{j,0}(r,\varphi,x_3) = \sqrt{2}\, L^0_j(2|k|r^2) e^{-|k|r^2} e^{ikx_3}$$

and the j summation is over $j = 0, \frac{1}{2}, 1, \frac{3}{2}, \cdots$.

Additional References for Chapter 7

Sect. 7.1: [JK 1], [Jo 1], [Jo 4-5], [Dix 2, 4-6], [Goo 3], [Hul 1].

Sect. 7.2: [Jo 25], [FP], [He 6], [Hö 5].

Sect. 7.3: [Jo 26], [JM], [Dix 9], [Fo 2], [FS 2], [Gav 1], [Goo 9].

Sect. 7.4: [JK 2], [Bony], [E], [BGGV], [HN 2], [HN 10], [Le], [MS], [Rot 1], [RSt].

CHAPTER 8. INFINITE-DIMENSIONAL LIE ALGEBRAS

In this chapter we study a selection of infinite-dimensional Lie algebras which have recently played a role in representation theory and in operator algebras. We shall construct Hermitian representations of these Lie algebras, and show that the representations "generate" C^*-algebras. In fact, we show that some recent constructions of simple C^*-algebras may be carried out more directly in the setting of infinite-dimensional Lie algebras. Our construction of the Lie algebras, and representations, uses extensions and cocycles heavily. Although the Lie algebras are infinite-dimensional, we shall show, using C^*-algebra theory, that they are paired naturally with Lie groups.

The Lie groups will be identified as extensions in the category of groups. The representations of the Lie algebra will be exponentiated to the Lie group, and we show that the corresponding representations of the group give rise to *covariant systems* for the underlying C^*-algebra.

8.1 Rotation Algebras

Let Γ be a (discrete) Abelian group, and let k be a field of characteristic zero. Let $L = k\cdot\Gamma$ be the free vector space over k based on Γ. Elements in L will be written in the form $\lambda = \sum x_\gamma \cdot \gamma$ where $x_\gamma \in k$, and the nonzero terms in the sum are indexed by a finite subset of Γ, the finite subset depending on λ.

Let $B : \Gamma \times \Gamma \longrightarrow k$ be a k-valued function satisfying:

$$B(\gamma,\zeta)B(\xi,\gamma+\zeta) + B(\zeta,\xi)B(\gamma,\zeta+\xi) + B(\xi,\gamma)B(\zeta,\xi+\gamma) = 0 \quad \text{for all } \xi,\gamma,\zeta \in \Gamma. \tag{8.1.1a}$$

$$B(\gamma,\xi) + B(\xi,\gamma) = 0 \qquad \gamma,\xi \in \Gamma. \tag{8.1.1b}$$

We now define a Lie bracket

$$[\cdot,\cdot] \quad L \times L \longrightarrow L$$

on the generators (*viz.*, Γ) for L as follows:

$$[\xi,\gamma] := B(\xi,\gamma)(\xi+\gamma) \qquad \xi,\gamma \in \Gamma; \tag{8.1.2}$$

extension to $L \times L$ by bilinearity. The following observation is immediate: The function B satisfies (8.1.1) if and only if the Lie bracket $[\cdot,\cdot]$ satisfies the Jacobi-identity, i.e.,

$$[\xi,[\gamma,\zeta]] + [\gamma,[\zeta,\xi]] + [\zeta,[\xi,\gamma]] = 0, \quad \xi,\gamma,\zeta \in L. \tag{8.1.3}$$

As for the proof, note that, if (8.1.1) is assumed, then we need only verify (8.1.3) for elements in the basis $\Gamma \subset L$. This is because of the bilinearity of $[\cdot,\cdot]$. First, B extends by bilinearity from $\Gamma \times \Gamma$ to $L \times L$, which means that $[\cdot,\cdot]$ becomes bilinear as well. But the verification of (8.1.3) for triples of elements in Γ amounts to just a re-writing of (8.1.1).

If, conversely, property (8.1.3) is assumed for the bracket $[\cdot,\cdot]$, defined by (8.1.2), then it follows, by substitution, that (8.1.1) holds.

For the second part of the construction, we impose the additional assumption that Γ is itself a free Abelian group, i.e., it has a set of generators, Ω say, such that

$$\Gamma = \mathbb{Z} \cdot \Omega. \tag{8.1.4}$$

This means that elements in Γ may be represented as finite sums

$$\gamma = \sum_{\omega\in\Omega} n_\omega \cdot \omega$$

where $n_\omega \in \mathbb{Z}$. and the summation is over a finite subset of Ω; the finite subset varies, of course, with the element γ in $\Gamma = \mathbb{Z} \cdot \Omega$.

Let V be the free vector space over k based on the set Ω, i.e., consisting of elements

$$v = \sum_{\omega\in\Omega} v_\omega \cdot \omega$$

with coordinates $v_\omega \in k$, $\omega \in \Omega$, and again finite summation over subsets of Ω. We shall equip V with the structure of an Abelian Lie algebra, defining the bracket, $[\cdot,\cdot] : V \times V \longrightarrow V$, to be identically zero:

$$[v,v'] = 0, \qquad v,v' \in V.$$

We are now ready to specify the short exact sequence

$$0 \longrightarrow L \longrightarrow g \longrightarrow V \longrightarrow 0,$$

defining the extension g of V by L as a *semidirect* sum, or rather product in the category of Lie algebras. This semidirect product will be obtained from a representation φ of V on the Lie algebra of derivations of L. A derivation of L is a linear mapping, $D : L \longrightarrow L$, satisfying

$$D([\lambda,\lambda']) = [D(\lambda),\lambda'] + [\lambda,D(\lambda')], \quad \text{for } \lambda,\lambda' \in L. \tag{8.1.5}$$

Since the commutator

$$[D_1,D_2] = D_1D_2 - D_2D_1$$

formed from two derivations D_1,D_2 is again a derivation, it follows that

$$\mathrm{Der}(L) := (\text{all derivations of } L)$$

is again a Lie algebra under the commutator bracket, i.e., the Jacobi identity is satisfied on $\mathrm{Der}(L)$ relatiave to the commutator bracket.

A representation φ is a Lie algebra homomorphism

$$\varphi : V \longrightarrow \mathrm{Der}(L).$$

In case V is Abelian, we get

$$[\varphi(v),\varphi(v')] = 0, \qquad v,v' \in V. \tag{8.1.6}$$

Definition 8.1.1. The representation φ is said to be *nondegenerate* if there are no nonzero solutions λ in L to the system

$$\varphi(v)\lambda = 0, \qquad v \in V.$$

In the special case when φ is given by some $F : \Gamma \longrightarrow V^*$ via (8.1.8) below, and $\Gamma = \mathbb{Z} \cdot \Omega$, $V = k\cdot\Omega$, then nondegeneracy of φ is equivalent to the statement that the matrix

$$\{F(\omega)(\omega')\}_{\omega,\omega'\in\Omega}$$

is nondegenerate.

Consider the dual vector space $V^* = \mathrm{Hom}(V,k)$ of all linear functionals on V, i.e., linear mappings from V to the ground field k. To every homomorphism

$$F : \Gamma \longrightarrow V^*,$$

we shall associate a representation

$$\varphi : V \longrightarrow \mathrm{Der}(L). \tag{8.1.7}$$

For $v \in V$, the derivation $\varphi(v)$ is defined by

$$\varphi(v)(\xi) = F(\xi)(v) \cdot \xi, \qquad \xi \in \Gamma, \tag{8.1.8}$$

and $\varphi(v)$ is extended by linearity from Γ to L. We check below that $\varphi(v)$ is a derivation, and that (8.1.6) holds. Using again extension by linearity, it is enough to verify Leibniz' formula

$$\varphi(v)([\xi,\gamma]) = [\varphi(v)\xi,\gamma] + [\xi,\varphi(v)\gamma]$$

for $\xi,\gamma \in \Gamma$. But we have

$$\begin{aligned}\varphi(v)([\xi,\gamma]) &= B(\xi,\gamma)\varphi(v)(\xi+\gamma)\\ &= B(\xi,\gamma)F(\xi+\gamma)(v) \cdot (\xi+\gamma)\\ &= B(\xi,\gamma)(F(\xi)(v) + F(\gamma)(v)) \cdot (\xi+\gamma)\\ &= [\varphi(v)\xi,\gamma] + [\xi,\varphi(v)\gamma],\end{aligned}$$

which is property (8.1.5).

We now verify formula (8.1.6):

$$\begin{aligned}[\varphi(v),\varphi(v')](\xi) &= \varphi(v)\varphi(v')\xi - \varphi(v')\varphi(v)\xi\\ &= \varphi(v)F(\xi)(v')\xi - \varphi(v')F(\xi)(v)\xi\\ &= F(\xi)(v')\varphi(v)\xi - F(\xi)(v)\varphi(v')\xi\\ &= F(\xi)(v')F(\xi)(v)\xi - F(\xi)(v)F(\xi)(v')\xi\\ &= 0, \qquad \text{for } v,v' \in V.\end{aligned}$$

Since L is spanned by the basis vectors ξ from Γ, formula (8.1.6) follows from this.

On the (vector) space $\mathfrak{g} = V \oplus L = \{(v,\lambda) : v \in V,\ \lambda \in L\}$, we define the following Lie bracket

$$[(v,\lambda),(v',\lambda')] := (0,[\lambda,\lambda']+\varphi(v)\lambda'-\varphi(v')\lambda). \tag{8.1.9}$$

We leave to the reader the verification that this bracket in $\mathfrak{g}$ does indeed satisfy the Jacobi-identity.

We shall identify L with an ideal in $\mathfrak{g}$ via the mapping $\lambda \longrightarrow (0,\lambda)$. The mapping $(v,\lambda) \longrightarrow v$ is a Lie homomorphism with L as kernel, and we have the short exact sequence

$$0 \longrightarrow L \longrightarrow \mathfrak{g} \longrightarrow V \longrightarrow 0$$

of Lie algebras, with $\mathfrak{g}/L \simeq V$.

Having the semidirect product $\mathfrak{g}$, formed from V and L, we may now study how V is embedded as a sub-Lie algebra of $\mathfrak{g}$ via the homomorphism

$$v \longrightarrow (v,0).$$

In this connection, we have

Proposition 8.1.2. Let $\varphi : V \longrightarrow \mathrm{Der}(L)$ be a given representation, and identify V with a Lie subalgebra of $\mathfrak{g}$. Then V is maximal Abelian in $\mathfrak{g}$ if and only if φ is nondegenerate.

Proof. Immediate!

We have shown how a given homomorphism $F : \Gamma \longrightarrow \mathrm{Hom}(V,k)$ determines an extension $\mathfrak{g} = \mathfrak{g}_F$ of V by L.

We now turn to the explicit construction of such homomorphisms in the free case, described above:

$$\Gamma = \mathbb{Z} \cdot \Omega, \qquad V = k \cdot \Omega.$$

But we shall assume that $k = \mathbb{R}$, or $\mathbb{C}$. In case $k = \mathbb{C}$, we shall define first

$$F : \Gamma \longrightarrow \mathrm{Hom}(\mathbb{R}\cdot\Omega, k), \tag{8.1.10}$$

and then extend (subsequently) to $\mathbb{C} \cdot \Omega$. We shall consider the one-torus $\mathbb{T} = \mathbb{R}/2\pi\mathbb{Z}$ as the dual of the group $\mathbb{Z}$, in short, $\mathbb{T} \simeq \hat{\mathbb{Z}}$ (= the one-dimensional unitary representations of the group $\mathbb{Z}$). For $t \in \mathbb{T}$, $n \in \mathbb{Z}$, we shall write $\langle t,n \rangle$ for the value of t at n. This is the familiar pairing of the dual group $\hat{\mathbb{Z}}$ with $\mathbb{Z}$. Let

$$\eta : \mathbb{R} \longrightarrow \mathbb{R}/2\pi\mathbb{Z}$$

be the canonical homomorphism onto the quotient. Let

$$v = \sum_{\omega \in \Omega} x_\omega \cdot \omega, \qquad x_\omega \in \mathbb{R},$$

and

$$\xi = \sum_{\omega \in \Omega} n_\omega \cdot \omega, \qquad n_\omega \in \mathbb{Z}.$$

We define

$$F(\xi)(v) = \prod_{\omega \in \Omega} \langle \eta(x_\omega), n_\omega \rangle \tag{8.1.11}$$

and note that there is at most a finite number of factors in the product ($\neq 1$) since n_ω is $\neq 0$ for at most a finite number of indices ω in Ω.

In the sequel, when reference is made to the semidirect product of the two Lie algebras V and L, it will always be understood (unless otherwise stated) that the action of V on L is specified by the homomorphism F determined in (8.1.11). This means that the Lie algebra $\mathfrak{g}$ depends on the basis set Ω, and on the Lie cocycle B in formula (8.1.2). This is because B determines uniquely the Lie bracket of L. In the sequel, we shall use the notation $\mathfrak{g} = V$ +) L to denote a semidirect product (in the category of Lie algebras). We shall not always make a distinction between the *external* semidirect product (described above) and the *internal* semidirect product. (The latter is defined for a given Lie algebra $\mathfrak{g}$, say, with two Lie subalgebras, V and L, say, satisfying conditions (i)-(iv) below:

(i) $$\mathfrak{g} = V + L,,$$

(ii) $$V \cap L = \{0\},$$

(iii) L is an ideal in g,

(iv) $g/L \simeq V$.

The notation $g = V +) L$ will also be used for the *internal* semidirect product (really sum) in the category of Lie algebras. The external/internal distinction will frequently be clear from the context, but it is not an important distinction.

We note that, if the semidirect product $g = V +) L$ is determined by the matrix function F in (8.1.11), then V is maximal Abelian in g. This follows immediately from Proposition 8.1.2 since the matrix from (8.1.11) is nondegenerate.

We now turn to the study of a class of examples of Lie cocycles B. They will be constructed from cocycles A of projective representations U of Γ.

Let $\mathfrak{A}$ be an associative $*$-algebra over $\mathbb{C}$ with unit. Recall, a $*$-operation on $\mathfrak{A}$ is an involutive antiautomorphism which is conjugate linear:

(i) $$a^{**} = a \qquad a \in \mathfrak{A},$$

(ii) $$(ab)^* = b^* a^*, \qquad a, b \in \mathfrak{A}.$$

(iii) $a \longrightarrow a^*$ is conjugate linear,

(iv) $I^* = I$ when I denotes the unit element.

An element u in $\mathfrak{A}$ is said to be *unitary* if it is invertible, and

$$u^* = u^{-1}.$$

Equivalently, $uu^* = u^*u = I$.

Definition 8.1.3. A mapping from Γ into the group of unitary elements in $\mathfrak{A}$ is said to be a *projective representation* if there is a function

$$A : \Gamma \times \Gamma \longrightarrow \mathbb{T}^1$$

such that

$$u_\xi u_\gamma = A(\xi, \gamma) u_{\xi+\gamma}, \qquad \xi, \gamma \in \Gamma. \tag{8.1.12}$$

Let $\mathcal{U}(\mathfrak{A})$ be the group of unitary elements in $\mathfrak{A}$. Then a projective representation may be described equivalently as a unitary representation of an extension of $\mathbb{T}^1$ by $\mathcal{U}(\mathfrak{A})$. After rewriting the defining property of projective representations, we arrive at the familiar cocycle identities for A:

$$A(\gamma,\zeta)A(\xi,\gamma+\zeta) = A(\xi+\gamma,\zeta)A(\xi,\gamma), \quad \text{for all triples } \xi,\gamma,\xi \in \Gamma. \tag{8.1.13}$$

A nontrivial example is obtained by taking $\Gamma = \mathbb{Z}^2 = \mathbb{Z} \times \mathbb{Z}$ and defining the cocycle A from a given "rotation angle" θ, $0 < \theta < 1$. This example is studied extensively [BEJ], [Con 3], [CR], [Eff], [PiVo], and the algebra $\mathfrak{A}$, after completion in a C^*-norm, is called the irrational rotation algebra (for irrational angle θ, that is). Moreover, $\mathfrak{A}$ is *simple* if and only if θ is *irrational*. The index θ describes the isomorphism classes [Ri 4], [Bla], and the K-groups of $\mathfrak{A}$ are known. For this example, we define

$$\begin{aligned} A((m,n),(m',n')) &= \exp(\sqrt{-1}\ 2\pi\ nm'\theta) \\ &= \rho^{nm'}, \end{aligned} \tag{8.1.14}$$

where $$\rho := \exp(\sqrt{-1}\ 2\pi\ \theta),$$

and $$(m,n) \in \mathbb{Z}^2, \qquad (m',n') \in \mathbb{Z}^2.$$

We leave to the reader the verification of the cocycle identity (8.1.13).

The next lemma gives a construction of Lie cocycles, starting with $\mathbb{T}^1$-valued cocycles at the group level.

Lemma 8.1.4. Let A be a cocycle on Γ given by (8.1.12), and define

$$B(\xi,\gamma) = A(\xi,\gamma) - A(\gamma,\xi), \qquad \xi,\gamma \in \Gamma. \tag{8.1.15}$$

Then it follows that B is a Lie cocycle, i.e., that (8.1.1) is satisfied.

Proof. We have

$$B(\gamma,\zeta)B(\xi,\gamma+\zeta) + B(\zeta,\xi)B(\gamma,\zeta+\xi) + B(\xi,\gamma)B(\zeta,\xi+\gamma)$$
$$= (A(\gamma,\zeta)-A(\zeta,\gamma))(A(\xi,\gamma+\zeta)-A(\gamma+\zeta,\xi))$$
$$+ (A(\zeta,\xi)-A(\xi,\zeta))(A(\gamma,\zeta+\xi)-A(\zeta+\xi,\gamma))$$
$$+ (A(\xi,\gamma)-A(\gamma,\xi))(A(\zeta,\xi+\gamma)-A(\xi+\gamma,\zeta)), \quad \xi,\gamma,\zeta \in \Gamma.$$

When the multiplication is completed we get a total of twelve terms that can be paired together into six pairs such that (8.1.13) applies to each pair and gives zero. As an example of a pair, we list:

$$A(\gamma,\zeta)A(\xi,\gamma+\zeta)-A(\xi,\gamma)A(\xi+\gamma,\zeta) = 0,$$

when (8.1.13) is applied. We leave it to the reader to identify the remaining five pairs.

Remark 8.1.5. The argument from the proof of Lemma 8.1.4 also shows that the function R, from $\Gamma \times \Gamma$ to $\mathbb{T}^1$, defined by

$$R(\xi,\gamma) = A(\xi,\gamma)A(\gamma,\xi)^{-1}, \qquad \xi,\gamma \in \Gamma, \tag{8.1.16}$$

satisfies

$$u_\xi u_\gamma = R(\xi,\gamma)u_\gamma u_\xi \tag{8.1.17}$$

whenever the given cocycle A is associated with a projective representation $\gamma \longrightarrow u_\gamma$ from Γ into $\mathcal{U}(\mathfrak{A})$, cf. formula (8.1.12).

We shall assume in the sequel that $\mathfrak{A}$ is generated by the set of unitaries $\{u_\gamma : \gamma \in \Gamma\}$, and that a cocycle A is given with corresponding projective representation, $\gamma \longrightarrow u_\gamma$.

We proceed to give a representation of L on $\mathfrak{A}$, and also a representation of V on $\mathfrak{A}$. We show that the resulting representation of the semidirect product $\mathfrak{g} = V +) L$ is implemented by a natural representation of $\mathfrak{g}$ in Hilbert space. Moreover, we give a covariant representation on the C^*-algebra obtained from a natural C^*-completion of $\mathfrak{A}$.

The separate constructions outlined above are made explicit in the following four observations.

Observation 8.1.6. Let $\gamma \longrightarrow u_\gamma$ be a given projective representation of Γ, and let A be the corresponding cocycle. Let

$$B(\xi,\gamma) = A(\xi,\gamma) - A(\gamma,\xi), \qquad \xi,\gamma \in \Gamma,$$

and define

$$\pi(\xi)u_\gamma = B(\xi,\gamma)u_{\gamma+\xi}. \tag{8.1.18}$$

Then π extends to a representation

$$\pi : L \longrightarrow \mathrm{Der}(\mathfrak{A})$$

where $\mathrm{Der}(\mathfrak{A})$ denotes the Lie algebra of derivations in the algebra $\mathfrak{A}$ generated by $\{u_\gamma : \gamma \in \Gamma\}$.

Proof. Since $L = k\Gamma$ where k is the ground field, it is enough to check that

$$\pi([\xi,\xi'])u_\gamma = \pi(\xi)\pi(\xi')u_\gamma - \pi(\xi')\pi(\xi)u_\gamma \tag{8.1.19}$$

holds for all ξ,ξ',γ in Γ. But when the formulas (8.1.18) and

$$[\xi,\xi'] = B(\xi,\xi')(\xi+\xi'),$$

are substituted, it follows that the desired conclusion (8.1.19) is equivalent to the known cocycle formula (8.1.1) for B, cf. Lemma 8.1.4.

Similarly, we must check that $\pi(\xi)$ is a derivation in the algebra $\mathfrak{A}$ generated by the u_γ's. Using generators, this reduces to the verification of the following formula:

$$\pi(\xi)(u_\gamma u_\zeta) = (\pi(\xi)u_\gamma)u_\zeta + u_\gamma\pi(\xi)u_\zeta, \qquad \text{for all } \xi,\gamma,\zeta \in \Gamma.$$

But this follows again from the cocycle formula for A, cf. also Lemma 8.1.4.

Observation 8.1.7. Each derivation $\{\pi(\lambda) : \lambda \in \Gamma\}$ is inner, in fact, if

$$\lambda = \sum_\xi x_\xi \cdot \xi \in L, \qquad x_\xi \in k,$$

the corresponding element

$$a = \sum_{\xi} x_{\xi} \cdot u_{\xi} \in \mathfrak{A}$$

implements the derivation, i.e.,

$$\pi(\lambda) = \text{ad } a.$$

Proof. By linearity, it is enough to verify the formula

$$\pi(\xi)u_{\gamma} = u_{\xi}u_{\gamma} - u_{\gamma}u_{\xi}$$

for all $\xi, \gamma \in \Gamma$. But

$$\begin{aligned}\pi(\xi)u_{\gamma} &= B(\xi,\gamma)u_{\gamma+\xi} \\ &= A(\xi,\gamma)u_{\gamma+\xi} - A(\gamma,\xi)u_{\gamma+\xi} \\ &= u_{\xi}u_{\gamma} - u_{\gamma}u_{\xi} \\ &= \text{ad}(u_{\xi})(u_{\gamma}),\end{aligned}$$

which completes the proof.

Definition 8.1.8. (a) A function

$$R : \Gamma \times \Gamma \longrightarrow \mathbb{T}$$

is said to be an *antisymmetric bicharacter* if

(i) $\xi \longrightarrow R(\xi,\eta)$ is a character on Γ for all $\gamma \in \Gamma$.

(ii) $R(\xi,\gamma) = R(\gamma,\xi)^{-1}$, $\xi,\gamma \in \Gamma$,

(b) Let G denote the dual group to Γ, i.e., the compact group $\hat{\Gamma}$ of all (unitary) characters on Γ. Then a given bicharacter gives rise to a homomorphism, $\Gamma \longrightarrow G \simeq \hat{\Gamma}$, defined by $\gamma \longrightarrow R(\cdot,\gamma)$. We say that R is *nondegenerate* if the kernel of this homomorphism is trivial.

Observation 8.1.9. There is a representation

$$\pi : G \longrightarrow \text{Aut}(\mathfrak{A})$$

which is determined uniquely by

(8.1.20) $$\pi(g)u_\xi = \langle g,\xi\rangle u_\xi, \qquad g \in G, \quad \xi \in \Gamma.$$

Proof. We have

$$\begin{aligned}\pi(g)(u_\xi u_\gamma) &= \pi(g)(A(\xi,\gamma)u_{\xi+\gamma}) \\ &= A(\xi,\gamma)\langle g,\xi\rangle\langle g,\gamma\rangle u_{\xi+\gamma} \\ &= \pi(g)(u_\xi)\pi(g)(u_\gamma)\end{aligned}$$

for $g \in G$, $\xi,\gamma \in \Gamma$. Similarly,

$$\begin{aligned}\pi(g)(u_\xi^*) &= \pi(g)(u_\xi^{-1}) \\ &= (\pi(g)(u_\xi))^{-1} \\ &= \langle g,\xi\rangle^{-1} u_\xi^{-1} \\ &= (\pi(g)(u_\xi))^*.\end{aligned}$$

Since $\mathfrak{A}$ is generated by the family $\{u_\gamma : \gamma \in \Gamma\}$ of unitaries, it follows that $\pi(g)$ is defined uniquely by (8.1.20) as a $*$-automorphism of $\mathfrak{A}$. It is immediate that $\pi(g_1g_2) = \pi(g_1)\pi(g_2)$ for all g_1 and g_2 in G, so π is a representation.

We shall need the following purely algebraic lemma.

Lemma 8.1.10. Let

$$\pi : G \longrightarrow \mathrm{Aut}(\mathfrak{A})$$

be a representation of a compact Abelian group G on a $*$-algebra $\mathfrak{A}$, and assume that the representation is diagonalized by a family of unitary elements $\{u_\gamma : \gamma \in \hat{G}\}$. That is, we have

$$\pi(g)u_\gamma = \langle g,\gamma\rangle u_\gamma \qquad g \in G, \quad \gamma \in \hat{G} = \Gamma.$$

Then it follows that the set $\{u_\gamma : \gamma \in \Gamma\}$ is linearly independent.

Proof. We prove by induction that we cannot have a nontrivial finite linear combination:

$$\sum_{\xi\in\Gamma} x_\xi u_\xi = 0. \tag{8.1.21}$$

The first step in the induction is clear. Pick a linear combination (8.1.21) with the smallest possible number of nonzero terms, if any. If $x_\gamma \neq 0$, then, upon normalization, we may assume that

$$u_\gamma + \sum_{\substack{\xi\in\Gamma \\ \xi\neq\gamma}} x_\xi u_\xi = 0. \tag{8.1.22}$$

Apply the automorphism $\pi(g)$ to (8.1.22). We get

$$\langle g,\gamma\rangle u_\gamma + \sum_{\xi\neq\gamma} \langle g,\xi\rangle u_\xi = 0. \tag{8.1.23}$$

Now multiply through in equation (8.1.22) with the number $\langle g,\gamma\rangle$, and subtract the resulting two equations. We get

$$\sum_{\xi\neq\gamma} x_\xi(\langle g,\xi\rangle - \langle g,\gamma\rangle) u_\xi = 0. \tag{8.1.24}$$

But if, in this sum, $x_\xi \neq 0$ for some ξ, then we may choose g in G such that $\langle g,\xi\rangle - \langle g,\gamma\rangle \neq 0$. It follows that (8.1.24) is a linear combination of the u_ξ's with a fewer number of terms than the original one (8.1.21). This is a contradiction, and we have proved linear independence.

Theorem 8.1.11. Let Γ be a discrete group, and let $\mathfrak{A}$ be a $*$-algebra with unit-element I. Let $\gamma \longrightarrow u_\gamma$ be a projective representation of Γ into the group of unitary elements in $\mathfrak{A}$. Let $A : \Gamma \times \Gamma \longrightarrow \mathbb{T}$ be the cocycle determined by (8.1.12), and let $R : \Gamma \times \Gamma \longrightarrow \mathbb{T}$ be given by

$$R(\xi,\gamma) = A(\xi,\gamma)A(\gamma,\xi)^{-1}, \qquad \xi,\gamma \in \Gamma.$$

Then it follows that R is an antisymmetric bicharacter of Γ.

Proof. For an arbitrary triple of elements ξ_1, ξ_2, γ in Γ we have the formula:

$$\begin{aligned} R(\xi_1+\xi_2,\gamma)u_\gamma u_{\xi_1+\xi_2} &= u_{\xi_1+\xi_2}u_\gamma \\ &= A(\xi_1,\xi_2)^{-1}\ u_{\xi_1}u_{\xi_2}u_\gamma \\ &= A(\xi_1,\xi_2)^{-1}\ R(\xi_2,\gamma)u_{\xi_1}u_\gamma u_{\xi_2} \\ &= A(\xi_1,\xi_2)^{-1}\ R(\xi_2,\gamma)R(\xi_1,\gamma)u_\gamma u_{\xi_1}u_{\xi_2} \\ &= R(\xi_1,\gamma)R(\xi_2,\gamma)u_\gamma u_{\xi_1+\xi_2}. \end{aligned}$$

Since the unitary elements in $\mathfrak{A}$ form a group, the element $u_\gamma u_{\xi_1+\xi_2}$ is unitary, and it follows that

$$R(\xi_1+\xi_2,\gamma) = R(\xi_1,\gamma)R(\xi_2,\gamma).$$

Since R is antisymmetric by definition, we conclude that R is an antisymmetric bicharacter.

The following result about the representation $\pi : G \longrightarrow \mathrm{Aut}(\mathfrak{A})$ will be included without proof. We refer the reader to [OPT] and [Sℓ] for the proofs.

Theorem 8.1.12. Let $\pi : G \longrightarrow \mathrm{Aut}(\mathfrak{A})$ be the representation defined by (8.1.20).

(a) Then the following two conditions are satisfied:

(i) For all $\xi \in \Gamma$, the set

$$\{a \in \mathfrak{A} : \pi(g)a = \langle g,\xi\rangle a,\ g \in G\}$$

consists only of the scalars times u_ξ.

(ii) The fixed-point algebra

$$\{a \in \mathfrak{A} : \pi(g)a = a,\ g \in G\}$$

consists of the scalars times I.
(We shall say that π is *ergodic*.)

(b) The following three conditions are equivalent:

(i) R is nondegenerate.

(ii) $\mathfrak{A}$ is simple, i.e., no two-sided ideals other than the two trivial extreme ones.

(iii) The center of $\mathfrak{A}$ consists of the scalars times I.

"Proof". We shall sketch the details for part (a).

By familiar results on ergodic actions of compact Abelian groups, it is enough to prove that the fixed-point subalgebra is one-dimensional.

Suppose $a \in \mathfrak{A}$ is in the fixed-point subalgebra. Let

$$a = \sum_{\xi} x_{\xi} u_{\xi}$$

be the Fourier series of a. Then

$$a = \pi(g)a = \sum_{\xi} x_{\xi} \langle g, \xi \rangle u_{\xi}.$$

We proved at the outset (Lemma 8.1.10) that the elements $\{u_{\xi} : \xi \in \Gamma\}$ are linearly independent. By comparison of Fourier coefficients, we get $x_{\xi} = 0$ for $\xi \in \Gamma \backslash \{0\}$. It follows that $a = x_0 \cdot I$ which is the desired conclusion.

As a corollary of the above construction we read off the (perhaps) surprising conclusion that every nondegenerate antisymmetric bicharacter cobounds.

Corollary 8.1.13. Let Γ be a discrete Abelian group, and let $R : \Gamma \times \Gamma \longrightarrow \mathbb{T}$ be a nondegenerate antisymmetric bicharacter. Then there is a cocycle $A : \Gamma \times \Gamma \longrightarrow \mathbb{T}$ such that

$$R(\xi, \eta) = A(\xi, \gamma) A(\gamma, \xi)^{-1}, \qquad \xi, \gamma \in \Gamma.$$

Proof. We have seen that there is a simple algebra $\mathfrak{A}$, spanned by a basis of unitary elements $\{u_{\gamma} : \gamma \in \Gamma\}$, and a representation $\pi : G \longrightarrow \mathrm{Aut}(\mathfrak{A})$ given by

$$\pi(g) u_{\xi} = \langle g, \xi \rangle u_{\xi}, \qquad g \in G, \quad \xi \in \Gamma,$$

where G is the dual compact Abelian group. We noted that for each $\xi \in \Gamma$, the spectral subspace

$$\mathfrak{A}(\xi) = \{a \in \mathfrak{A} : \pi(g)a = \langle g,\xi\rangle a,\ g \in G\}$$

is one-dimensional. It follows that $u_\gamma u_\xi \in \mathfrak{A}(\gamma+\xi)$, $\gamma,\xi \in \Gamma$. As a result, there is a number (which we shall denote $A(\gamma,\xi) \in \mathbb{T}$) such that $u_\gamma u_\xi = A(\gamma,\xi)u_{\gamma+\xi}$. It follows from this that A satisfiess (8.1.13), and that

$$\begin{aligned} R(\gamma,\xi)u_\xi u_\gamma &= u_\gamma u_\xi \\ &= A(\gamma,\xi)u_{\gamma+\xi} \\ &= A(\gamma,\xi)A(\xi,\gamma)^{-1}\ u_\xi u_\gamma . \end{aligned}$$

We obtain the desired solution

$$R(\gamma,\xi) = A(\gamma,\xi)A(\xi,\gamma)^{-1}$$

upon cancellation of the unitary element $u_\xi u_\gamma$.

We note that the corollary may be regarded as a converse to Theorem 8.1.11.

8.2 Completions of $\mathfrak{U}_{\mathbb{C}}(L)$

A. The Trace Completion

In Section 8.1, we constructed the infinite-dimensional Lie algebra L over a given discrete Abelian group Γ. As a complex Lie algebra, L is the free vector space $L = \mathbb{C}\cdot\Gamma$ equipped with the Lie bracket, defined by

$$(8.2.1) \qquad [\xi,\gamma] = B(\xi,\gamma)(\xi+\gamma), \qquad \text{for } \xi,\gamma \in \Gamma,$$

where $B : \Gamma\times\Gamma \longrightarrow \mathbb{C}$ is assumed to satisfy the Lie algebraic cocycle identity (8.1.1). In fact, the bracket, defined in (8.2.1), extends by linearity to L, and the cocycle identity (8.1.1) is equivalent with the Jacobi-identity for the bracket.

We considered the case where L carries a representation

$$\pi : G \longrightarrow \mathrm{Aut}(L)$$

of the dual compact Abelian group $G = \hat{\Gamma}$ which is diagonalized by $\{\gamma : \gamma \in \Gamma\} \subset L$, i.e.,

(8.2.2) $$\pi(g)\gamma = \langle g,\gamma\rangle\gamma, \qquad g \in G, \quad \gamma \in \Gamma.$$

Let $\mathfrak{U}_{\mathbb{C}}(L)$ be the universal enveloping algebra over the complex numbers $\mathbb{C}$. Since

(8.2.3) $$\pi : G \longrightarrow \mathrm{Aut}(L)$$

is a group of automorphisms on L, it has a canonical extension, which we shall denote by $\tilde{\pi}$ for the moment:

(8.2.4) $$\tilde{\pi} : G \longrightarrow \mathrm{Aut}(\mathfrak{U}_{\mathbb{C}}(L)).$$

Note that, in (8.2.3), we have a group of Lie automorphisms, while, in (8.2.4), we have a group of automorphisms of the associative unital algebra $\mathfrak{U}_{\mathbb{C}}(L)$. We shall introduce the notation

(8.2.5) $$\mathfrak{U} := \mathfrak{U}_{\mathbb{C}}(L),$$

and consider the spectral subspaces

(8.2.6) $$\mathfrak{U}(\xi) = \{a \in \mathfrak{U} : \tilde{\pi}(g)a = \langle g,\xi\rangle a, \ g \in G\},$$

defined for $\xi \in \Gamma$. In the special case $\xi = 0$, the spectral subspace is the fixed-point subalgebra:

(8.2.7) $$\mathfrak{U}(0) = \mathfrak{U}^G = \{a \in \mathfrak{U} : \tilde{\pi}(g)a = a, \ g \in G\}$$

We imposed the assumption that $\mathfrak{U}^G$ be *one-dimensional*. It follows then from general theory that each space $\mathfrak{U}(\xi)$, $\xi \in \Gamma$, is one-dimensional. Note that ξ plays two roles: It is a unitary character on the compact group G, and it is an element in L, and in $\mathfrak{U}$, since L may be identified as a Lie subalgebra of $\mathfrak{U}$. To avoid ambiguity, we shall write u_ξ, in place of ξ, when ξ is viewed as an element in $\mathfrak{U}$.

It follows that each spectral subspace $\mathfrak{U}(\xi)$ is spanned by the single element u_ξ. Since $u_\xi u_\gamma \in \mathfrak{U}(\xi+\gamma)$ for pairs of elements ξ,γ in Γ, there is a function $A : \Gamma \times \Gamma \longrightarrow \mathbb{C}$ such that

(8.2.8) $$u_\xi u_\gamma = A(\xi,\gamma)u_{\xi+\gamma}.$$

We shall impose the added assumption that A takes values in the circle group $\mathbb{T}$.

If we define

$$R(\xi,\gamma) = A(\xi,\gamma)A(\gamma,\xi)^{-1}$$

then we get the formula

$$u_\xi u_\gamma = R(\xi,\gamma)u_\gamma u_\xi, \qquad \xi,\gamma \in \Gamma. \tag{8.2.9}$$

Indeed,

$$\begin{aligned} u_\xi u_\gamma &= A(\xi,\gamma)u_{\xi+\gamma} \\ &= A(\xi,\gamma)A(\gamma,\xi)^{-1}\, u_\gamma u_\xi \end{aligned}$$

where the formula (8.2.8) has been used twice.

We further proved, in Theorem 8.1.11, that R is an anti-symmetric bicharacter on Γ.

We now define a $*$-operation on $\mathfrak{A}$. First define

$$u_\gamma^* = A(\gamma,\gamma)u_{-\gamma}, \qquad \gamma \in \Gamma, \tag{8.2.10}$$

on the basis elements $\{u_\gamma : \gamma \in \Gamma\}$. We claim that the definition (8.2.10) extends to the enveloping algebra $\mathfrak{A}$ and yields a $*$-operation on $\mathfrak{A}$ relative to which we have

$$\tilde{\pi}(g)(a^*) = (\tilde{\pi}(g)(a))^* \quad \text{for all} \quad g \in G,\ a \in \mathfrak{A}. \tag{8.2.11}$$

As a result, G is represented as a group of $*$-automorphisms on $\mathfrak{A}$.

To prove (8.2.11), we need the observation that B is a coboundary; indeed,

$$B(\xi,\gamma) = A(\xi,\gamma)-A(\gamma,\xi). \tag{8.2.12}$$

The proof of (8.2.12) is similar to that of (8.2.9). We have

$$\begin{aligned} B(\xi,\gamma)u_{\xi+\gamma} &= [u_\xi,u_\gamma] \\ &= u_\xi u_\gamma - u_\gamma u_\xi \\ &= A(\xi,\gamma)u_{\xi+\gamma} - A(\gamma,\xi)u_{\gamma+\xi} \\ &= (A(\xi,\gamma)-A(\gamma,\xi))u_{\xi+\gamma}, \end{aligned}$$

which proves (8.2.12).

If $u_\gamma \longrightarrow u_\gamma^*$ is extended to L by linearity, we get

$$[\lambda,\mu]^* = -[\mu^*,\lambda^*], \qquad \lambda,\mu \in L. \tag{8.2.13}$$

By the universal property of $\mathfrak{A} = \mathfrak{U}_{\mathbb{C}}(L)$, we then get

$$(ab)^* = b^*a^*, \qquad a,b \in \mathfrak{A}. \tag{8.2.14}$$

It is enough to verify (8.2.13) on the basis elements. The proof goes as follows:

$$\begin{aligned}
[u_\gamma^*, u_\xi^*] &= [A(\gamma,\gamma)u_{-\gamma}, A(\xi,\xi)u_{-\xi}] \\
&= A(\gamma,\gamma)A(\xi,\xi)[u_{-\gamma},u_{-\xi}] \\
&= A(\gamma,\gamma)A(\xi,\xi)B(-\gamma,-\xi)u_{-\gamma-\xi} \\
&= A(\gamma,\gamma)A(\xi,\xi)(A(-\gamma,-\xi)-A(-\xi,-\gamma))u_{-\gamma-\xi} \\
&= A(\xi+\gamma,\xi+\gamma)A(\xi,\gamma)^{-1}\, u_{-\gamma-\xi} - A(\xi+\gamma,\xi+\gamma)A(\gamma,\xi)^{-1}u_{-\gamma-\xi} \\
&= -[u_\gamma,u_\xi]^*, \qquad \gamma,\xi \in \Gamma.
\end{aligned}$$

Before stating our first result, we note that L and $\mathfrak{U}_{\mathbb{C}}(L)$ are isomorphic as vector spaces. This is because they both have the vectors $\{u_\gamma : \gamma \in \Gamma\}$ as a basis. Indeed, we proved in Lemma 8.1.10 the linear independence in $\mathfrak{U}_{\mathbb{C}}(L)$. It will therefore be convenient to denote the two representations of G on L, resp., on $\mathfrak{U}_{\mathbb{C}}(L)$, by the same letter, π; i.e., π is a $*$-representation of G on L, and on $\mathfrak{U}_{\mathbb{C}}(L)$.

We shall now assume that the (antisymmetric bicharacter) R:

$$R(\xi,\gamma) = A(\xi,\gamma)A(\gamma,\xi)^{-1}, \qquad \xi,\gamma \in \Gamma,$$

is nondegenerate. We saw in Theorem 8.1.12 that $\mathfrak{A}$ is then a *simple* algebra.

Definitions 8.2.1. A linear functional f on $\mathfrak{A}$ is said to be *positive* if

$$f(a^*a) \geq 0, \qquad a \in \mathfrak{A}.$$

It is said to be *normalized* if

$$f(u_0) = 1.$$

Recall that u_0 is the unit-element in $\mathfrak{U}$. A normalized positive linear functional is called a *state*.

A state f is said to be a *trace* if $f(ab) = f(ba)$, $a,b \in \mathfrak{U}$.

Lemma 8.2.2. The algebra $\mathfrak{U}$ has a unique trace.

Proof. Elements in $\mathfrak{U}$ may be expanded in the basis $\{u_\xi : \xi \in \Gamma\}$,

$$a = \sum_\xi x_\xi u_\xi \qquad (\xi \in \Gamma,\ x_\xi \in \mathbb{C}).$$

If we define

$$f(a) = x_0, \tag{8.2.15}$$

it is easy to verify that f is a trace. Details are left to the reader.

The uniqueness is proved below. Assume f is a given trace. We will prove the formula (8.2.15). Note that

$$f(a) = f(u_\gamma a u_\gamma^{-1}),$$

and

$$f(a) = \sum_\xi x_\xi f(u_\xi).$$

It follows that

$$\sum_\xi x_\xi f(u_\xi) = \sum_\xi x_\xi R(\gamma,\xi) f(u_\xi). \tag{8.2.16}$$

But, if $f(u_\xi) \neq 0$ for some $\xi \neq 0$, then we may pick $\gamma \in \Gamma$ such that $R(\gamma,\xi) \neq 1$. This contradicts (8.2.16), and it follows that $f(u_\xi) = 0$ whenever $\xi \in \Gamma\backslash\{0\}$. Since f is normalized, by definition, we have $f(u_0) = 1$. The uniqueness is proved.

We shall now apply the Gelfand-Naimark-Segal (GNS) construction to the trace state f. The reader is referred to [BR], and Section 2.8 above for the standard facts on the GNS construction.

The following known lemma will be useful:

Lemma 8.2.3. Let f be the trace state on $\mathfrak{A}$. Then the kernel of f is trivial. (We say that f is faithful.)

Proof. The assertion of the lemma amounts to the statement that the only solution to

$$f(a^*a) = 0 \tag{8.2.17}$$

is $a = 0$. Let $\mathcal{K}$ denote the solutions to (8.2.17). By the Schwarz inequality

$$|f(ba)|^2 \leq f(bb^*)f(a^*a), \tag{8.2.18}$$

it follows that $\mathcal{K}$ is a left-ideal. Since $f(a^*a) = f(aa^*)$, it is also a right-ideal. But u_0 is not in $\mathcal{K}$, and it follows from the simplicity of $\mathfrak{A}$ that $\mathcal{K} = \{0\}$.

The completion of $\mathfrak{A}$, in the norm $a \longrightarrow f(a^*a)$, will be denoted $\mathcal{H}_2$, or $\mathcal{L}^2(\mathfrak{A}, f)$. Every automorphism of $\mathfrak{A}$ implements a unitary operator on $\mathcal{H}_2$.

Let π be the representation of L on $\mathfrak{A}$ described in Observation 8.1.7. Since $\mathfrak{A}$ is the enveloping algebra, π extends uniquely to a representation, also denoted π, of $\mathfrak{A}$. In fact, it follows from 8.1.7 that

$$\pi(a)b = ab, \qquad a \in \mathfrak{A}, \quad b \in \mathfrak{A}.$$

It follows from (8.2.18) that each operator $\pi(a)$ extends uniquely to a bounded operator on $\mathcal{H}_2$, and that π is a Hermitian representation.

We also get, from the GNS-construction, the existence of a familiar bijection between, on the one hand, $\mathrm{Aut}(\mathfrak{A})$, and, on the other, the group of unitary operators U on $\mathcal{H}_2$ satisfying

$$Uu_0 = u_0, \quad \text{and} \quad U\pi(\mathfrak{A})U^* = \pi(\mathfrak{A}). \tag{8.2.19}$$

Indeed, if $\alpha \in \mathrm{Aut}(\mathfrak{A})$ is given, then the corresponding unitary operator U_α is determined by

$$U_\alpha(a) = \alpha(a), \qquad a \in \mathfrak{A}. \tag{8.2.20}$$

We verify that U_α, defined by (8.2.20), is unitary, and that the properties (8.2.19) hold. Indeed,

$$\begin{aligned}\|U_\alpha a\|_2^2 &= \|\alpha(a)\|_2^2 = f(\alpha(a)^*\alpha(a)) \\ &= f(\alpha(a^*a)) = f(a^*a) \\ &= \|a\|_2^2\end{aligned}$$

where uniqueness of the trace was used. It follows from this that U_α is unitary.

We have

$$U_\alpha u_0 = \alpha(u_0) = \alpha(1) = 1 = u_0.$$

and

$$\begin{aligned}U_\alpha \pi(a)b &= U_\alpha(ab) \\ &= \alpha(ab) \\ &= \alpha(a)\alpha(b) \\ &= \pi(\alpha(a))U_\alpha b, \qquad a,b \in \mathfrak{A},\end{aligned}$$

from which we derive the covariance formula

$$U_\alpha \pi(a) U_\alpha^* = \pi(\alpha(a)). \tag{8.2.21}$$

We are now ready to prove the main result which states that every nondegenerate antisymmetric bicharacter gives rise to a natural construction of a representation of the pair Γ, G on the trace Hilbert space $\mathcal{H}_2$.

First note that we have a representation of $G \times \Gamma$ on $\mathcal{H}_2$ which is given as follows. The representation will be denoted by α. If $g \in G$, then $\alpha(g)$ is the automorphism of $\mathfrak{A}$, given by

$$\alpha(g)u_\xi = \langle g,\xi\rangle u_\xi, \qquad g \in G, \quad \xi \in \Gamma. \tag{8.2.22}$$

If $\gamma \in \Gamma$, then an automorphism $\alpha(\gamma)$ is determined by

$$\alpha(\gamma)(u_\xi) = R(\gamma,\xi)u_\xi = u_\gamma u_\xi u_\gamma^*. \tag{8.2.23}$$

This is indeed a representation of $G \times \Gamma$ since $\alpha(g)\alpha(\gamma) = \alpha(\gamma)\alpha(g)$ for all $g \in G$, $\gamma \in \Gamma$.

For each $(g,\gamma) \in G \times \Gamma$, let $U(g,\gamma)$ be the unitary operator on $\mathcal{H}_2$ which is determined from the automorphism $\alpha(g,\gamma)$ via the GNS-construction.

Theorem 8.2.4. The set of unitary operators $\{U(g,\gamma) : (g,\gamma) \in G \times \Gamma\}$ induced from the representation

$$\alpha \quad : G \times \Gamma \longrightarrow \mathrm{Aut}(\mathfrak{A})$$

forms a unitary *representation* of $G \times \Gamma$ which is simultaneously diagonalized by the *orthonormal* basis $\{u_\xi : \xi \in \Gamma\}$ for $\mathcal{H}_2$, i.e., the von Neumann algebra generated by $\{U(g,\gamma)\}$ is maximal Abelian.

Proof. It follows from the properties of the trace that $\{u_\xi : \xi \in \Gamma\}$ is an orthonormal basis for $\mathcal{H}_2$. Specifically, $f(u_\xi^* u_\gamma)$ = the Kronecker delta function in $\xi,\gamma \in \Gamma$. Generally, a representation on $\mathfrak{A}$ induces only a projective representation. But the theorem states that, in this case, the induced projective representation is a representation.

We have, for $g \in G$, $\gamma,\xi \in \Gamma$, the formula:

$$\begin{aligned} U(g,\gamma)u_\xi &= \alpha(g,\gamma)u_\xi \\ &= \langle g,\xi\rangle R(\gamma,\xi)u_\xi, \end{aligned}$$

which proves that the operator family $\{U(g,\gamma) : (g,\gamma) \in G \times \Gamma\}$ is simultaneously diagonalized by the orthonormal basis $\{u_\xi\}$. It follows, in particular, that the operators $U(e,\gamma)$ and $U(g,0)$ are mutually commuting for $\gamma \in \Gamma$ and $g \in G$.

We now prove that the von Neumann algebra generated by $\{U(g,\gamma) : (g,\gamma) \in G\times\Gamma\}$ is maximal Abelian. It follows from the fact that the eigenspaces are one-dimensional. For all $\xi \in \Gamma$, the joint eigenspace is

$$\mathcal{H}_2(\xi) = \{a \in \mathcal{H}_2 : U(g,\gamma)a = \langle g,\xi\rangle R(\gamma,\xi)a,\ (g,\gamma) \in G \times \Gamma\},$$

and we show that it is spanned by u_ξ.

Let $a \in \mathcal{H}_2(\xi)$, and consider the Fourier series decomposition of a relative to the action of $U(g) = U(g,0)$, $g \in G$. We have

$$a = \sum_{\zeta \in \Gamma} \hat{a}(\zeta)$$

where

$$\hat{a}(\zeta) = \int_G \overline{\langle g,\zeta\rangle} U(g)a\, dg.$$

It follows that $\hat{a}(\zeta) = 0$ for $\zeta \neq \xi$, and

$$a = \int_G \langle g,\xi\rangle U(g)a\, dg. \tag{8.2.24}$$

When the orthogonal expansion

$$a = \sum_{\zeta \in \Gamma} f(au_\zeta^*)u_\zeta \tag{8.2.25}$$

is substituted into (8.2.24), we arrive at the conclusion

$$\sum_\zeta f(au_\zeta^*)u_\zeta = f(au_\xi^*)u_\xi$$

which, in turn, yields

$$f(au_\zeta^*) = \begin{cases} f(au_\xi^*), & \zeta = \xi, \\ 0 \quad , & \zeta \neq \xi. \end{cases}$$

Substitution back into (8.2.25) yields $a = f(au_\xi^*)u_\xi$, which proves that the eigenspace $\mathcal{H}_2(\xi)$ is spanned by the single basis vector u_ξ.

We now read off from this that the commutant $\{U(g,\gamma) : (g,\gamma) \in G \times \Gamma\}'$ is Abelian. For suppose $X \in \mathcal{B}(\mathcal{H}_2)$ is in the commutant. Then it must leave invariant each of the eigenspaces $\mathcal{H}_2(\xi)$, $\xi \in \Gamma$. It follows that Xu_ξ is proportional to u_ξ, for all $\xi \in \Gamma$. The commutant is therefore isomorphic to $\ell^{\infty}(\Gamma)$ = all bounded sequences on Γ. In particular, it is Abelian. This completes the proof of Theorem 8.2.4.

Remark 8.2.5. The second half of the proof yields the stronger information, namely, that the commutant of $\{U(g,0) : g \in G\}$ is Abelian. We show below that the commutant of $\{U(e,\gamma) : \gamma \in \Gamma\}$ is also Abelian. In fact we have

$$\begin{aligned} \{U(e,\gamma) : \gamma \in \Gamma\}' &= \{U(g,0) : g \in G\}' \\ &= \{U(g,\gamma) : (g,\gamma) \in G \times \Gamma\}' \\ &\simeq \ell^{\infty}(\Gamma). \end{aligned}$$

The only part of these assertions which is not already proved is that

$$\{U(e,\gamma) : \gamma \in \Gamma\}' \simeq \ell^{\infty}(\Gamma). \tag{8.2.26}$$

The proof below follows the idea from the proof of Theorem 8.2.4. We must show that the eigenspace

$$\mathcal{H}_2^e(\xi) = \{a \in \mathcal{H}_2 : U(e,\gamma)a = R(\gamma,\xi)a, \ \gamma \in \Gamma\}$$

is one-dimensional for all $\xi \in \Gamma$. Once we have proved this, the desired conclusion (8.2.26) is immediate. The proof that $\mathcal{H}_2^e(\xi)$ is one-dimensional is based on the orthogonal expansion

$$a = \sum_{\zeta\in\Gamma} f(au_\zeta^*)u_\zeta. \tag{8.2.27}$$

Substitution of this into

$$U(e,\gamma)a = R(\gamma,\xi)a, \qquad \gamma \in \Gamma,$$

yields

$$\sum_{\zeta} (R(\gamma,\zeta)-R(\gamma,\xi))f(au_\zeta^*)u_\zeta = 0.$$

If $f(au_\zeta^*) \neq 0$ for some $\zeta \neq \xi$, then we may pick γ such that $R(\gamma,\zeta-\xi) \neq 1$. It follows that

$$R(\gamma,\zeta)-R(\gamma,\xi) = R(\gamma,\xi)(R(\gamma,\zeta-\xi)-1) \neq 0,$$

which is a contradiction. Therefore, the expansion (8.2.27)

can have only one term, viz.,

$$a = f(au_\xi^*)u_\xi.$$

This completes the proof.

B. The C^**-norm Completion*

In the remainder of this section we recall the C^*-algebraic approach to the rotation algebras. Our present purpose is to exponentiate the representation of the Lie algebra $\mathfrak{g}$ from the short exact sequence

$$0 \longrightarrow L \longrightarrow \mathfrak{g} \longrightarrow V \longrightarrow 0$$

of Lie algebras.

We have obtained a representation π of the Lie algebra extension $\mathfrak{g}$ by operators on the trace Hilbert space $\mathcal{H}_2$, and we now address the question of exactness for this representation, i.e., finding a representation of some ∞-dimensional Lie (type) group such that the corresponding infinitesimal representation agrees with the given representation π of $\mathfrak{g}$.

The starting point is a discrete Abelian group Γ, and a nondegenerate antisymmetric bicharacter on Γ. We denote by $\hat{\Gamma}$ the dual compact Abelian group, and we let $\mathfrak{A}(R)$ be the $*$-algebra described above (which is spanned linearly by the basis elements $\{u_\gamma : \gamma \in \Gamma\}$).

We shall describe below the C^*-algebra completion $A(R)$ of $\mathfrak{A}(R)$, and we shall denote by $\mathcal{U}(A(R))$ the group of all unitary elements in $A(R)$. The infinite-dimensional group $G(R)$ with Lie algebra equal to $\mathfrak{g}$ will be constructed in the next section as an extension of $\hat{\Gamma}$ by $\mathcal{U}(A(R))$, i.e., given by a short exact sequence

$$(e) \longrightarrow \mathcal{U}(A(R)) \longrightarrow G(R) \longrightarrow \hat{\Gamma} \longrightarrow (e)$$

in the category of groups.

Recall the representation π of $\mathfrak{A}(R)$ given by

$$\pi(a)b = ab, \qquad a \in \mathfrak{A}(R), \quad b \in \mathfrak{A}(R).$$

We saw that $\pi(a)$ extends uniquely to a bounded operator on $\mathcal{H}_2$ satisfying

$$\langle \pi(a)h_1, b_2\rangle = \langle h_1, \pi(a^*)h_2\rangle$$

for $h_1, h_2 \in \mathcal{H}_2$, and $a \in \mathfrak{A}(R)$. The inner product $\langle\cdot,\cdot\rangle$ is the trace-inner product on $\mathcal{H}_2$.

Now $\pi(a)$ is a bounded operator on $\mathcal{H}_2$, and we shall denote its norm $\|\pi(a)\|$.

Lemma 8.2.6. If $\|\pi(a)\| = 0$ for some $a \in \mathfrak{A}(R)$, then $a = 0$.

Proof. The kernel of

$$\pi : \mathfrak{A}(R) \longrightarrow \mathfrak{B}(\mathcal{H}_2)$$

is a two-sided ideal in $\mathfrak{A}(R)$. Since $\pi(1) = I$, this ideal must be zero; the lemma is proved.

Definition 8.2.7. We define the C^*-norm on $\mathfrak{A}(R)$ by

$$\|a\| := \|\pi(a)\|. \tag{8.2.28}$$

It follows from the lemma that it is a norm. It is also immediate from (8.2.28) that it is a C^*-norm, i.e., that

$$\|a^*a\| = \|a\|^2, \qquad a \in \mathfrak{A}(R). \tag{8.2.29}$$

For

$$\begin{aligned}\|a^*a\| &= \|\pi(a^*a)\| \\ &= \|\pi(a)^*\pi(a)\| \\ &= \|\pi(a)\|^2 \\ &= \|a\|^2.\end{aligned}$$

It follows from Theorem 8.2.4 that the representations π of $\mathfrak{g}$, and α of $G \times \Gamma$, on $\mathfrak{A}(R)$, extend naturally to representations on the new completion $A(R)$. We shall denote the extended representations by the same symbol, i.e.,

$$\alpha : G \times \Gamma \longrightarrow \mathrm{Aut}(A(R)).$$

Our main objective is to construct groups of automorphisms on $A(R)$, and unitary representations on $\mathcal{H}_2$ which form covariant systems: We say that a representation α of some group G,

$$\alpha : G \longrightarrow \mathrm{Aut}(A(R)),$$

and a unitary representation U of G on $\mathcal{H}_2$, form a covariant system if

$$U_g\ \pi(a)U_g^* = \pi(\alpha_g(a)), \qquad g \in G, \quad a \in A(R). \tag{8.2.30}$$

Since we have Lie groups, the representations will be constructed from one-parameter subgroups.

We shall therefore be especially interested in the covariant systems (8.2.30) when the group G is the real line. In this case, we talk about *one-parameter groups of automorphisms* and *unitary one-parameter groups*. The corresponding infinitesimal generator will be studied.

If

$$\alpha : \mathbb{R} \longrightarrow \mathrm{Aut}(A(R))$$

is a strongly continuous one-parameter group of automorphisms, then the *infinitesimal generator* δ is defined by

$$\delta(a) = \frac{d}{dt}\,\alpha(t)(a)\Big|_{t=0} \tag{8.2.31}$$

on elements a in $A(R)$ such that the limit $\lim\limits_{t\to 0} t^{-1}(\alpha(t)(a)-a)$ exists. We shall view δ as an operator in $A(R)$, and denote its domain $D(\delta)$.

The following result [BR, v.I, Thm. 3.1.19], [Yo, IX.8 Thm. Phillips-Lumer] follows from classical Hille-Yosida theory.

Proposition 8.2.8. An operator δ in $A(R)$ is a generator if and only if:

(i) The domain of δ is dense.

(ii) δ is a closed linear operator.

(iii) δ is a derivation.

(iv) $\delta(a^*) = \delta(a)^*$, $\quad a \in D(\delta) = D(\delta)^*$.

(v) The two operators $\pm\,\delta$ are both dissipative in $A(R)$.

(vi) $\{a \pm \delta(a) : a \in D(\delta)\} = A(R)$.

We have the following result:

Theorem 8.2.9. Suppose δ satisfies (i)-(iv), and (a) and (b) below:

(a) $$\mathfrak{A}(R) \subset D(\delta),$$

(b) $$\{\delta(a) : a \in \mathfrak{A}(R)\} \subset \mathfrak{A}(R).$$

Then we may conclude that δ is a generator.

The reader is referred to [Bra 7, Lemma 2.9.4] for more details, and a proof.

As a *corollary*, we note that every derivation δ of $\mathfrak{A}(R)$ which satisfies

(8.2.32) $$\delta(a^*) = \delta(a)^*, \qquad a \in \mathfrak{A}(R),$$

is closable, and $\overline{\delta}$ is a generator.

We saw in Theorem 8.2.4 that a given antisymmetric bicharacter R gives rise to a unitary representation U of $G \times \Gamma$ on the trace Hilbert space $\mathcal{H}_2$, where $G = \hat{\Gamma}$. If $\{u_\xi : \xi \in \Gamma\}$ denotes the orthonormal basis described before the statement of Theorem 8.2.4, then the representation U is determined by the identity $U(g,\gamma)u_\xi = \langle g,\xi\rangle R(\gamma,\xi)u_\xi$, $g \in G$, $\gamma,\xi \in \Gamma$. The elements u_ξ are viewed as vectors in $\mathcal{H}_2$. Also, recall that the C^*-algebra $A(R)$ is generated by $\{u_\xi : \xi \in \Gamma\}$, and that the u_ξ's are now viewed as unitary operators on $\mathcal{H}_2$. They satisfy

(8.2.33) $$u_\gamma u_\xi = R(\gamma,\xi)u_\xi u_\gamma. \qquad \gamma,\xi \in \Gamma.$$

The theorem below may be viewed as a converse of Theorem 8.2.4.

We show that the C^*-algebra $A(R)$ and the unitary generators $\{u_\xi\}$ may be *reconstructed* from a given unitary representation U of $G \times \Gamma$ which is assumed to have the properties which are listed for U in Theorem 8.2.4.

We also saw, in Corollary 8.1.13, that any nondegenerate antisymmetric bicharacter automatically cobounds, i.e., it may be represented in the form

$$R(\gamma,\xi) = A(\gamma,\xi)A(\xi,\gamma)^{-1}$$

for some cycle A. This cycle, in turn, defines a central extension of $\mathbb{T}$ by Γ.

Let Γ_A denote the extension:

$$(0) \longrightarrow \Gamma \longrightarrow \Gamma_A \longrightarrow \mathbb{T} \longrightarrow (1).$$

Recall that elements in Γ_A are pairs (γ,t) with $\gamma \in \Gamma$ and $t \in \mathbb{T}$. The "product" in Γ_A is defined by

$$(8.2.34) \qquad (\gamma,t) \cdot (\gamma',t') = (\gamma+\gamma',A(\gamma,\gamma')tt'),$$

while the homomorphisms $\Gamma \longrightarrow \Gamma_A$ and $\Gamma_A \longrightarrow \mathbb{T}$ are given by the respective mappings, $\gamma \longrightarrow (\gamma,1)$, respectively, $(\gamma,t) \longrightarrow t$. It follows that the latter two mappings are indeed homomorphisms when the product in Γ_A is given by (8.2.34).

The study of the C^*-algebra $A(R)$ may be approached from three distinct viewpoints, but the result is the same: The three avenues are:

(a) The functor $R \longrightarrow A(R)$ from "*antisymmetric bicharacters on discrete Abelian groups*" into C^*-algebras. In fact. this is a functor into C^*-*algebra dynamical systems with* <u>*ergodic*</u> *actions of compact Abelian groups.*" Starting with a discrete Abelian group Γ, the corresponding compact group G is $\hat{\Gamma}$.

The previous discussion shows that this is indeed a functor,

$$(R,\Gamma) \longrightarrow (A(R),G):$$

A morphism $(R,\Gamma) \xrightarrow{\varphi} (R',\Gamma')$ maps to a morphism (by the functor)

$$(A(R),G) \xrightarrow{\varphi^*} (A(R'),G)$$

of C^*-algebra dynamical systems.

The respective morphisms are given as follows: $\varphi : \Gamma \longrightarrow \Gamma'$

is a homomorphism of groups, and the diagram

$$\begin{array}{ccc} \Gamma & \xrightarrow{\varphi} & \Gamma' \\ R \downarrow & & \downarrow R' \\ \hat{\Gamma} & \xrightarrow{\hat{\varphi}} & \hat{\Gamma}' \end{array}$$

is commutative. Similarly, morphisms $(A(R),G) \longrightarrow (A(R'),G')$ are given by homomorphisms of C^*-algebras $A(R) \longrightarrow A(R')$ which intertwine the respective compact group actions.

(b) The second approach is to study *projective representations* of discrete Abelian groups Γ. If $A : \Gamma \times \Gamma \longrightarrow \mathbb{T}$ is a cocycle of some projection representation, then the bicharacter R is given by

$$R(\gamma,\xi) = A(\gamma,\xi)A(\xi,\gamma)^{-1}, \qquad \gamma,\xi \in \Gamma.$$

(c) The third approach is to study unitary representations π of central extensions

$$0 \longrightarrow \Gamma \longrightarrow \Gamma_A \longrightarrow \mathbb{T} \longrightarrow (1).$$

A central extension is given by a cocycle A. If π is a unitary representation of Γ_A, the mapping $\gamma \longrightarrow \pi((\gamma,1))$ then defines a projective representation of Γ, and *vice versa*.

We saw in Theorem 8.2.4 that there is a natural unitary representation U of $\hat{\Gamma} \times \Gamma$ associated to (a), or equivalently (b) or (c); if R is the bicharacter, then U is given by

$$U(g,\gamma)u_\xi = \langle g,\xi\rangle R\langle\gamma,\xi\rangle u_\xi, \quad g \in \hat{\Gamma}, \quad \gamma,\xi \in \Gamma,$$

and moreover, the representation

$$g \longrightarrow U(g,0) \tag{8.2.35}$$

is unitarily equivalent to the regular representation of G on $\mathcal{L}^2(G)$. Elements in $\Gamma \simeq \hat{G}$ may be viewed as an orthonormal basis for $\mathcal{L}^2(G)$ since

$$\int_G \langle g,\gamma\rangle \overline{\langle g,\xi\rangle} dg = \begin{cases} 1 & \text{if } \gamma = \xi \\ 0 & \text{if } \gamma \neq \xi. \end{cases}$$

The intertwining mapping

$$W : \mathcal{H}_2(\text{trace}) \longrightarrow \mathcal{L}^2(G)$$

is determined by

(8.2.36) $$Wu_\xi = \xi, \qquad \xi \in \Gamma,$$

and satisfies

(8.2.37) $$WU(g,0)W^*\xi = \langle g,\xi\rangle\xi.$$

We now give a necessary and sufficient condition for a unitary representation of $\hat{\Gamma} \times \Gamma$ to be associated with an antisymmetric bicharacter.

Theorem 8.2.10. Let U be a unitary representation of $G \times \Gamma$ on $\mathcal{L}^2(G)$ where $G = \hat{\Gamma}$, and assume

$$U(g,0)\xi = \langle g,\xi\rangle\xi, \qquad g \in G, \quad \xi \in \Gamma.$$

Let S be the shift-representation of Γ defined by

$$S(\gamma)\xi = \gamma+\xi, \qquad \gamma \in \Gamma, \quad \xi \in \Gamma.$$

Then $U\big|_{\{1\}\times\Gamma}$ is given by an antisymmetric bicharacter if and only if

(8.2.38) $$U(1,\gamma)S(\xi)U(1,-\gamma)S(-\xi) = S(\gamma)U(1,\xi)S(-\gamma)U(1,-\xi).$$

Moreover, the bicharacter is nondegenerate if and only if the representation $U\big|_{\{1\}\times\Gamma}$ is faithful.

Proof. Since $\gamma \longrightarrow U(1,\gamma)$ is a homomorphism, the function $R : \Gamma \times \Gamma \longrightarrow \mathbb{T}$ defined by

$$U(1,\gamma)\xi = R(\gamma,\xi)\xi, \qquad \gamma,\xi \in \Gamma,$$

is a homomorphism in γ, when ξ is fixed. Note that $U(1,\gamma)S(\xi)U(1,-\gamma)S(-\xi)$ is a commutator, and that it commutes with $\{U(g,0) : g \in G\}$. It follows that it is multiplication by a scalar. A similar conclusion holds for the right hand side of equation (8.2.38). When the two sides are compared we get

$$R(\gamma,\xi) = R(-\xi,\gamma) = R(\xi,\gamma)^{-1}$$

which shows that R is antisymmetric. Since it is a character in the first variable, it is also in the second variable.

We leave the proof of part two to the reader.

8.3 Extensions of $\hat{\Gamma}$ by $\mathcal{U}(A(R))$

Let R be an antisymmetric bicharacter of a given discrete Abelian group Γ, and let $\hat{\Gamma}$ be the dual compact Abelian group. We shall assume that R is *nondegenerate*, so the associated C^*-algebra $A(R)$ is simple. Let $\mathcal{U}(A(R))$ be the group of unitary elements in $A(R)$.

Let $\alpha : \hat{\Gamma} \longrightarrow \mathrm{Aut}(A(R))$ be the canonical action of $\hat{\Gamma}$ on $A(R)$ which is determined by

$$\alpha(k)(u_\xi) = \langle k,\xi\rangle u_\xi, \qquad k \in \hat{\Gamma}, \quad \xi \in \Gamma. \tag{8.3.1}$$

It is immediate that α restricts to an action of $\hat{\Gamma}$ on $\mathcal{U}(A(R))$. This is just because an automorphism maps unitaries to unitaries.

We shall now construct the extension of $\hat{\Gamma}$ by $\mathcal{U}(A(R))$. The extension will be denoted $G_0(R)$, and it consists of pairs (k,u), $k \in \hat{\Gamma}$, $u \in \mathcal{U}(A(R))$, with the group product defined by

$$(k,u)(k',u') = (kk',u\alpha(k)(u')). \tag{8.3.2}$$

The two homomorphisms of the short exact sequence

$$(e) \longrightarrow \mathcal{U}(A(R)) \longrightarrow G_0(R) \longrightarrow \hat{\Gamma} \longrightarrow (e)$$

are given by

$$u \longrightarrow (1,u), \qquad u \in \mathcal{U}(A(R)),$$

and

$$(k,u) \longrightarrow k, \qquad k \in \hat{\Gamma},$$

respectively.

We have a natural representation of $G_0(R)$ by $*$-automorphisms of the C^*-algebra: Every element $u \in \mathcal{U}(A(R))$ defines an inner automorphism of $A(R)$, and we shall denote it by $\mathrm{Ad}(u)$. It is the adjoint representation:

$$\mathrm{Ad}(u)(a) = uau^{-1}, \qquad a \in A(R). \tag{8.3.3}$$

We define the representation φ of $G_0(R)$ as follows:

$$(8.3.4) \qquad \varphi(k,u) = \mathrm{Ad}(u)\alpha_k, \qquad k \in \hat{\Gamma}, \quad u \in \mathcal{U}(A(R)).$$

This is a representation since

$$\begin{aligned}\varphi((k,u)\cdot(k',u')) &= \varphi((kk',u\alpha(k)(u'))) \\ &= \mathrm{Ad}(u\alpha(k)(u'))\alpha(kk') \\ &= \mathrm{Ad}(u)\mathrm{Ad}(\alpha(k)u')\alpha(k)\alpha(k') \\ &= \mathrm{Ad}(u)\alpha(k)\mathrm{Ad}(u')\alpha(k)^{-1}\alpha(k)\alpha(k') \\ &= \mathrm{Ad}(u)\alpha(k)\mathrm{Ad}(u')\alpha(k') \\ &= \varphi((k,u))\varphi((k',u')).\end{aligned}$$

We shall be interested in constructing a Lie (type) group which has as Lie algebra $\mathfrak{g}$, the Lie algebra of all *derivations* of the algebra $\mathfrak{A}(R)$. A linear mapping

$$\delta : \mathfrak{A}(R) \longrightarrow \mathfrak{A}(R)$$

is said to be a derivation of $\mathfrak{A}(R)$ if

$$\text{(i)} \qquad \delta(ab) = \delta(a)b + a\delta(b), \qquad a,b \in \mathfrak{A}(R).$$

and

$$\text{(ii)} \qquad \delta(a^*) = \delta(a)^*, \qquad a \in \mathfrak{A}(R).$$

Because of certain properties of the *exponential mapping* from the Lie algebra to the Lie group (to be discussed below), we have included property (ii) into the *definition* of a derivation.

In the special case when Γ is finitely generated, we will show that every derivation, $\delta : \mathfrak{A}(R) \longrightarrow \mathfrak{A}(R)$, exponentiates. But the exponential mapping is not given by a power series expansion. We shall denote it by "exp" nonetheless. The Lie algebra of all derivations in $\mathfrak{A}(R)$ will be denoted $\mathfrak{g}$, and we prove

$$(8.3.5) \qquad \exp : \mathfrak{g} \longrightarrow G(R).$$

In the more general case, when Γ is not finitely generated, the exponential mapping is still defined. Theorem 8.2.9 is applied for this purpose.

Since the representation $\varphi((k,u)) = (\mathrm{Ad}\ u)\circ\alpha(k)$ is faithful, we may identify $G_0(R)$ with the group of automorphisms $\varphi(G_0(R))$. The group $G(R)$ which we shall need will be a completion of $\varphi(G_0(R))$. We shall denote by $G_1(R)$ the set of automorphisms which can be approximated by elements form $\varphi(G_0(R))$. More precisely, we say that $\tau \in \mathrm{Aut}(A(R))$ belongs to $G_1(R)$ if there is a net $\{u_\lambda\} \subset \mathcal{U}(A(R))$ and $k \in \Gamma$ such that

$$\tau = \lim_\lambda \mathrm{Ad}(u_\lambda)\alpha(k).$$

The limit is in the weak topology, i.e., for all $a \in A(R)$, and vectors $h_1, h_2 \in \mathcal{H}_2$, we have

$$\lim_\lambda \langle \mathrm{Ad}(u_\lambda)\alpha(k)(a)h_1, h_2\rangle = \langle \tau(a)h_1, h_2\rangle.$$

Note that $G_1(R)$ is (presumably) not a group. The desired group $G(R)$ will satisfy

$$G_0(R) \subset G(R) \subset G_1(R), \tag{8.3.6}$$

and we now proceed to construct this group.

The construction of $G(R)$ is similar to that of $G_0(R)$. For $G_0(R)$, we take an extension of $\hat{\Gamma}$ by the group of unitary elements in the C^*-algebra $A(R)$. We have identified $\mathfrak{A}(R)$ with an algebra of bounded operators on the trace Hilbert space $\mathcal{H}_2$, and $A(R)$ is the norm completion (inside $\mathcal{B}(\mathcal{H}_2)$) relative to the C^*-norm induced from $\mathcal{B}(\mathcal{H}_2)$. The construction of $G(R)$ will be the same, with the single exception that we shall work with the group of unitary elements in the *von Neumann algebra* generated by $\mathfrak{A}(R)$, rather than the C^*-algebra. Recall that this von Neumann algebra, denoted $M(R)$, is the completion of $\mathfrak{A}(R)$ in the weak, or equivalently, strong operator topology. By the double commutant theorem, we also have

$$M(R) = \mathfrak{A}(R)''.$$

We have

$$(1) \longrightarrow \mathcal{U}(M(R)) \longrightarrow G(R) \longrightarrow \hat{\Gamma} \longrightarrow (1).$$

To make explicit the group product in $G(R)$, we need to know that the action (alias representation)

$$\alpha : \hat{\Gamma} \longrightarrow \mathrm{Aut}(A(R)) \tag{8.3.7}$$

extends to a representation (also denoted by α)

$$\alpha : \hat{\Gamma} \longrightarrow \mathrm{Aut}(M(R)). \tag{8.3.8}$$

But this follows from Theorem 8.2.4 where we showed that the unitary representation

$$U : \hat{\Gamma} \longrightarrow (\text{unitary operators on } \mathcal{H}_2),$$

given by

$$U_g(u_\xi) = \langle g, \xi \rangle u_\xi, \qquad \xi \in \Gamma, \tag{8.3.9}$$

is strongly continuous, and satisfies

$$U_g \, a \, U_g^* = \alpha(g)(a), \quad \text{for} \quad g \in \hat{\Gamma} \quad \text{and} \quad a \in A(R). \tag{8.3.10}$$

It follows from (8.3.10) that $a \longrightarrow \alpha(g)(a)$ is continuous on $A(R)$, in the weak operator topology. It therefore extends to the completion in this topology, and this completion is $M(R)$. The extended representation also satisfies (8.3.10), but now for $a \in M(R)$, the larger of the two algebras.

As a result, the definition of the product in $G_0(R)$, namely (8.3.2), carries over to $G(R)$: If $u' \in \mathcal{U}(M(R))$, then $\alpha(k)(u') \in \mathcal{U}(M(R))$ for all $k \in \hat{\Gamma}$. The product (8.3.2) is therefore well defined on

$$G(R) = \{(k,u) : k \in \hat{\Gamma}, \ u \in \mathcal{U}(M(R))\}, \tag{8.3.11}$$

and it is given by the generalized formula (8.3.2) above.

In the remainder of this section, we will show that the familiar techniques from finite-dimensional Lie groups carry over to the infinite-dimensional mapping $\exp : \mathfrak{g} \longrightarrow G(R)$, and prove a generalized Campbell-Baker-Hausdorff formula.

We first show that the pair, Lie algebra/Lie group, $\mathfrak{g}$, resp., $G(R)$, comes equipped with a natural exponential mapping. Recall, elements δ in $\mathfrak{g}$ are derivations, defined on $\mathfrak{A}(R) = \text{span}\{u_\xi : \xi \in \Gamma\}$, and mapping into $\mathfrak{A}(R)$. They satisfy

$$\delta(a^*) = \delta(a)^*, \qquad a \in \mathfrak{A}(R).$$

We noted in Theorem 8.2.9 that each such δ is closable as an operator in $A(R)$, and that $\overline{\delta}$ generates a one-parameter group of $*$-automorphisms in $A(R)$. We now proceed to analyze the automorphism group

$$t \longrightarrow \exp(t\overline{\delta}) : \mathbb{R} \longrightarrow \text{Aut}(A(R)),$$

and to show that the exponential mapping, defined this way, maps into $G(R)$, i.e.,

$$\exp : \mathfrak{g} \longrightarrow G(R). \tag{8.3.12}$$

Our discussion of (8.3.12) separates into two cases. For the first case, an added assumption on Γ is in force, namely that Γ is a *finitely generated* group. This assumption will then be removed in step two. Most of the details will be supplied for step one, while step two will be rather sketchy.

In the study of Lie theory for $G(R)$ the following lemma will be used without mention:

Lemma 8.3.1. There is a one-to-one correspondence between continuous one-parameter subgroups $\{(k(t)),u(t)) : t \in \mathbb{R}\} \subset G(R)$ and pairs $\{k(t)\} \subset \hat{\Gamma}$, $\{u(t)\} \subset \mathcal{U}(M(R))$ where $\{k(t))$ is a continuous one-parameter subgroup of $\hat{\Gamma}$ and $\{u(t)\}$ is a cocycle, i.e.,

$$u(t+s) = u(t)\, \alpha_{k(t)}(u(s)), \qquad t,s \in \mathbb{R}.$$

For a given $\delta \in \mathfrak{g}$, there exists by [BEJ, Theorem 1.1] a continuous one-parameter subgroup $\{k(t) : t \in \mathbb{R}\} \subset \hat{\Gamma}$, and an element $p \in \mathfrak{A}(R)$, $p^* = -p$ such that

$$\delta(a) = \delta_k(a) + pa - ap, \qquad a \in \mathfrak{A}(R), \tag{8.3.13}$$

where

(8.3.14) $$\delta_k(a) = \frac{d}{dt}\Big|_{t=0} \alpha(k(t))(a).$$

(We have used here the assumption on Γ : a finite number of generators.) The decomposition (8.3.13) is unique. Using the simplicity of $A(R)$, we may also note that if p is chosen subject to the condition

$$\mathrm{trace}(p) = 0,$$

then p is also unique. We shall use the terminology $\mathrm{ad}(p)$ for the inner derivation given by

(8.3.15) $$a \longrightarrow pa - ap.$$

The condition $p^* = -p$ yields

$$\mathrm{ad}(p)(a^*) = (\mathrm{ad}\ p(a))^*, \qquad a \in \mathfrak{A}(R),$$

and $e^{t\ \mathrm{ad}\ p}(a) = e^{tp}\ ae^{-tp}$ where $\{e^{tp} : t \in \mathbb{R}\}$ is a one-parameter subgroup of $\mathcal{U}(A(R))$. Define

$$\alpha(t) = e^{t\delta_k}, \qquad t \in \mathbb{R}.$$

It is a one-parameter group of automorphisms, and it is determined by

$$\alpha(t)(u_\xi) = \langle k(t),\xi\rangle u_\xi, \qquad \xi \in \Gamma,$$

where $\{k(t) : t \in \mathbb{R}\} \subset \hat{\Gamma}$ is the subgroup determined, in (8.3.13), by the given element $\delta \in \mathfrak{g}$.

We now define a cocycle $E(t)$, depending on p and $\{\alpha(t)\}$, with values in $\mathcal{U}(A(R))$. It is familiar from perturbation theory, and it allows us to write down a formula for the one-parameter group $\{e^{t\delta} : t \in \mathbb{R}\} \subset \mathrm{Aut}(A(R))$, and to identify it as a subgroup of $G_0(R)$. Define the following time-ordered exponential expansion:

(8.3.16) $$E(t) = 1 + \sum_{n=1}^{\infty} \int\limits_{t\geq t_1\geq t_2\geq\cdots\geq t_n\geq 0} dt_1 \int dt_2 \cdots \int dt_n\ \alpha(t_n)(p)\cdots\alpha(t_1)(p).$$

It follows, of course, that, in the special case where the subgroup $\{k(t) : t \in \mathbb{R}\} \subset \hat{\Gamma}$ reduces to a single point, then $E(t) = \exp(tp)$ is just the unitary one-parameter group generated by p. (The general formula for $E(t)$ is also valid, *mutatis mutandis* when $t \leq 0$.) It is known that $E(t) \in A(R)$, [BR, v. II, Proposition 5.4.1]. In fact, $E(t)$ is unitary, and satisfies the differential equations:

$$\frac{dE(t)}{dt} = \delta_k(E(t)) + pE(t) \tag{8.3.17}$$

and

$$\frac{dE(t)}{dt} = E(t)\alpha(t)(p). \tag{8.3.18}$$

In particular, $E(t) \in D(\delta_k)$ for all $t \in \mathbb{R}$. Formula (8.3.17) is from [BJ 3, Lemma 2.1]. Finally, $E(t)$ satisfies the cocycle identity

$$E(t+s) = E(t)\alpha(t)(E(s)), \qquad t,s \in \mathbb{R}, \tag{8.3.19}$$

cf. also Lemma 8.3.1 above.

We now define

$$\beta(t)(a) = E(t)\alpha(t)(a)E(t)^*, \qquad a \in A(R),$$

or equivalently,

$$\beta(t) = \mathrm{Ad}(E(t)) \circ \alpha(t), \qquad t \in R. \tag{8.3.20}$$

It follows from (8.3.19) that $\{\beta(t) : t \in \mathbb{R}\}$ is a one-parameter group of automorphisms, and from (8.3.18) that $\delta = \delta_k + \mathrm{ad}(p)$ is the infinitesimal generator of this one-parameter group. It is immediate from (8.3.16) that $\{\beta(t) : t \in \mathbb{R}\}$ is, in fact, a one-parameter subgroup of $G_0(R)$. Since the original derivation, $\delta \in \mathfrak{g}$, was arbitrary, we have defined an exponential mapping

$$\exp : \mathfrak{g} \longrightarrow G_0(R),$$

$$\exp(t\delta) := \beta(t), \qquad t \in \mathbb{R}, \tag{8.3.21}$$

with $\beta(t)$ and $E(t)$ given by (8.3.16)-(8.3.20).

We recall that there is a one-to-one correspondence between the continuous one-parameter subgroups

$$\{k(t) : t \in \mathbb{R}\} \subset \hat{\Gamma}$$

and the real vector space $\mathrm{Hom}(\Gamma,\mathbb{R})$. Let $v : \Gamma \longrightarrow \mathbb{R}$ be a homomorphism, then there is a unique subgroup $\{k(t)\} \subset \hat{\Gamma}$ such that

$$\langle k(t),\xi\rangle = \exp(it\, v(\xi)). \tag{8.3.22}$$

If conversely $\{k(t)\}$ is given, then the left hand side of (8.3.22) determines some $v \in \mathrm{Hom}(\Gamma,\mathbb{R})$.

Let $\{k(t)\}$ be a one-parameter group, and let v be the corresponding homomorphism. We shall use the notation:

$$\alpha_v(t) = \alpha(k(t))$$

$$\delta_v(a) = \frac{d}{dt}\,\alpha_v(t)(a)\Big|_{t=0}.$$

By virtue of (8.3.22), we have

$$\delta_v(u_\xi) = i\, v(\xi)u_\xi, \qquad \xi \in \Gamma. \tag{8.3.23}$$

Since we have a duality between the two vector spaces $\mathbb{R}\cdot\Gamma$ and $\mathrm{Hom}(\Gamma,\mathbb{R})$, the function $\xi \longrightarrow v(\xi)$ will be denoted $\langle v,\xi\rangle = v(\xi)$, and (8.3.23) then takes the form

$$\delta_v(u_\xi) = i\langle v,\xi\rangle u_\xi, \qquad \xi \in \Gamma.$$

We now return to the group

$$G_0(R) = \{(k,u) : k \in \hat{\Gamma},\ u \in \mathcal{U}(A(R))\}$$

and its Lie algebra

$$\mathfrak{g} = \{(v,\lambda) : v \in \mathrm{Hom}(\Gamma,\mathbb{R}),\ \lambda \in L_{\mathbb{R}}\}.$$

Recall that $L_{\mathbb{R}}$ is the real Lie algebra of all inner derivations. We saw that there is an isomorphism between $L_{\mathbb{R}}$ and the Lie algebra of elements p in $\mathfrak{A}(R)$ satisfying $p^* = -p$, and $\mathrm{trace}(p) = 0$. This means that we may identify $\mathfrak{g}$ with

$$\{(v,p) : v \in \mathrm{Hom}(\Gamma,\mathbb{R}),\ p \text{ skew Hermitian, and trace-less}\}$$

with the Lie bracket

$$[(v,p),(v',p')] = (0,[p,p']+\delta_v(p')-\delta_{v'}(p)).$$

Using formulas (8.3.1), (8.3.4), and (8.3.11), the adjoint

representation of $G_0(R)$ in $\mathfrak{g}$ then takes the form:

(8.3.24) $$\mathrm{Ad}_{G_0(R)}(k)(v,p) = (v,\alpha_k(p)), \qquad k \in \hat{\Gamma},$$

(8.3.25) $$\mathrm{Ad}_{G_0(R)}(u)(v,p) = (v,upu^{-1}-\delta_v(u)u^{-1}), \quad u \in \mathcal{U}(A(R)).$$

We remark that $\mathrm{Ad}_{G_0(R)}(u)$ is not defined on $\mathfrak{g}$ for all unitary elements in the C^*-algebra $A(R)$, but only for elements such that $u \in D(\delta_v)$ for all $v \in \mathrm{Hom}(\Gamma,\mathbb{R})$. This means that u must be a unitary element in the algebra $A_1(R)$ of C^1-elements for the $\hat{\Gamma}$-action on $A(R)$. We have

$$A_1(R) = \{a \in A(R) : (k \longrightarrow \alpha_k(a)) \in C^1(\hat{\Gamma},A(R))\}.$$

As a vector space, $\mathfrak{g}$ is just the direct sum of the two vector spaces $\mathrm{Hom}(\Gamma,\mathbb{R})$ and $\{p : p^* = -p,\ \mathrm{trace}(p) = 0\}$, it follows that the two operators

$$\mathrm{Ad}_{G_0(R)}(k) \quad \text{and} \quad \mathrm{Ad}_{G_0(R)}(u)$$

may be written as 2×2 matrices relative to the decomposition (v,p) of $\mathfrak{g}$. The respective matrices are

$$\mathrm{Ad}_{G_0(R)}(k) = \begin{bmatrix} I & 0 \\ 0 & \alpha_k \end{bmatrix}, \qquad k \in \hat{\Gamma},$$

$$\mathrm{Ad}_{G_0(R)}(u) = \left[\begin{array}{c|c} I & 0 \\ \hline -\delta_{\bullet}(u)u^{-1} & \mathrm{Ad}_u \end{array}\right]$$

where the entry $-\delta_{\bullet}(u)u^{-1}$ indicates the linear dependence on v. Recall that

$$\delta_v = d\alpha(v)$$

where $\alpha : \hat{\Gamma} \longrightarrow \mathrm{Aut}(A(R))$ is the canonical action of the compact group $\hat{\Gamma}$. The Lie algebra of this group $\mathrm{Hom}(\Gamma,\mathbb{R})$ is finite-dimensional if and only if Γ is finitely generated. But, even if it is finite-dimensional, $\mathfrak{g}$ is still infinite-dimensional.

Since the Lie algebra $\mathfrak{g}$ is infinite-dimensional, continuity properties of the adjoint representation

$$\mathrm{Ad}_{G_0(R)} \quad \text{of} \quad G_0(R) \quad \text{on} \quad \mathfrak{g}$$

must be examined. In this connection we have

Lemma 8.3.2. The automorphism

$$\mathrm{Ad}_{G_0(R)}(k,u) : \mathfrak{g} \longrightarrow \mathfrak{g}$$

is continuous for all $k \in \hat{\Gamma}$, and $u \in \mathcal{U}(A_1(R))$.

Proof. The continuity follows form an inspection of formulas (8.3.24)-(8.3.25), of the corresponding matrix representations. The nontrivial diagonal entries are α_k, and Ad_u; both automorphisms relative to the C^*-norm. Both are isometric relative to this norm. The off diagonal term is clearly continuous in v if $\mathrm{Hom}(\Gamma,\mathbb{R})$ is finite-dimensional, i.e., if Γ has a finite number of generators. In the general case, we may pick an ℓ^1-norm, defined relative to the natural basis for $\mathrm{Hom}(\Gamma,\mathbb{R})$. Then $v \longrightarrow \delta_v(u)u^{-1}$ will also be continuous relative to the respective norms, and the lemma follows. The reader is referred to (8.1.4)-(8.1.11) for details on $\mathrm{Hom}(\Gamma,\mathbb{R})$, and the "natural" basis.

Despite the lemma, it appears that the norm-bounded operators $\mathrm{Ad}_{G_0(R)}(k,u)$ are not strongly continuous on the group $\hat{\Gamma} \times \mathcal{U}(A(R))$ where they are defined.

Fortunately, it suffices, for the study of Lie theory, to have the strong continuity along one-parameter groups.

Lemma 8.3.3. Let $(v,p) \in \mathfrak{g}$, and let

$$(k(t),e^{tq}) \subset \hat{\Gamma} \times \mathcal{U}(A(R))$$

be a one-parameter group, $q = -q^{*} \in \mathfrak{X}(R)$. Then it follows that

$$\lim_{t\to 0}\|(v,p) - \mathrm{Ad}_{G_0(R)}(k(t),e^{tq})(v,p)\| = 0,$$

when g is given the norm topology from Lemma 8.3.2.

Proof. Each of the three terms in (8.3.23)-(8.3.24), but $\delta_v(u)u^{-1}$, is clearly strongly continuous on $\hat{\Gamma} \times \mathcal{U}(A(R))$ with the specified topology. We must show, for fixed $v \in \mathrm{Hom}(\Gamma,\mathbb{R})$, that the nonlinear mapping $u \longrightarrow \delta_v(u)u^{-1}$ is continuous on one-parameter subgroups $u(t) = e^{tq}$, $t \in \mathbb{R}$.

But

$$\delta_v(u(t)) = \int_0^t ds\; u(s)\delta_v(q)u(t-s) \tag{8.3.26}$$

by virtue of [BR, vol.I, Lemma 3.2.31]. (The result (8.3.26) is a standard fact about closed derivations.)

It follows that

$$\delta_v(u(t))u(t)^{*} = \int_0^t ds\; u(s)\delta_v(q)u(s)^{*},$$

and the problem is reduced to considering the term $u(s)\delta_v(q)u(s)^{*}$ under the integral sign. We may simplify further using that $q \in \mathfrak{X}(R)$, i.e., it is a *finite* combination of the basis elements $\{u_\xi : \xi \in \Gamma\}$. But we have $\delta_v(u_\xi) = i\langle v,\xi\rangle u_\xi$, so $u(s)\delta_v(q)u(s)^{*}$ is a finite linear combination of terms $u(s)u_\xi\, u(s)^{*}$. These terms are continuous in s, since $u(s) = e^{sq}$, and q is a bounded operator on $\mathcal{H}_2$. When the different terms are combined, it follows that

$$\lim_{t\to t_0}\|\delta_v(u(t))u(t)^{*}-\delta_v(u(t_0))u(t_0)^{*}\| = 0,$$

for all $t_0 \in \mathbb{R}$. The strong continuity along one-parameter groups follows.

Remark 8.3.4. The conclusion of Lemma 8.3.3 also holds when we consider continuity along one-parameter subgroups of $G_0(R)$ of the form $(k(t),E(t))$ where $E(t)$ is a cocycle, cf. Lemma 8.3.1. The proof depends on a modification of formula (8.3.26) using the cocycle properties of $E(t)$, and (8.3.16)-(8.3.17) above. We shall omit details on this technical point.

We shall use the two lemmas (and the remark) to prove a generalized Campbell-Baker-Hausdorff formula for the Lie algebra g with exponential mapping,

$$\exp : g \longrightarrow G_0(R),$$

defined by (8.3.20)-(8.3.21). We have proved that, for all $\delta_1,\delta_2 \in g$, the automorphism $\exp\delta_1 \exp\delta_2$ is, up to winding number, of the form $\exp\delta_3$ where δ_3 is given by the familiar Campbell-Hausdorff formula.

Theorem 8.3.5. Let $\delta_1,\delta_2 \in g$ be given, and let

$$\psi(z) = \frac{z \log z}{z-1}$$

be defined on the Riemann surface for the complex logarithm. Then the endomorphism, $g \longrightarrow g$:

$$\psi(\mathrm{Ad}_{G_0(R)}(\delta_1)\mathrm{Ad}_{G_0(R)}(t\delta_2))$$

is defined, up to winding number, by functional calculus applied to

$$\mathrm{Ad}_{G_0(R)}(\delta_1)\mathrm{Ad}_{G_0(R)}(t\delta_2).$$

The integral

$$I(\delta_1,\delta_2) = \int_0^1 \psi(\mathrm{Ad}_{G_0(R)}(\delta_1)\mathrm{Ad}_{G_0(R)}(t\delta_2))(\delta_2)dt$$

is norm-convergent in g, and

$$\delta_3 := \delta_1 + I(\delta_1,\delta_2)$$

is a solution to the equation

$$\delta_3 = \log(e^{\delta_1} e^{\delta_2}).$$

It is determined by the familiar C-B-H formula obtained by expanding $\psi(z)$ in a Laurent series.

One of the simple applications of the C-B-H formula to the Lie theory in finite dimensions is a proof of the "Trotter-product formulas"

$$(8.3.27) \qquad \exp t(\delta_1+\delta_2) = \lim_{n\to\infty}\left[\exp\frac{t\delta_1}{n}\exp\frac{t\delta_2}{n}\right]^n$$

and

$$(8.3.28) \quad \exp t^2[\delta_1,\delta_2] = \lim_{n\to\infty}\left[\exp\frac{t\delta_1}{n}\exp\frac{t\delta_2}{n}\exp\frac{-t\delta_1}{n}\exp\frac{-t\delta_2}{n}\right]^{n^2} .$$

These formulas are still valid in the present context, and they may be derived directly as an application of the Trotter-Chernoff convergence theorem [Ne 5, I.8, Thms. 2-5] and [Che] for strongly continuous one-parameter groups of isometries. In formulas (8.3.27)-(8.2.28), we start with a given pair of elements $\delta_1,\delta_2 \in g$. They generate one-parameter groups of automorphisms of $A(R)$. Since $A(R)$ is a C^*-algebra, these one-parameter groups are isometric and strongly continuous. But the new elements $\delta_1+\delta_2$ and $[\delta_1,\delta_2]$ are also in g, and therefore they themselves generate one-parameter groups. Formulas (8.3.27)-(8.3.28) therefore follow from the Trotter-Chernoff theorem.

We conclude this section with some examples of associated Lie algebras where the Lie theory of the present section is available.

Let Γ be a discrete Abelian group and let R be an anti-symmetric, nondegenerate bicharacter on Γ. Let $\mathfrak{A}(R)$ be the associated simple $*$-algebra, with C^*-norm completion $A(R)$. As above, we denote by

$$\alpha : \hat{\Gamma} \longrightarrow \mathrm{Aut}(A(R))$$

the canonical ergodic action of the dual compact group $\hat{\Gamma}$. It will be assumed first that Γ is finitely generated, and that, for each $\gamma \in \Gamma$, the function $\xi \longrightarrow (R(\xi,\gamma)-1)^{-1}$ is of at most polynomial growth on Γ. We let $A_\infty(R)$ be the algebra of

C^∞-elements for the action α on $\hat{\Gamma}$, and we consider the Lie algebra $\mathfrak{g}_\infty$ of all derivations defined on $A_\infty(R)$ and mapping $A_\infty(R)$ into itself. (The condition $\delta(a^*) = \delta(a)^*$, $a \in A_\infty(R)$, $\delta \in \mathfrak{g}_\infty$ is implicit in the definition.) We showed in [BEJ] that $\mathfrak{g}_\infty$ is an extension of the Lie algebra $\mathrm{Hom}(\Gamma,\mathbb{R})$ by the inner derivations on $A_\infty(R)$, i.e., that every $\delta \in \mathfrak{g}_\infty$ decomposes uniquely into a sum

$$\delta = \delta_v + \mathrm{ad}\ p$$

with $v \in \mathrm{Hom}(\Gamma,\mathbb{R})$, $p \in A_\infty(R)$. With the side-conditions:

$$p^* = -p, \qquad \mathrm{trace}(p) = 0,$$

the element p is unique as well. This means that the earlier results of the present section carry over to the Lie algebra $\mathfrak{g}_\infty$.

In a very general setting, the Lie algebra $\mathfrak{g}_\infty$ of all derivations in the ring $A_\infty(R)$ is quite closely connected to an algebra of derivations defined only on the smaller ring $\mathfrak{A}(R)$. The following result is valid *also when* Γ *is not assumed finitely generated*, and the polynomial growth condition on $\xi \longrightarrow (R(\xi,\gamma)-1)^{-1}$ may not be satisfied. We do place the following restriction on Γ: *Every finite rank subgroup of* Γ *is assumed to be finitely generated*. This assumption is satisfied if Γ is of the form

$$\Gamma = \text{finite} \oplus \bigoplus_{\Omega} \mathbb{Z},$$

i.e., Γ is the direct sum of a finite Abelian group, and a free Abelian group. We showed in [BEJ] that the converse assertion also holds.

We proved in [BEJ, Corollary 5.12] that every derivation δ, defined on $\mathfrak{A}(R)$, extends by closure to $A_\infty(R)$, and $\bar{\delta}$ is sequentially continuous from $A_\infty(R)$ to $A(R)$ when $A_\infty(R)$ is equipped with its natural inductive-limit-Fréchet (LF) topology, and $A(R)$ is equipped with the norm topology coming from the C^*-norm.

If it is further assumed that $\delta(\mathfrak{A}(R)) \subset A_\infty(R)$, then it follows that

$$\overline{\delta}(A_\infty(R)) \subset A_\infty(R),$$

see [BGJ, Corollary 5.13].

When the two results are combined, it follows that the study of derivations

$$\delta : A_\infty(R) \longrightarrow A_\infty(R)$$

is equivalent to the study of derivations

$$\delta : \mathfrak{A}(R) \longrightarrow A_\infty(R).$$

The latter type of derivations are in one-to-one correspondence with cocycles on Γ with values in $A_\infty(R)$.

A function

$$c : \Gamma \longrightarrow A_\infty(R)$$

is said to be a cocycle if

(i) $$c(\gamma+\xi) = u_\xi^* \, c(\gamma)u_\xi + c(\xi),$$

(ii) $$c(\gamma)^* = u_\gamma^* \, c(-\gamma)u_\gamma,$$

(iii) $$c(\gamma)^* = -c(\gamma),$$

for all $\xi,\gamma \in \Gamma$. For a given cocycle c, i.e., a function on Γ as specified with properties (i)-(iii), we may define $\delta \in \mathfrak{g}_\infty$ by the basis condition

$$\delta(u_\xi) = u_\xi \, c(\xi), \qquad \xi \in \Gamma.$$

If, conversely, $\delta \in \mathfrak{g}_\infty$ is given, then the function c, defined by

$$c(\xi) = u_\xi^* \, \delta(u_\xi), \qquad \xi \in \Gamma,$$

is a cocycle as specified above. The reader is referred to [BGJ, Proposition 2.3] for a proof of (a more general result!) the equivalence between $\mathfrak{g}_\infty$ and the space of cocycles.

In the decomposition of δ, $\delta = \delta_V +$ (approximately inner), the approximately inner component is given directly in terms of the cocycle. The reader is referred to [BEJ, Theorem 2.1] for details on the general case. In this general case, a variety of limit arguments are combined with Fourier series approximations, and kernel theory. We shall comment, here, only on the simpler special case where the approximately inner component is known to be inner. (Two different classes of examples of this phenomenon are mentioned above.) If c is a cocycle, and

$$c(\gamma) = \sum_{\xi} \hat{c}(\gamma,\xi) u_{\xi}$$

is the corresponding Fourier series,

$$\hat{c}(\gamma,\xi) = u_{\gamma}^{*} \int_{\hat{\Gamma}} dk \overline{\langle k,\xi \rangle}\, \alpha_k(c(\gamma)),$$

then the element p in the decomposition (8.3.13) is given by

$$p = \sum_{\substack{\xi \in \Gamma \\ \xi \neq 0}} d(\xi) u_{\xi}$$

where the numbers $d(\xi)$ are obtained as a solution to the system of equations:

$$\hat{c}(\gamma,\xi) = d(\xi)(R(\xi,\gamma)-1).$$

Existence, and essential uniqueness, for this system is treated thoroughly in [BEJ].

Since g_{∞} is a Lie algebra with the Lie bracket, given by

$$[\delta_1,\delta_2] = \delta_1\delta_2 - \delta_2\delta_1,$$

it follows that the space $\mathscr{C}(\Gamma, A_{\infty}(R))$ of cocycles acquires a structure as a Lie algebra. Since the bijection

$$\delta \longrightarrow c : g_{\infty} \longrightarrow \mathscr{C}(\Gamma, A_{\infty}(R))$$

is a linear isomorphism, the familiar bracket on g_{∞} induces a new Lie bracket on $\mathscr{C}(\Gamma, A_{\infty}(R))$, and it is automatic that the

Jacobi-identity is valid for this bracket. We have proved that the Lie algebra $\mathfrak{g}_\infty$ is an extension of $\mathrm{Hom}(\Gamma,\mathbb{R})$ by the approximately inner derivations in $A_\infty(R)$, and it follows as a corollary that the (isomorphic) Lie algebra $\mathcal{C}(\Gamma,A_\infty(R))$ has the same properties. This point of view is developed in [Jo 23, Thm. 3.1] where a formula is also given for the Lie bracket on $\mathcal{C}(\Gamma,A_\infty(R))$.

We recall the formula for this Lie bracket below:

Let $\delta_1,\delta_2 \in \mathfrak{g}_\infty$, and let $c_1,c_2 \in \mathcal{C}(\Gamma,A_\infty(R))$ be the respective cocycles. Let $[\cdot,\cdot]$ denote the commutator-bracket in $A_\infty(R)$. Then the cocycle c determined by

$$[\delta_1,\delta_2] = \delta_1\delta_2-\delta_2\delta_1$$

is

$$c(\gamma) = [c_1(\gamma),c_2(\gamma)] + \delta_1(c_2(\gamma)) - \delta_2(c_1(\gamma)), \qquad \gamma \in \Gamma.$$

It is possible, of course, to give the formula for c purely in terms of c_1 and c_2, i.e., without reference to the respective derivations δ_1 and δ_2. Since we shall not need the formula in the present context, details are omitted; but they may be found in [Jo 23].

We conclude with a remark on the modifications required in Theorem 8.3.5 for treating algebras built over discrete Abelian groups Γ which are not necessarily finitely generated. We will have to assume that every finite rank subgroup of Γ is finitely generated. It is not known, in this generality, whether every derivation

$$\delta : \mathfrak{A}(R) \longrightarrow \mathfrak{A}(R)$$

is automatically a generator.

We shall therefore not be able to consider the Lie algebra of *all* derivations in $\mathfrak{A}(R)$ but only the subalgebra of derivations with the following restriction.

We shall consider only the subalgebra:

$$\mathfrak{g}_c = \{\delta \in \mathfrak{g} : \delta \text{ has an associated pair } (E,v)\}$$

where $v \in \mathrm{Hom}(\Gamma,\mathbb{R})$, with one-parameter subgroup $\{k(t)\} \subset \hat{\Gamma}$, given by

$$\langle k(t),\xi\rangle = \exp(it\langle v,\xi\rangle), \qquad t \in \mathbb{R}, \quad \xi \in \Gamma,$$

and

$$E : \mathbb{R} \longrightarrow \mathcal{U}(M(R))$$

is a cocycle, i.e., it satisfies

$$E(t+s) = E(t)\ \alpha(k(t))(E(s)), \qquad t,s \in \mathbb{R},$$

where α is the canonical action of $\hat{\Gamma}$ on the von Neumann algebra $M(R) = \mathfrak{A}(R)''$.

It follows form the results, developed in the beginning of the present section, that $\mathfrak{g}_c$ is indeed a Lie algebra, and that the exponential mapping of $\mathfrak{g}_c$ takes values in the group $G(R)$ described in (8.3.6).

8.4 Extensions of $u(n,1)$

In the previous section we described a functorial construction of representations of infinite-dimensional extensions of Abelian Lie algebras. We saw that the representations are determined by an *ergodic* action of a certain compact Abelian group on a $*$-algebra. We found that the C^*-completion plays an important role in the study of the representations.

In this section, we shall sketch a similar program for functorial representations of a class of infinite-dimensional extensions of non-Abelian Lie algebras. While, in Section 8.1, we had the short exact sequence,

$$0 \longrightarrow L \longrightarrow \mathfrak{g} \longrightarrow V \longrightarrow 0,$$

with V Abelian, we shall concentrate here on the case where V is a copy of the reductive Lie algebra $u(n,1)$ of $(n+1) \times (n+1)$ complex matrices.

Our treatment will follow closely the outline in the previous sections, as well as the research papers [Voi 1] and [BEvGJ].

We first recall the Lie group $U(n,1)$ and its Lie algebra $u(n,1)$. The group $U(n,1)$ is defined to be the group of invertible $(n+1) \times (n+1)$ complex matrices g which leave invariant the quadratic form

$$Q(z,z) = z_0\bar{z}_0 - \sum_{i=1}^{n} z_i\bar{z}_i \tag{8.4.1}$$

defined for $z = (z_0, z_1, z_2, \cdots, z_n) \in \mathbb{C}^{n+1}$. The action gz is defined by matrix multiplication, with z viewed as a column vector. The invariance property is

$$Q(gz,gz) = Q(z,z), \qquad z \in \mathbb{C}^{n+1}.$$

Let x be a $(n+1) \times (n+1)$ complex matrix, and let $g(t) = \exp(tx) = \sum_{k=0}^{\infty} \frac{t^k}{k!} x^k$. We note that

$$Q(g(t)z, g(t)z) = Q(z,z), \qquad z \in \mathbb{C}^{n+1}, \quad t \in \mathbb{R},$$

if and only if

$$Q(xz,z) + Q(z,xz) = 0, \qquad z \in \mathbb{C}^{n+1}. \tag{8.4.2}$$

It follows that the Lie algebra $u(n,1)$ of the group $U(n,1)$ consists of matrices x satisfying (8.4.2.), and moreover that $u(n,1)$ is an extension of $u(n)$ by the Abelian Lie algebra $\mathbb{C}^n$:

$$0 \longrightarrow \mathbb{C}^n \longrightarrow u(n,1) \longrightarrow u(n) \longrightarrow 0.$$

The Lie algebra $u(n)$ is determined by the quadratic form

$$\langle z,z \rangle = Q'(z,z) = \sum_{i=1}^{n} z_i\bar{z}_i, \quad z = (z_1, \cdots, z_n) \in \mathbb{C}^n.$$

It consists of $n \times n$ complex matrices satisfying

$$Q'(xz,z) + Q'(z,xz) = 0, \qquad z \in \mathbb{C}^n. \tag{8.4.3}$$

The form Q' is obtained from Q by specializing to $z_0 = 0$. It follows from (8.4.3) that $u(n)$ is indeed the familiar Lie algebra of all skew-Hermitian operators in the complex n dimensional Hilbert space $\mathbb{C}^n$.

First recall that $u(n)$ is a central extension of $su(n)$ by one real dimension. For the action of $su(n)$ on $\mathbb{C}^n$, we take the familiar matrix multiplication. This identifies $su(n,1)$ as an extension

$$0 \longrightarrow \mathbb{C}^n \longrightarrow su(n,1) \longrightarrow su(n) \longrightarrow 0.$$

The final step in the extension construction for $u(n,1)$ is determined by a sesquilinear mapping from $\mathbb{C}^n$ into $u(n)$. Let $\xi,\eta \in \mathbb{C}^n$ be a pair of vectors, and let $\xi\rangle\langle\eta$ be the partial isometry with initial space $\mathbb{C}\eta$ and final space $\mathbb{C}\xi$. Then $(\xi\rangle\langle\eta)^* = \eta\rangle\langle\xi$. (This is the Dirac formalism for rank one operators.)

The quadratic form

$$\mathbb{C}^n \times \mathbb{C}^n \longrightarrow u(n)$$

is defined by

$$\xi,\eta \longrightarrow \xi\rangle\langle\eta - \eta\rangle\langle\xi,$$

where we have used Dirac's notation for rank one operators on $\mathbb{C}^n$. Recall that $\xi\rangle\langle\eta$ is proportional to the partial isometry in $\mathbb{C}^n$ with initial space $\mathbb{C}\eta$ and final space $\mathbb{C}\xi$.

When the different extension steps are combined, we arrive at the following parametrization of the elements in $u(n,1)$:

(8.4.4) $$x = (a,A,\xi)$$

with $$a \in i\mathbb{R},$$

$$A^* = -A,$$

and $$\xi \in \mathbb{C}^n,$$

and Lie bracket

$$[x,x'] = (\langle\xi\xi'\rangle-\langle\xi'\xi\rangle,[A,A']+\xi\rangle\langle\xi'-\xi'\rangle\langle\xi,\ a'\xi-a\xi').$$

We form the tensor algebra over the finite-dimensional Hilbert space $H = \mathbb{C}^n$:

(8.4.5) $$T(H) = \sum_{m=0}^{\infty}{}^{\oplus}\ H^{\otimes m}$$

where

$$H^{\otimes 0} = \mathbb{C}\Omega,$$

with a specified vacuum vector Ω,

$$H^{\otimes 1} = H,$$

and inductively

$$H^{\otimes(m+1)} = (H^{\otimes m}) \otimes H.$$

This algebra $T(H)$ carries again an inner product, obtained by tensoring the inner product on H for each n, and then taking orthogonal direct sums in the category of Hilbert spaces. This is the infinite-particle Boltzmann-statistics Hilbert space built on H when H is viewed as one particle space. We have the familiar second quantization functor Γ from the group of unitaries on H to the group of unitaries on $T(H)$. It is defined by

(8.4.6) $$\Gamma(A)(\xi_1 \otimes \cdots \otimes \xi_m) = A\xi_1 \otimes \cdots \otimes A\xi_m, \quad A \in U(n), \xi_i \in H, \quad i = 1, \cdots, m, \quad m = 1, 2, \cdots,$$

and $$\Gamma(A)\Omega = \Omega, \qquad A \in U(n).$$

We shall also need the annihilation operators $\{a(\xi) : \xi \in H\}$ defined by

(8.4.7) $$a(\xi)(\eta_1 \otimes \cdots \otimes \eta_m) = \sum_{i=1}^{m} \langle \eta_i, \xi \rangle \eta_1 \otimes \cdots \hat{i} \cdots \otimes \eta_m$$

where the symbol $\hat{i}$ indicates that the vector η_i is omitted. Each operator $a(\xi)$ has dense domain in $T(H)$, and the adjoint operator $a(\xi)^* = a^*(\xi)$ is the creation operator which is now given by

(8.4.8) $$a^*(\xi)(\eta_1 \otimes \cdots \otimes \eta_m) = \sum_{i=0}^{m} \eta_1 \otimes \cdots \otimes \eta_i \otimes \xi \otimes \eta_{i+1} \otimes \cdots \otimes \eta_m$$

where it is understood that the first term in this sum is

$$\xi \otimes \eta_1 \otimes \cdots \otimes \eta_m,$$

while the last term is

$$\eta_1 \otimes \cdots \otimes \eta_m \otimes \xi .$$

In particular, we have

$$a^*(\xi) : H^{\otimes m} \longrightarrow H^{\otimes(m+1)},$$

$$a(\xi) : H^{\otimes m} \longrightarrow H^{\otimes(m-1)},$$

and the vacuum condition:

(8.4.9) $$a(\xi)\Omega = 0, \qquad \xi \in H.$$

An elementary calculation shows that the vectors in $H^{\otimes m}$, $m = 0,1,\cdots,$ are analytic for each of the operators $a^*(\xi)$ and $a(\xi)$, $\xi \in H$. We may therefore apply the result in [GJ 2, Theorem 3.1], see also the Appendix below, to the operator Lie algebra generated by

(8.4.10) $$\{a^*(\xi)-a(\xi) : \xi \in H\}.$$

But this Lie algebra (over $\mathbb{R}$) is obtained by applying a Hermitian representation to the initial Lie algebra $u(n,1)$. Recall that elements x in $u(n,1)$ are parametrized:

$$x = (a,A,\xi),$$

$$a \in i\mathbb{R}, \quad A \in u(n), \quad \xi \in H.$$

The representation ρ of $u(n,1)$ is determined by the conditions:

(8.4.11) $$\rho(0,0,\xi) := a^*(\xi) - a(\xi), \qquad \xi \in H,$$

and

$$\rho(0,A,0)(\eta_1 \otimes \cdots \otimes \eta_m) = \sum_{i=1}^{m} \eta_1 \cdots \otimes \eta_{i-1} \otimes A\eta_i \otimes \eta_{i+1} \otimes \cdots \eta_m .$$

In fact, on $u(n)$, we have

$$\rho = d\Gamma,$$

where Γ is the representation of $U(n)$ which is determined in (8.4.6).

The representation ρ of $u(n,1)$ is exact, and there is a unique unitary representation W of the Lie group $U(n,1)$ such that

$$\rho = dW \qquad \text{on} \quad \mathfrak{u}(n,1). \tag{8.4.12}$$

We should note that [GJ 2, Theorem 3.1], strictly speaking, only gives the existence of a unitary representation W of the universal covering group $\widetilde{U(n,1)}$. But since we have the constraint that

$$W\Big|_{U(n)} = \Gamma,$$

i.e., the restriction of W is *predetermined*, it follows that the integrated representation W is trivial on the kernel of the covering mapping

$$\widetilde{U(n,1)} \longrightarrow U(n,1).$$

It follows that W passes to the quotient and defines a genuine unitary representation of U(n,1), and that (8.4.12) holds.

We proceed to describe a covariant system on T(H) for this representation.

Since T(H) is an algebra it carries the natural left-regular representation π defined by

$$\pi(a)b = ab, \qquad a,b \in T(H), \tag{8.4.13}$$

when the product ab is taken in T(H). But H is naturally embedded into T(H), so, for $\xi \in H \subset T(H)$, (8.4.13) specializes to

$$\pi(\xi)\eta_1 \otimes\cdots\otimes \eta_m = \xi \otimes \eta_1 \otimes\cdots\otimes \eta_m. \tag{8.4.14}$$

We shall denote by $\mathfrak{A}(H)$ the smallest $*$-algebra of operators on T(H) which contains the identity operator I and the operators $\pi(\xi)$, $\xi \in H$, from (8.4.14). Since $\mathfrak{A}(H)$ contains the adjoint operator $\pi(\xi)^* = \pi^*(\xi)$, we note the formula

$$\pi^*(\xi)(\eta_1 \otimes\cdots\otimes \eta_m) = \langle \eta_1, \xi\rangle \eta_2 \otimes\cdots\otimes \eta_m$$

for this operator.

It follows that

$$\pi^*(\xi)\pi(\xi) = \|\xi\|^2 \cdot I, \qquad \xi \in H,$$

and, in particular, that $\pi(\xi)$ is an isometry if ξ is a unit-vector in one-particle space H, i.e., $\|\xi\|^2 = \langle \xi, \xi \rangle = Q'(\xi,\xi) = 1$. Even when ξ is not a unit-vector, the operator $\pi(\xi)$ is bounded, and, for the operator norm, we have

$$\|\pi(\xi)\|^2 = \|\pi^*(\xi)\pi(\xi)\| = \|\xi\|^2.$$

The norm-completion of $\mathfrak{A}(H)$ inside $\mathfrak{B}(T(H))$ will be denoted C(H). It is a C^*-algebra, of course.

Theorem 8.4.1. There is a representation

$$\alpha : U(n,1) \longrightarrow \mathrm{Aut}(C(H))$$

satisfying

$$\alpha_g(a) = W_g \, a \, W_g^*, \qquad a \in C(H), \quad g \in U(n,1). \tag{8.4.15}$$

Proof. Suppose (8.4.15) holds. Let $x \in u(n,1)$, and set $g(t) = \exp(tx)$, $t \in \mathbb{R}$. Then a formal differentiation of $W_{g(t)} \, a \, W^*_{g(t)}$, $a \in C(H)$, yields:

$$\begin{aligned} d\alpha_x(a) &= \frac{d}{dt}\Big|_{t=0} \alpha_{g(t)}(a) \\ &= \frac{d}{dt}\Big|_{t=0} W_{g(t)} \, a \, W_{g(-t)} \\ &= \left[\frac{d}{dt}\Big|_{t=0} W_{g(t)}\right] a - a\left[\frac{d}{dt}\Big|_{t=0} W_{g(t)}\right] \\ &= dW(x)a - a\, dW(x) \\ &= [dW(x), a] \\ &= [\rho(x), a]. \end{aligned}$$

It follows that, for x in the Lie algebra, the operator $d\alpha(x)$ is given by a commutator which is defined in terms of the original representation ρ of $u(n,1)$. The representation ρ extends naturally to the universal enveloping algebra $\mathfrak{U}_{\mathbb{C}}(u(n,1))$ and the extended representation, also denoted by ρ, is Hermitian relative to the $*$-operation on $\mathfrak{U}_{\mathbb{C}}(u(n,1))$ and the inner product on the tensor algebra T(H), alias the Boltzmann-statistics Fock space.

It follows from the discussion of $u(n,1)$ that the algebra $\rho(\mathfrak{U}_{\mathbb{C}}(u(n,1))$ is generated by the operators

$$\rho(0,0,\xi) = a^*(\xi)-a(\xi), \qquad \xi \in H.$$

By taking real and imaginary parts, it follows that $\rho(\mathfrak{U}_{\mathbb{C}}(u(n,1))$ is also generated by the annihilation, resp., creation operators $a(\xi)$, resp., $a^*(\xi)$, $\xi \in H$.

This discussion shows that the infinitesimal version of the covariance formula (8.4.15) amounts to the statement that the commutator brackets

$$[a^*(\xi),\cdot] \text{ and } [a(\xi),\cdot]$$

normalize the algebra $\mathfrak{A}(H)$. Since $\mathfrak{A}(H)$ is a $*$-algebra, it is enough to verify one of the two conditions. But $\mathfrak{A}(H)$ is generated by the operators $\pi(\eta)$, $\eta \in H$, so it is enough to verify that

$$[a^*(\xi),\pi(\eta)] \in \mathfrak{A}(H)$$

for all $\xi,\eta \in H$. However, this is immediate from the commutation relations

$$\begin{aligned}[a^*(\xi),\pi(\eta)] &= \pi(\xi)\pi(\eta)\\ &= \pi(\xi \otimes \eta).\end{aligned}$$

We also have

$$[a(\xi),\pi(\eta)] = \langle\eta,\xi\rangle I$$

where I is the identity operator on $T(H)$. It follows that

$$\begin{aligned}\text{(8.4.16)} \qquad d\alpha(0,0,\xi)(\pi(\eta)) &= [a^*(\xi),\pi(\eta)] - [a(\xi),\pi(\eta)]\\ &= \pi(\xi)\pi(\eta) - \langle\eta,\xi\rangle I.\end{aligned}$$

We have still not proved (8.4.15), but only an infinitesimal version of it. But using again the result [GJ 2, Thm. 3.1], this time for the C^*-algebra $C(H)$, rather than the Hilbert space $T(H)$, we now proceed to exponentiate the infinitesimal covariance condition. To use the cited result, we must verify that the elements in $\mathfrak{A}(H) \subset C(H)$ are analytic for the derivations $d\alpha(0,0,\xi)$. These derivations generate the Lie algebra of all derivations $d\alpha(u(n,1))$. But the analytic elements for a family of derivations is an algebra (i.e., it is closed under multiplication), so it is enough to check that the generating elements $\pi(\eta)$, $\eta \in H$, in $\mathfrak{A}(H)$ are analytic.

The formula (8.4.16) above may readily be iterated, and we arrive at the desired conclusion that each of the elements $\pi(\eta)$, $\eta \in H$, is analytic for the derivation $D_\xi = d\alpha(0,0,\xi)$. Specifically, we have

$$D_\xi \pi(\eta) = \pi(\xi)\pi(\eta) - \langle \eta, \xi \rangle I, \tag{8.4.17}$$

and

$$\begin{aligned} D_\xi^2 \pi(\eta) &= D_\xi(\pi(\xi)\pi(\eta)) \\ &= D_\xi(\pi(\xi))\pi(\eta) + \pi(\xi)D_\xi(\pi(\eta)) \\ &= 2\pi(\xi)^2\pi(\eta) - \|\xi\|^2\pi(\eta) - \langle \eta, \xi \rangle \pi(\xi). \end{aligned}$$

It follows that

$$\|D_\xi \pi(\eta)\| \le 2\|\xi\|\|\eta\|$$

and

$$\|D_\xi^2 \pi(\eta)\| \le 4\|\xi\|^2\|\eta\|.$$

Now make the inductive assumption

$$\|D_\xi^k \pi(\eta)\| \le (2\|\xi\|)^k\|\eta\|k!. \qquad 1 \le k \le m. \tag{8.4.18}$$

We then get

$$\begin{aligned} D_\xi^{m+1}\, \pi(\eta) &= D_\xi^m(\pi(\xi)\pi(\eta)) \\ &= \sum_{k=0}^{m} \binom{m}{k} D_\xi^k\, \pi(\xi) D_\xi^{m-k}\, \pi(\eta), \end{aligned}$$

and

$$\begin{aligned} \|D_\xi^{m+1}\, \pi(\eta)\| &\le \sum_{k=0}^{m} \binom{m}{k} (2\|\xi\|)^k (2\|\xi\|)^{m-k} \|\xi\|\|\eta\| \times k!(m-k)! \\ &= (m+1)m!(2\|\xi\|)^m\|\xi\|\|\eta\| \\ &\le (m+1)!(2\|\xi\|)^{m+1}\, \|\eta\|, \end{aligned}$$

which proves that (8.4.18) also holds for the next step $m+1$.

We have proved that the radius of convergence for the power

series

$$\sum_{m=0}^{\infty} \frac{t^m}{m!} \|D_\xi^m \pi(\eta)\|$$

is at least $(2\|\xi\|)^{-1}$, which is, of course, a half when ξ is a unit-vector in H. This means that the radius of convergence depends only on ξ and not on the vector η.

We have a faithful state f on $C(H)$ which is simply given by the Fock vacuum vector Ω, that is,

$$f(a) = \langle a\Omega, \Omega\rangle, \qquad a \in C(H).$$

Since $W_g\Omega = \Omega$, $g \in G = U(n,1)$, and therefore

$$dW(x)\Omega = 0, \qquad x \in \mathfrak{g} = \mathfrak{u}(n,1),$$

we conclude that

$$f(d\alpha(x)(a)) = 0, \qquad x \in \mathfrak{g}, \quad a \in \mathfrak{A}(H). \tag{8.4.19}$$

Indeed,

$$\begin{aligned} f(d\alpha(x)(a)) &= \langle d\alpha(x)(a)\Omega, \Omega\rangle \\ &= \langle (dW(x)a - a\, dW(x))\Omega, \Omega\rangle \\ &= \langle dW(x)a\Omega, \Omega\rangle \\ &= -\langle a\Omega, dW(x)\Omega\rangle \\ &= 0, \end{aligned}$$

which is the assertion (8.4.19).

The generator theorem [BR, vol.I, Thm. 3.2.50] now applies to the derivation $d\alpha(x)$ for $x = (0,0,\xi) \in \mathfrak{g} = \mathfrak{u}(n,1)$, and we conclude that each derivation $d\alpha(x)$ generates a strongly continuous one-parameter group of $*$-automorphisms of the C^*-algebra $C(H)$, i.e.,

$$t \longrightarrow \alpha(tx) : \mathbb{R} \longrightarrow \mathrm{Aut}(C(H))$$

is a representation of $\mathbb{R}$ with

$$d\alpha(x)(a) = \left.\frac{d}{dt}\right|_{t=0} \alpha(tx)(a)$$

for a in the domain of the infinitesimal generator.

We have verified the assumptions in [GJ 2, Thm. 3.1], and

we conclude that there is a representation

$$\alpha : G \longrightarrow \mathrm{Aut}(C(H)) \tag{8.4.20}$$

such that

$$d\alpha(x) = \frac{d}{dt}\Big|_{t=0} \alpha(\exp(tx)), \qquad x \in \mathfrak{g}. \tag{8.4.21}$$

Note that we get (8.4.21) as an identity on the algebra of C^∞-vectors for the action (8.4.20), i.e., on

$$C^\infty(G) = \{a \in C(H) : (g \longrightarrow \alpha_g(a)) \in C^\infty(G, C(H))\}.$$

Also note that the identity (8.4.21) is automatically valid for all $x \in \mathfrak{g} = \mathfrak{u}(n,1)$, not just for the special Lie algebra elements x of the form $(0,0,\xi)$, $\xi \in H$.

Strictly speaking, [GJ 2, Thm. 3.1] only yields a representation

$$\alpha : \tilde{G} \longrightarrow \mathrm{Aut}(C(H))$$

of the universal covering group of $U(n,1)$. But this representation is trivial on the kernel of the covering mapping $\tilde{G} \xrightarrow{\Phi} G$ from simply connected $\tilde{G}$. This is because any possible loop in G must be contained in $U(n) \times S^1$ since the vector component of the Cartan decomposition for $U(n,1)$ is trivially simply connected. It is just a copy of $\mathbb{C}^n$, i.e,

$$G/U(n) \times S^1 \simeq \mathbb{C}^n, \tag{8.4.22}$$

and the representation α in (8.4.15) is already specified on $U(n) \times S^1$: On the $U(n)$ component, we have

$$\alpha(k)(a) = \Gamma(k) a \Gamma(k)^*, \qquad a \in C(H), \quad k \in U(n),$$

and, for k in S^1, the automorphism $\alpha(k)$ is trivial.

It follows that

$$\alpha : \tilde{G} \longrightarrow \mathrm{Aut}(C(H))$$

does indeed pass to the quotient

$$\tilde{G}/\mathrm{kernel}(\Phi) \simeq G \tag{8.4.23}$$

and yields a genuine representation of G. We shall refer to this representation as the *canonical representation* of the Lie group $U(n,1)$ on the C^*-algebra $C(H)$.

We now recall the definition of the simple C^*-algebra $\mathcal{O}_n$, also called the Cuntz-algebra. The C^*-algebra $C(H)$ described above is, in fact, an extension of $\mathcal{O}_n$ by the compact operators $\mathcal{K}$ on the Fock space $T(H)$, i.e., we have a short exact sequence

(8.4.24) $$0 \longrightarrow \mathcal{K} \longrightarrow C(H) \longrightarrow \mathcal{O}_n \longrightarrow 0$$

of C^*-algebra. In the case $n = 1$ we do *not* get an extension of a simpel C^*-algebra, but rather an extension of the Abelian C^*-algebra of all continuous functions on the circle S^1, i.e.,

$$0 \longrightarrow \mathcal{K} \longrightarrow C(H) \longrightarrow C(S^1) \longrightarrow 0.$$

In this case, $C(H)$ is, of course, the C^*-algebra generated by a single isometry with multiplicity one. The operator $V = \pi(\xi)$, $\|\xi\| = 1$, $\xi \in \mathbb{C}^1$, is just the unilateral shift. This is because $T(H)$ is just $\ell^2(\mathbb{N})$ when H is one-dimensional. Pick an orthonormal basis $\{e_n : n = 0,1,2,\cdots\}$. Then V is defined by

$$Ve_n = e_{n+1}, \qquad n = 0,1,\cdots.$$

The algebra $C(H)$ was studied first in this case by Coburn, [Cob]. (The algebra $C(H)$ is called the Toeplitz algebra, also when $\dim H > 1$.)

In the general case, $1 < n < \infty$, we proceed to describe the short exact sequence (8.4.24).

The representation π from (8.4.13)-(8.4.14) satisfies the following properties:

$$\pi^*(\xi)\pi(\eta) = \langle \eta, \xi \rangle I,$$

(8.4.25) $$\pi(\eta)\pi^*(\xi)(\zeta_1 \otimes \cdots \otimes \zeta_m) = \langle \zeta_1, \xi \rangle \eta \otimes \zeta_2 \otimes \cdots \otimes \zeta_m,$$

and

$$\pi(\eta)\pi^*(\xi)\Omega = 0, \qquad \xi, \eta \in H.$$

If $e_1, \cdots, e_n$ is an orthonormal basis for H, and $V_i = \pi(e_i)$, then we have

$$V_i^* V_j = \delta_{ij} I, \qquad i \leq i,j \leq n,$$

and

$$\sum_{i=1}^{n} V_i V_i^* < I,$$

i.e., the V_i's are isometries with orthogonal range projections. The sum $\sum_1^n V_i V_i^*$ is the projection onto the orthogonal complement of the vacuum. This follows from (8.4.25). Indeed,

$$\left[\sum_1^n V_i V_i^*\right]\zeta_1 \otimes \cdots \otimes \zeta_m = \sum_1^n \langle \zeta_1, e_i \rangle e_i \otimes \zeta_2 \otimes \cdots \otimes \zeta_m$$

$$= \zeta_1 \otimes \cdots \otimes \zeta_m.$$

Moreover, the ideal in $\mathcal{X}(H)$ generated by $P = I - \sum_1^n V_i V_i^*$, i.e.,

$$\mathcal{X}(H) P \, \mathcal{X}(H),$$

consists of all finite rank operators in the Hilbert space $T(H)$. Let $\mathcal{K}$ be the corresponding closed ideal, and let

$$\varphi : C(H) \longrightarrow C(H)/\mathcal{K}$$

be the quotient mapping. Let

$$S_i = \varphi(V_i), \qquad 1 \le i \le n. \tag{8.4.26}$$

Then

$$S_i^* S_j = \delta_{ij} I \tag{8.4.27a}$$

and

$$\sum_1^n S_i S_i^* = I. \tag{8.4.27b}$$

Cuntz [Cu] showed that this C^*-algebra is *simple*. It is called the Cuntz algebra, and is denoted $\mathcal{O}_n$.

It follows from Theorem 8.4.1 that the representation

$$\alpha : G \longrightarrow \mathrm{Aut}(C(H))$$

passes to the quotient $\mathcal{O}_n = C(H)/\mathcal{K}$ since the compact operators $\mathcal{K}$ are invariant under conjugation by the unitary representation W from (8.4.15). We shall denote the resulting action of $G = U(n,1)$ on $\mathcal{O}_n$ also by α and refer to it as the canonical action. The image of $\mathfrak{X}(H)$ under the quotient homomorphism $\varphi : C(H) \longrightarrow \mathcal{O}_n$ will be denoted $\mathcal{D}_n$. It is a dense subalgebra of $\mathcal{O}_n$, and it is simple as well. It is generated by the isometries S_i, $i = 1,\cdots,n$.

We shall be interested in the Lie algebra $\mathcal{L}_n$ of all derivations in $\mathcal{D}_n$, i.e., all linear endomorphisms of $\mathcal{D}_n$ satisfying the two conditions:

$$\delta(ab) = \delta(a)b + a\delta(b)$$

and

$$\delta(a^*) = \delta(a)^*, \qquad a,b \in \mathcal{D}_n.$$

It is an infinite-dimensional Lie algebra over $\mathbb{R}$ under the commutator bracket, and it contains the canonical copy of $u(n,1)$ which is obtained by differentiation of the Lie group action $\alpha : U(n,1) \longrightarrow \mathrm{Aut}(\mathcal{O}_n)$, i.e., the Lie algebra

$$\{d\alpha(x) : x \in u(n,1)\} \subset \mathcal{L}_n$$

with $d\alpha$ given by (8.4.21).

We now record the following three results from [BEvGJ] and [BEv 2].

Theorem 8.4.2. (a) Every δ in $\mathcal{L}_n$ is closable and $\overline{\delta}$ generates a strongly continuous representation of $\mathbb{R}$, i.e., $e^{t\overline{\delta}}$ is a one-parameter group of $*$-automorphisms of $\mathcal{O}_n$.

(b) Elements $d\alpha(x) \neq 0$ in the subalgebra $d\alpha(u(n,1))$ are non-approximately inner.

(c) If an element δ in $\mathcal{L}_n$ is approximately inner, and invariant under the S^1-action, then it is inner.

Remarks 8.4.3. Part (a) is the generator result, and the reference is [BE 2]. The idea of the proof is based on a contribution by Thomsen [Tho 2]. The Thomsen paper, in turn, was an appendix to [BGJ]. Parts (b) and (c) are contained in [BEvGJ]. Part (c) is the main result in [BEvGJ] and it is called the Dichotomy Theorem.

The Lie algebra $\mathcal{L}_n$ shares property (a) with the corresponding infinite-dimensional Lie algebra studied in Section 8.3. But it is not true that every δ in $\mathcal{L}_n$ is an inner perturbation of an element from $d\alpha(u(n,1))$ = (the Lie algebra of the canonical action).

The reader is referred to [Bra 7, Section 2.9] for additional results on $\mathcal{L}_n$, and on derivations in C^*-algebras with properties similar to those of $\mathcal{O}_n$. The generator result, viz., property (a) in Theorem 8.4.2 above, is Lemma 2.9.4 in [Bra 7].

The group of all automorphisms of $\mathcal{O}_n$, i.e., $\mathrm{Aut}(\mathcal{O}_n)$ is not particularly well understood. But the subgroup of $\mathrm{Aut}(\mathcal{O}_n)$ generated by $\exp(\mathcal{L}_n) = \{\exp \delta : \delta \in \mathcal{L}_n\}$ is better understood. While it does not appear that Theorem 8.3.5 (from the previous section) carries over to $\exp \mathcal{L}_n$, the corresponding Trotter-approximation formulas (8.3.27)-(8.3.28) are valid. The proof is identical to the one given in Section 8.3.

Based on the results from Sections 8.2-8.3, one might perhaps expect that the quotient of $\mathcal{L}_n$ by the ideal of inner derivations is isomorphic to the canonical $u(n,1)$-Lie algebra. This is not the case. In fact, the quotients $\mathcal{L}_n/\{\text{inners}\}$ is infinite-dimensional.

This follows from the following theorem. First, two definitions: Let $\mathcal{D}_n^{sk}$ denote the skew-Hermitian elements in $\mathcal{D}_n$. Let σ be the shift of $\mathcal{O}_n$, given by

(8.4.28) $$\sigma(a) = \sum_{i=1}^{n} S_i a S_i^*.$$

Define $\Lambda : \mathcal{L}_n \longrightarrow \mathcal{D}_n^{sk}$ by $\Lambda(\delta) = \sum_i \delta(S_i)S_i^*$, and define $\Gamma : \mathcal{D}_n^{sk} \longrightarrow \mathcal{L}_n$ by $\Gamma(\ell)$ = the unique derivation δ which is determined on the generators S_i by

$$\delta(S_i) = \ell S_i, \qquad 1 \leq i \leq n.$$

Then the two functions are each other's inverses:

(8.4.29) $$\mathcal{L}_n \underset{\Gamma}{\overset{\Lambda}{\rightleftarrows}} \mathcal{D}_n^{sk}.$$

To show, e.g., that $\Gamma(\Lambda(\delta)) = \delta$ we note that

$$\left[\sum_i \delta(S_i)S_i^*\right]S_k = \sum_i \delta(S_i)S_i^*S_k$$
$$= \delta(S_k), \qquad 1 \leq k \leq n,$$

by virtue of (8.4.27a).

Theorem 8.4.4. (a) Define the bracket in $\mathcal{D}_n^{sk}$ as follows:

$$[\ell_1,\ell_2]_{sk} = [\ell_2,\ell_1] + \Gamma(\ell_1)(\ell_2) - \Gamma(\ell_2)(\ell_1).$$

Then $\mathcal{D}_n^{sk}$ is a Lie algebra, and it is isomorphic to $\mathcal{L}_n$.

(b) Under the isomorphism in (a), the inner derivations in $\mathcal{L}_n$ are mapped onto the set

$$\{b-\sigma(b) \ : \ b \in \mathcal{D}_n^{sk}\} \subset \mathcal{D}_n^{sk}.$$

Proof. Let $\Gamma(\ell_1) = \delta_1$ and $\Gamma(\ell_2) = \delta_2$, then

$$[\delta_1,\delta_2](S_i) = \delta_1(\delta_2(S_i)) - \delta_2(\delta_1(S_i))$$
$$= \delta_1(\ell_2 S_i) - \delta_2(\ell_1 S_i)$$
$$= \delta_1(\ell_2)S_i + \ell_2\delta_1(S_i) - \delta_2(\ell_1)S_i - \ell_1\delta_2(S_i)$$
$$= (\delta_1(\ell_2)-\delta_2(\ell_1) + [\ell_2,\ell_1])S_i, \qquad 1 \leq i \leq n.$$

It follows that

$$\Lambda([\delta_1,\delta_2]) = \delta_1(\ell_2) - \delta_2(\ell_1) + [\ell_1,\ell_2],$$

which proves (a).

As for (b), suppose $\ell = b-\sigma(b)$, and let $\delta = \Gamma(\ell)$. Then

$$\begin{aligned}\delta(S_k) &= bS_k - \sum_i S_i b S_i^* S_k \\ &= bS_k - S_k b \\ &= (\mathrm{ad}\ b)(S_k), \qquad 1 \leq k \leq n.\end{aligned}$$

Since $\mathfrak{D}_n$ is generated by the S_k's, it follows that $\delta = \mathrm{ad}\ b$, i.e., δ is inner. Since the argument works both ways, (b) is proved.

To study the quotient $\mathcal{L}_n/\{\text{inner derivations in } \mathcal{L}_n\}$, or equivalently, $\mathfrak{D}_n^{sk}/\{(I-\sigma)\mathfrak{D}_n^{sk}\}$, a class of states on $\mathcal{O}_n$ were introduced by the authors of [Cu] and [BEvGJ]. We recall the states. They are often called the Cuntz states, and they are indexed by unit-vectors ξ in one-particle space $H = \mathbb{C}^n$. Let $\xi = (\xi_1,\cdots,\zeta_n) \in \mathbb{C}^n$ and define f_ξ as follows. On monomials of the form $P = S_{i_1}\cdots S_{i_h} S_{j_1}^* \cdots S_{j_k}^*$, we set

$$f_\xi(P) = \xi_{i_1}\cdots\xi_{i_h}\bar{\xi}_{j_1}\cdots\bar{\xi}_{j_k}, \tag{8.4.30}$$

or equivalently,

$$f_\xi(\varphi(\eta_1)\cdots\varphi(\eta_h)\varphi(\zeta_1)^*\cdots\varphi(\zeta_k)^*) = \prod_{r=1}^{h}\langle\eta_r,\xi\rangle \prod_{s=1}^{k}\langle\xi,\zeta_s\rangle. \tag{8.4.31}$$

We show in [BEvJG, Proposition 2.1] that f_ξ is well defined, and is a state when $\|\xi\|_2^2 = \sum_{i=1}^{n} \xi_i\bar{\xi}_i = 1$.

The proof (in [BEvGJ] of a more general result) is not particularly difficult. But the part of it that we need is a special case of a result on completely positive mappings in

C^*-algebras, and the latter subject is beyond the scope of the present book. We shall simply work with the states f_ξ in the form (8.4.30) or (8.4.31). Note that, if $P \in \mathcal{O}_n$, then $f_\xi(P)$ may be viewed as the symbol of P in the weak sense of symbols of pseudodifferential operators. It is immediate that the states are shift invariant, i.e., that

$$f_\xi(\sigma(P)) = f_\xi(P), \qquad p \in \mathcal{O}_n.$$

It follows that $f_\xi(\ell) = 0$ for $\ell \in (I-\delta)\mathfrak{D}_n$, so each f_ξ passes to the quotient

$$\mathfrak{D}_n^{sk}/\{(I-\sigma)\mathfrak{D}_n^{sk}\}.$$

However, there are elements, not in $(I-\sigma)\mathfrak{D}_n$, but which are nonetheless in the kernel of the states $\{f_\xi : \xi \in \mathbb{C}_1^n\}$; for example, $\ell = S_1(S_1S_1^*-\sigma(S_1S_1^*))$ is not in $(I-\sigma)\mathfrak{D}_n$, although $f_\xi(\ell) = 0$, $\xi \in \mathbb{C}_1^n$.

We shall need the following observation about the image, under $\Lambda : \mathcal{L}_n \longrightarrow \mathfrak{D}_n^{sk}$, of the canonical Lie algebra $d\alpha(u(n,1))$.

Proposition 8.4.5. For $A \in u(n)$, set $\delta_A = d\alpha(0,A,0)$, and, for $\xi \in \mathbb{C}^n$, $\delta_\xi = d\alpha(0,0,\xi)$.

Then

$$\Lambda(\delta_A) = \sum_{i,j} S_iA_{ji}S_j^* \tag{8.4.32}$$

and

$$\Lambda(\delta_\xi) = \varphi(\xi) - \varphi(\xi)^* \tag{8.4.33}$$

where the A_{ji}'s are the matrix entries of A, and φ is the canonical homomorphism $\varphi : C(H) \longrightarrow \mathcal{O}_n$.

(We have used the shorthand notation $\varphi(\xi)$ rather than the more precise one $\varphi(\pi(\xi))$. Similarly, $S_i = \varphi(\pi(e_i)) = \varphi(V_i)$,

$1 \leq i \leq n$, where $\{e_i\}$ is the canonical basis for $\mathbb{C}^n$.)

It follows that

$$f_\eta(\Lambda(\delta_A)) = \sum_{i,j} \eta_i A_{ji} \bar{\eta}_j = \langle A\eta, \eta\rangle$$

and

$$f_\eta(\Lambda(\delta_\xi)) = \langle \xi, \eta\rangle - \langle \eta, \xi\rangle$$
$$= 2i \operatorname{Im}\langle \xi, \eta\rangle$$

for arbitrary unit-vectors $\eta = (\eta_1, \cdots, \eta_n) \in \mathbb{C}^n$. As a result, we have the:

Corollary 8.4.6. If $A \neq 0$, then δ_A is *non*-approximately inner; and if $\xi \neq 0$, then δ_ξ is *non*-approximately inner.

Proofs. The proofs of the two results 8.4.5 and 8.4.6 are quite straightforward, using just the formulas for the derivations δ_A, resp., δ_ξ, which we derived in the proof of Theorem 8.4.1.

We also used the following easy fact to get the formulas (8.4.32)-(8.4.33). For a given derivation δ, and a given element ℓ in $\mathcal{O}_n$, we have : $\Lambda(\delta) = \ell$ if and only if

$$\delta(S_i) = \ell S_i, \qquad 1 \leq i \leq n. \tag{8.4.34}$$

Indeed, if (8.4.34) holds, then

$$\ell = \sum_i \ell S_i S_i^* = \sum_i \delta(S_i) S_i^* = \Lambda(\delta).$$

Conversely, the element $\Lambda(\delta)$ satisfies (8.4.34).

While the quotient $\mathcal{L}_n/\{\text{inner derivations}\}$ is infinite-dimensional, and contains an isomorphic copy of $u(n,1)$, there is not decomposition result which is a direct analogy to the decomposition (8.3.13) for the rotation algebras, cf. [BEJ].

We have $\Lambda(d\alpha(u(n,1)) \subset \mathcal{D}_n^{sk}$, and we let $\mathcal{M}_n$ denote subspace of $\mathfrak{D}_n$ spanned by elements of the form $S_{m_1}\Lambda(d\alpha(x))S^*_{m_2}$ where $x \in u(n,1)$ and S_{m_1}, resp., S_{m_2}, are multi-indexed monomials in the generators S_i. Since

$$f_\eta(S_{m_1}\Lambda(d\alpha(x))S^*_{m_2}) = \eta_{m_1} f_\eta(\Lambda(d\alpha(x)))\bar{\eta}_{m_2}$$

for unit-vectors $\eta = (\eta_1,\cdots,\eta_n) \in \mathbb{C}^n$, it follows that the derivations generated by the elements $S_{m_1}\Lambda(d\alpha(x))S^*_{m_2}$ are non-approximately inner when $x \neq 0$.

It can be shown that the quotient $\mathfrak{D}_n/(I-\sigma)\mathfrak{D}_n$ is spanned by these elements, and it follows from this that the quotient $\mathcal{L}_n/\{\text{inner derivations}\}$ is generated, in this sense, by the canonical Lie algebra $d\alpha(u(n,1))$.

This appears to be a substitute, of some kind, for the decomposition theorem for derivations in the rotation algebras.

Remark 8.4.7. Let $n \in \mathbb{N}$, $n > 1$, be given, and let M denote the ring of $n\times n$ complex matrices. Let $T(M) = \sum_{m=0}^{\infty} M^{\otimes m}$ be the one-sided tensor algebra, i.e., the one-sided infinite *algebraic* tensor product. Let $\sigma : T(M) \longrightarrow T(M)$ be the shift defined by

$$\sigma(1) = 1, \quad \sigma(a_1\otimes\cdots\otimes a_m) = 1 \otimes a_1 \otimes\cdots\otimes a_m,$$

$a_i \in M$, $m = 1,2,\cdots$, and let $\|\cdot\|$ denote the unique C^*-algebra norm on $T(M)$.

Then the result Theorem 8.4.2(c) above may be reformulated as follows: *The range of* $I-\sigma$ *in* $T(M)$ *is relatively closed with respect to* $\|\cdot\|$.

Stated this way, the result is of some interest since the corresponding statement about the one-sided shift in the category of Abelian C^*-algebras is false, as noted in [BEvGJ].

8.5 Vector Fields on the Circle

In Section 8.1 we constructed Lie algebras based on discrete Abelian groups. Let Γ be such a group, and let

$$B : \Gamma \times \Gamma \longrightarrow \mathbb{C}$$

be a cocycle, i.e., a function in two variables satisfying the Lie cocycle identity (8.1.1). This means that we may define a Lie bracket on the free vector space V over Γ by

$$[\xi,\xi'] = B(\xi,\xi')(\xi+\xi'), \tag{8.5.1}$$

and extension by linearity. We saw that the irrational rotation algebras may be constructed this way when $\Gamma = \mathbb{Z}^2$.

However, the simplest case is $\Gamma = \mathbb{Z}$. If we define the cocycle B by

$$B(n,n') = n'-n, \qquad n,n' \in \mathbb{Z}, \tag{8.5.2}$$

then formula (8.5.1) reads

$$[n,n'] = (n'-n)(n+n').$$

The resulting Lie algebra will be denoted $\mathfrak{d}$. We get a representation ρ of $\mathfrak{d}$ by setting

$$\rho(n) = -ie^{in\theta}\frac{d}{d\theta}. \qquad n \in \mathbb{Z}, \tag{8.5.3}$$

where $\rho(n)$ is interpreted as a vector field (i.e., a derivation) on the circle S^1.

Since the formula

$$[\rho(n),\rho(n')] = (n'-n)\rho(n+n') \tag{8.5.4}$$

can be checked by calculus, it follows that a Lie cocycle is indeed defined with formula (8.5.4).

Let C be a cocycle on $\mathfrak{d}$, as defined in Section 2.5, i.e.,

$$C(x,[y,z])+C(y,[z,x])+C(z,[x,y]) = 0, \qquad c,y,z \in \mathfrak{d}.$$

Then C defines a central extension of $\mathfrak{d}$, and ρ extends to a representation of this central extension $\mathfrak{d}_C$, or, alternatively, as noted in Section 2.5, ρ extends to a projective representation of $\mathfrak{d}$ itself.

Recall that $\mathfrak{d}_C$ consists of pairs $\{(x,a) : x \in \mathfrak{d},\ a \in \mathbb{C}\}$, with Lie bracket

$$(8.5.5) \qquad [(x,a),(x',a')] = ([x,x'],C(x,x')),$$

and the extended representation, denoted by ρ_C, is given by

$$(8.5.6) \qquad \rho_C((x,a)) = \rho(x) + aI,$$

where I is the identity, or rather the constant function one.

If we take

$$(8.5.7) \qquad C(n,n') = \delta(n+n')n(n^2-1)/12,$$

then the resulting Lie algebra $\mathfrak{d}_C$ is called the Virasoro algebra. (The term $\delta(n+n')$ in (8.5.7) refers to the Dirac delta function, i.e.,

$$\delta(n+n') = \begin{cases} 1 & \text{if } n' = -n \\ 0 & \text{if } n' \neq -n \end{cases}.)$$

It is known that the function C defined in (8.5.7) is a cocycle, i.e., that it satisfies (2.5.3) but this can also be verified directly.

There have recently been various papers on projective representations of the vector field Lie algebra $\mathfrak{d}$. This program is motivated by string theory, or, more precisely, the study of the, so-called, loop groups. In this study, the group of all orientation preserving diffeomorphisms of the circle S^1 serves as a Lie group for the Lie algebra $\mathfrak{d}$, and it is the projective unitary representations of this group which have occupied researchers in loop group theory, and in representation theory, as well. For a general discussion, the reader is referred to the recent book by Pressley and Segal [PS, Chs. 3-4]. One approach to such projective unitary representations is based on the Virasoro algebra, $\mathfrak{d}_C$. One looks for Hermitian representations of $\mathfrak{d}_C$, and tries to exponentiate these representations to get *unitary* representations of the corresponding group. If G denotes the group of orientation preserving diffeomorphisms of S^1, then we are looking for a central extension of G with $\mathfrak{d}_C$ as Lie algebra. But central extensions of G are given by cocycles on G. The appropriate cocycle is known; see, for instance, [PS]. It is obtained by exponentiation of the Lie algebra cocycle C.

If $\mathfrak{d}$ is identified with the Lie algebra of all polynomial vector fields on S^1,

$$x = f(e^{i\theta})\,\frac{d}{d\theta}, \qquad y = h(e^{i\theta})\,\frac{d}{d\theta},$$

then the formula (8.5.7) for C may be rewritten in the form

$$(8.5.8) \qquad C(x,y) = \frac{i}{24\pi}\int_0^{2\pi} d\theta \left\{\frac{d^2 f}{d\theta^2}(e^{i\theta}) + f(e^{i\theta})\right\} \frac{dh}{d\theta}(e^{i\theta}).$$

The group G is known to be simple, see [Her], but the exponential mapping

$$(8.5.9) \qquad \exp : \mathfrak{d} \longrightarrow G$$

lacks some of the familiar pleasant properties of the exponential mapping for finite-dimensional Lie groups. We mentioned in Section 2.10 that, in the finite-dimensional case, the image under the exponential mapping generates the connected component of the identity element in G. This fails badly for (8.5.9). In fact, the subgroup generated by the exponential mapping is nowhere dense in G, see [PS].

More specifically, let $\mathfrak{d}_0$ be the Lie algebra of all C^∞ real valued vector fields, i.e., vector fields of the form

$$(8.5.10) \qquad x = f(e^{i\theta})\frac{d}{d\theta},$$

with $f : S^1 \longrightarrow \mathbb{R}$ a given smooth function. Since S^1 is compact, every such vector field exponentiates to a flow on S^1, i.e., to a one-parameter group of diffeomorphisms on S^1. This follows from standard facts on differential equations, see, e.g., [L] and [CL, Ch. 17]. It is the subgroup of all diffeomorphisms, which is generated by the one-parameter subgroups, which is nowhere dense.

Consider the finite-dimensional case:

$$\exp : \mathfrak{g} \longrightarrow G \tag{8.5.11}$$

with G a given Lie group, and exponential mapping (8.5.11) from the Lie algebra $\mathfrak{g}$. Then a unitary representation U of G on a Hilbert space $\mathcal{H}$ is determined, essentially uniquely, by its value on one-parameter subgroups. That is to say, U is determined by the one-parameter groups

$$\{(U(\exp(tx))\}_{t\in\mathbb{R}} \tag{8.5.12}$$

when x ranges over the Lie algebra $\mathfrak{g}$. Once the Gårding space is specified, the one-parameter groups (8.5.12) are determined in turn by the derived representation dU of $\mathfrak{g}$. By Stone's theorem [RN], the one-parameter group (8.5.12) is determined by its skew-adjoint infinitesimal generator. But this generator is essentially skew-adjoint on the Gårding space (by a result of I.E. Segal [Se 7]), and it follows that U can be reconstructed from the derived representation dU. (If G is not simply connected, this must be modified slightly as noted in Section 8.4 above.)

If $\mathfrak{g}$ is not finite-dimensional, the program of exponentiating the infinitesimal Lie algebra of operators must be radically altered. While we outline a general reconstruction program for exponentiating operator Lie algebras in the finite-dimensional case in the Appendix below, each of the known infinite-dimensional Lie algebras must be treated on its own merit.

Recent papers by Roe Goodman and Norlan Wallach [GW 1-2] have carried out this program for positive energy representations of the Virasoro algebra.

We shall take a closer look at the highest weight (Hermitian) representations of the Virasoro algebra.

First, a bit of terminology: Let $V = \mathbb{C}\cdot\mathbb{Z}$ = the free complex vector space over $\mathbb{Z}$. We turn V into a Lie algebra via (8.5.1)-(8.5.2), and consider a central extension with generators $\{x_n : n \in \mathbb{Z}\}$ and a single central element y. The commutation relations (8.5.5) (with (8.5.7)) then take the form

(8.5.13) $$[x_n, x_m] = (m-n)x_{n+m} + \delta(n+m)\,\frac{n(n^2-1)}{12}\,y, \quad n,m \in \mathbb{Z}.$$

For efficiency we shall denote this Virasoro algebra by the single letter $\mathfrak{v}$. It follows that the elements $\{x_n,\ n \in \mathbb{Z}\} \cup \{y\}$ form a set of generators for the universal enveloping algebra $\mathfrak{U}(\mathfrak{v})$. We now turn this algebra into a $*$-algebra by defining a $*$-operation on the generators:

(8.5.14) $$x_n^* = x_{-n}, \qquad n \in \mathbb{Z}, \quad y^* = y.$$

The reader is encouraged to compare formulas (8.5.14) and (8.2.10).

We shall consider Hermitian representations of $\mathfrak{U}(\mathfrak{v}) \equiv \mathfrak{U}$. Recall from Section 5.1 that a Hermitian representation ρ of $\mathfrak{U}$ on a Hilbert space $\mathcal{H}$ is a representation ρ satisfying

(8.5.15) $$\langle \rho(x)a, b\rangle = \langle a, \rho(x^*)b\rangle \quad \text{for all} \quad x \in \mathfrak{U}, \quad a,b \in \mathcal{D}(\rho).$$

The linear space $\mathcal{D}(\rho)$ is called the domain of ρ, and it is assumed dense in $\mathcal{H}$, and invariant under each operator $\rho(x)$, $x \in \mathfrak{U}$.

Following [GW 2], we make the

Definition 8.5.1. A Hermitian representation ρ of $\mathfrak{U}$ with dense domain $\mathcal{D}(\rho)$ in a Hilbert space $\mathcal{H}$ is said to be a *highest weight* representation if

(1) the operator $\rho(x_0)$ diagonalizes on $\mathcal{D}(\rho)$ with eigenvalues of the form $\{h-n : n = 0,1,2,\cdots\}$,

(2) each of the corresponding eigenspaces is finite-dimensional,

(3) $\rho(y) = qI$, for some $q \in \mathbb{R}$,

and

(4) $\dim\{a \in \mathcal{D}(\rho) : \rho(x_0)a = h\,a\} = 1$.

A unit-vector in the one-dimensional eigenspace will be chosen once and for all, and it will be denoted by Ω.

For efficiency, the additional notation

(8.5.16) $$X_n = \rho(x_n), \qquad n \in \mathbb{Z}, \quad Y = \rho(y),$$

is introduced.

We note that the highest weight representations are parametrized with the pair h,q of real numbers. Using (8.5.13), we arrive at the operator commutation relations:

$$(8.5.17) \qquad [X_n, X_m] = (m-n)X_{n+m} + \delta(n+m)\frac{n(n^2-n)}{12} q, \qquad n,m \in \mathbb{Z}.$$

For $n = 1,2,\cdots$, consider the three operators, X_n, $X_{-n,}$ and X_0. They satisfy the pair of relations:

$$[X_0, X_{\pm n}] = \pm nX_{\pm n}$$

and

$$[X_{+n}, X_{-n}] = -2nX_0 + \frac{n(n^2-1)q}{12} I.$$

Let $n \in \mathbb{N}$ be fixed, and define

$$A_{\pm} = (1/n)X_{\pm n}$$

and

$$A_0 = (1/n)X_0 + \frac{q(1-n^2)}{24n} I.$$

Then the above relations translate into:

$$[A_+, A_-] = -2A_0,$$

and

$$[A_0, A_{\pm}] = \pm A_{\pm}.$$

But these are just the familiar generators for a representation of $s\ell_2(\mathbb{R})$. The operator A_0 is already diagonalized, and the Casimir operator is:

$$(8.5.18) \qquad C = \frac{1}{2}\{A_+, A_-\} - A_0^2.$$

It can be checked, using simple algebra, that the Casimir operator diagonalizes. Now Theorem 5.5.1 implies that the representation generated by the triple $\{A_{\pm}, A_0\}$ is exact.

This means that there is a unitary representation U_n of the simply connected covering group $\widetilde{SL_2(\mathbb{R})}$ such that:

$$
(8.5.19) \quad \begin{cases} dU_n(x_{\pm n}) = \rho(x_{\pm n}) \\ dU_n(x_0) = \rho(x_0) \\ dU_n(y) = \rho(y) = qI \end{cases} .
$$

It follows, in particular, that the representations U_n, $n = 1,2\cdots$, agree on the two basis elements x_0, y.

Recall that the Gårding space is identical with the space of C^∞-vectors. Using (8.5.18-19), in conjunction with Theorem 5.5.1 above, we conclude that the distinct unitary representations U_n have the same space of Gårding vectors, independent of the value of the index n. This common space is just the space of C^∞-vectors for the single operator X_0.

It is known [GW 2, Lemma 1.1] that the possibilities for the variables h,q are as follows:

(1) $h = q = 0$: This is a trivial representation.

(2) $h \leq 0$, and $q > 0$.

We shall omit discussion of possibility (1), and assume (2) in the sequel. Since the spectrum of X_0 is $\{h-n : n = 0,1,\cdots\}$ it follows that $X_0 \leq 0$, i.e., X_0 is a semibounded operator.

For each n, let ρ_n denote the restriction of ρ to the algebra generated by the three elements $\{x_{\pm n}, x_0 + \frac{q(1-n^2)}{24} y\}$. Then it follows from the proof of Theorem 5.5.1 that ρ_n is an essentially selfadjoint representation, and exact, and that

$$
\mathcal{D}(\bar{\rho}_n) = \mathcal{D}(\bar{\rho}_0).
$$

Since this holds for all n, it follows further that the original representation ρ is essentially selfadjoint, i.e., that

$$
\rho^* = \bar{\rho}
$$

where ρ^*, resp., $\bar{\rho}$, are defined in Section 5.2.

Let A be the operator $A = I-X_0$. Using now (5.5.5), we conclude that, for every $x \in \mathfrak{d}$, there is a constant C, depending on x, such that

$$\|\rho(x)a\| \leq C\|Aa\|, \qquad a \in \mathcal{D}(\rho),$$

and similarly,

$$\|A^{t-1}\rho(x)a\| \leq C_t\|A^t a\|, \qquad a \in \mathcal{D}(\rho). \tag{8.5.20}$$

If $\mathcal{H}_t$ is the domain of A^t, $t \in \mathbb{R}$, with graph norm:

$$a \longrightarrow \|A^t a\|,$$

then (8.5.20) states that $\rho(x)$ extends by completion to a bounded operator from $\mathcal{H}_t$ to $\mathcal{H}_{t-1}$.

It follows further from (8.5.20) that there is a constant c, independent of n, such that

$$\|X_{\pm n}a\| \leq c\{n-h+q(n^2-1)\}\|a\|, \quad n = 1,2,\cdots, \quad a \in \mathcal{D}(\rho), \tag{8.5.21}$$

satisfying $X_0 a = (h-n)a$. When the two results (8.5.20) and (8.5.21) are combined we conclude that the given representation ρ of $\mathfrak{v}$ extends by completion to the central extension of the Lie algebra of all smooth vector fields on S^1. The cocycle C is determined by (8.5.8) on the larger algebra of *all smooth* vector fields. The Lie bracket in the completed algebra $\tilde{\mathfrak{v}}$ is also given by formula (8.5.5), but now for all smooth vector fields x and x'.

Let $\tilde{\mathfrak{d}}$ be the Lie algebra of all smooth vector fields on S^1, and let $\tilde{\mathfrak{v}}$ be the corresponding central extension. The point of the larger Lie algebra of all smooth vector fields, rather than just the polynomial vector fields, is that the group G of orientation preserving diffeomorphisms on S^1 acts by conjugation on $\tilde{\mathfrak{d}}$, but *not* on $\mathfrak{d}$.

Concerning the cocycle C from (8.5.8) we have the following:

The extension of the given cocycle representation from $\mathfrak{d}$ to the Fréchet space completion $\tilde{\mathfrak{d}}$ = (all smooth vector fields on S^1) is a construction which is almost identical to the extension constructions for the Hermitian representations ρ which we considered in Section 6.3 above.

Lemma 8.5.2. Let the action of G on $\tilde{\mathfrak{d}}$ be given by $g, x \longrightarrow x^g$, where the vector field x^g, evaluated on a C^∞-function φ, is given by

$$x^g(\varphi) := (x(\varphi \circ g^{-1})) \circ g. \tag{8.5.22}$$

Then there is a continuous linear functional A_g on $\tilde{\mathfrak{d}}$ such that

$$C(x^g, z^g) = C(x,z) - A_g([x,z]), \qquad x, z \in \tilde{\mathfrak{d}}, \quad g \in G. \tag{8.5.23}$$

Proof. The reader is referred to [GW 1, Lemma 7.7] and [GW 2, Corollary 3.6] for details.

Corollary 8.5.3. For every $g \in G$, the two representations $x \longrightarrow \rho(x)$, and $x \longrightarrow \rho(x^g)$ are unitarily equivalent.

Proof. We first note that $x \longrightarrow \rho(x^g)$ is indeed a projective representation of $\tilde{\mathfrak{d}}$, or equivalently a representation of $\tilde{\mathfrak{d}}$. Set $\rho^g(x) := \rho(x^g)$, $x \in \tilde{\mathfrak{d}}$. Then

$$\begin{aligned} [\rho^g(x), \rho^g(z)] &= [\rho(x^g), \rho(z^g)] \\ &= \rho([x^g, z^g]) + qC(x^g, z^g)I \\ &= \rho([x,z]^g) + q\{C(x,z) - A_g([x,z])\}I \\ &= \rho^g([x,z]) + qC_g(x,z)I. \end{aligned}$$

It follows that the cocycle C_g defined by

$$C_g(x,z) := C(x,z) - A_g([x,z])$$

is associated with the representation ρ^g. Since the two cocycles are equivalent, it can be checked that the representations ρ and ρ^g are equivalent.

It is known, but highly nontrivial [GW 2, Theorem 4.2], that there is a unitary cocycle representation W of G which implements the equivalence, i.e.,

$$\rho^g(x) = W_g \, \rho(x) W_g^*, \qquad g \in G, \quad x \in \mathfrak{d}. \tag{8.5.24}$$

The complete details of proof are beyond the scope of the present monograph. The reader is referred to [GW 2].

We note that formula (8.5.22) takes a more explicit form with the following identifications: The vector field x is given by (8.5.10) for some function f on S^1. We define a function F on $\mathbb{R}$ by

$$F(\theta) = f(e^{i\theta}), \qquad \theta \in \mathbb{R},$$

and identify x with the vector field

$$x = F(\theta)\frac{d}{d\theta} \qquad \text{on } \mathbb{R}.$$

Elements g in the group of all orientation preserving diffeomorphisms of S^1 may be identified with functions, also denoted by g, on $\mathbb{R}$ which are smooth, strictly monotone increasing, and satisfying

$$g(\theta+2\pi) = g(\theta) + 2\pi, \qquad \theta \in \mathbb{R}.$$

Substitution into (8.5.22), and an application of the chain rule, yields

(8.5.25) $$x^g = \left(\frac{F\circ g}{g'}\right)\frac{d}{d\theta}$$

where $g' = \frac{dg}{d\theta}$. Note that the function $F\circ g$ is periodic since F is. Indeed, $(F\circ g)(\theta+2\pi) = F(g(\theta+2\pi)) = F(g(\theta)+2\pi) = F(g(\theta)) = (F\circ g)(\theta)$, $\theta \in \mathbb{R}$. Since x is a vector field (alias, a derivation), we have

$$x(\varphi\circ g^{-1}) = (\varphi'\circ g^{-1})x(g^{-1})$$

for smooth functions φ on S^1. It follows that

$$x(\varphi\circ g^{-1})\circ g = \varphi'(x(g^{-1})\circ g).$$

We have $(x(g^{-1})\circ g)(\theta) = \frac{F(g(\theta))}{g'(\theta)}$ for $\theta \in \mathbb{R}$, and formula (8.5.25) results upon substitution.

It follows from (8.5.25) that the mapping

(8.5.26) $$g,x \longrightarrow x^g : G \times \tilde{\delta} \longrightarrow \tilde{\delta}$$

is a representation, i.e., that

(8.5.27) $$[x^g, z^g] = [x,z]^g$$

and

(8.5.28) $$(x^g)^h = x^{gh} \qquad \text{for all } x,z \in \tilde{\mathfrak{d}}, \text{ and } g,h \in G.$$

Indeed, for smooth functions φ on S^1, we have

$$\begin{aligned}
[x^g, z^g](\varphi) &= x^g(z^g\varphi) - z^g(x^g\varphi) \\
&= x(z^g\varphi)\circ g^{-1}) \circ g - z((x^g\varphi)\circ g^{-1}) \circ g \\
&= x(z(\varphi\circ g^{-1})) \circ g - z(x(\varphi\circ g^{-1})) \circ g \\
&= ([x,z](\varphi\circ g^{-1})) \circ g,
\end{aligned}$$

proving (8.5.27). Similarly, we have

$$\begin{aligned}
(x^g)^h\varphi &= (x^g(\varphi\circ h^{-1})) \circ h \\
&= (x((\varphi\circ h^{-1})\circ g^{-1})) \circ gh \\
&= (x(\varphi\circ(gh)^{-1})) \circ gh \\
&= x^{gh}(\varphi),
\end{aligned}$$

proving (8.5.28).

We need to study extensions of the representation (8.5.26) from $\tilde{\mathfrak{d}}$ to $\tilde{\mathfrak{v}}$. The following lemma shows that such extensions are parametrized by representations of G on $(\tilde{\mathfrak{d}})^*$ = (all linear functionals on $\tilde{\mathfrak{d}}$).

Lemma 8.5.4. Let y be the central element in $\tilde{\mathfrak{v}}$, and define $y^g = y$. When the representation (8.5.26) is extended from $\tilde{\mathfrak{d}}$ to $\tilde{\mathfrak{v}}$ by linearity, the difference

$$[x_1+\lambda_1 y, x_2+\lambda_2 y]^g - [x_1^g+\lambda_1 y, x_2^g+\lambda_2 y]$$

is a scalar times y. This scalar is of the form $A_g([x_1,x_2])$ where $g \longrightarrow A_g$ is a representation of G in $\tilde{\mathfrak{d}}^*$.

Proof. We may assume without loss of generality that $\lambda_1 = \lambda_2 = 0$. Then

$$[(x_1,0),(x_2,0)] = ([x_1,x_2], C(x_1,x_2)) \quad \text{for } x_1,x_2 \in \tilde{\mathfrak{d}}.$$

The difference between the two elements is:

$$(0, C(x_1, x_2) - C(x_1^g, x_2^g)).$$

This proves the first part of the lemma. It follows from general theory [PS, Sections 4.2-4.3] that the difference has the stated representation, i.e., that it is a linear functional on the commutator $[x_1, x_2]$ in $\tilde{\delta}$.

The existence proof for the projective representation W from (8.5.24) is different from the corresponding finite-dimensional analogue in the step where the local homomorphism is extended to a global one. As noted in [GW 1], a version of the Nash-Moser theorem is employed for the infinite-dimensional problem.

Let

$$G_1 = \{g \in G : g(1) = 1\}.$$

Then G_1 is known to be contractible. This means that a given local homomorphism W from G_1 to $\mathcal{PU}(\mathcal{H}) := \mathcal{U}(\mathcal{H})/\mathbb{T}$ extends automatically to a homomorphism from G_1 to $\mathcal{PU}(\mathcal{H})$, i.e., to a projective representation. This means that W is determined by its local properties, i.e., it is enough to specify it as a local representation defined only on a neighborhood of the origin in G_1. The contractibility of G_1, and the Nash-Moser theorem, are from Hamilton's paper [Ha]. The identification of W as a local representation is carried out in [GW 1-2].

The given representation ρ is a highest weight representation, cf. Definition 8.5.1. This means, in particular, that the space

$$\{a \in \mathcal{H} : \rho(x_0)a = ha\}$$

is one-dimensional. The construction of the desired local representation W depends on the fact that the representation ρ^g, defined by $\rho^g(x) = \rho(x^g)$, is also a highest weight representation for g in a neighborhood of the origin in G_1.

Moreover, it is possible to pick a section, $g \longrightarrow \Omega(g)$, of vectors satisfying

$$\rho(x_0^g)\Omega(g) = h(g)\Omega(g),$$

with both the eigenvector $\Omega(g)$ and the eigenvalue $h(g)$, depending continuously on g, such that

$$\|\Omega(g)\| = 1,$$

and

$$(8.5.29) \qquad \langle\rho(x)\Omega(g),\Omega(g)\rangle = \langle\rho(x^g)\Omega,\Omega\rangle \quad \text{for } g \text{ in the neighborhood, and } x \in \tilde{\mathfrak{D}}.$$

That this can be done is the content of [GW 1, Theorem 3.5]. The representation W_g may be defined in terms of this section by the familiar Gelfand-Naimark-Segal (GNS) construction as follows:

$$W_g\ \rho(x)\Omega(g) := \rho(x^g)\Omega.$$

Each W_g will be a unitary operator by virtue of (8.5.29) since

$$\begin{aligned} \|W_g\ \rho(x)\Omega(g)\|^2 &= \|\rho(x^g)\Omega\|^2 \\ &= \langle\rho((x^*x)^g)\Omega,\Omega\rangle \\ &= \langle\rho(x^*x)\Omega(g),\Omega(g)\rangle \\ &= \|\rho(x)\Omega(g)\|^2, \qquad x \in \tilde{\mathfrak{D}}. \end{aligned}$$

Hence, W_g is isometric on a dense space of vectors in $\mathcal{H}$. The set of vectors of the form $\{\rho(x)\Omega(g) : x \in \mathfrak{D}\}$ is dense since $\Omega(g)$ is cyclic.

The covariance identity (8.5.24) will be satisfied by virtue of the very construction, i.e., the GNS-construction. Indeed, the identity

$$\begin{aligned} (W_g\ \rho(x))\rho(z)\Omega(g) &= W_g\ \rho(xz)\Omega(g) \\ &= \rho((xz)^g)\Omega \\ &= \rho(x^g)\rho(z^g)\Omega \\ &= (\rho(x^g)W_g)\rho(z)\Omega(g) \end{aligned}$$

holds for $x \in \tilde{\mathfrak{D}}$, and $z \in \mathfrak{U}(\tilde{\mathfrak{D}})$.

Using again cyclicity of the vector $\Omega(g)$, we conclude that

$$W_g \ \rho(x) = \rho(x^g)W_g,$$

or equivalently,

$$\rho(x^g) = W_g \ \rho(x)W_g^*, \qquad x \in \tilde{\mathfrak{d}},$$

which is the desired formula.

To prove the existence of a continuous section

$$g \longrightarrow (\Omega(g), h(g)),$$

defined in a neighborhood of the origin in G_1, and having the specified property (8.5.29), is an infinite-dimensional version of Rellich's theorem, cf. [Kat 2; VII.5, Thm. 3.9, p. 392, and VIII.5, Thm. 3.15, p. 462]. We need to consider the resolvent operator $\lambda, g \longrightarrow (\lambda I - \rho(x_0^g))^{-1}$. It follows from the properties of the representation ρ that the resolvent is compact, defined for $\lambda \in \mathbb{C}$, $\mathrm{Re}\,\lambda > 0$, and continuous in the g-variable. But the Rellich-Kato theorem applies only for values of g in finite-dimensional submanifolds of G_1. However, a suitable infinite-dimensional version of the R-K theorem may be obtained as an application of the Nash-Moser implicit function theorem. The reader is referred to [GW 1] and [Ha] for details on this important technical point.

8.6 Noncommutative Tori

In the previous section, we considered the Lie algebra $\mathfrak{d}$ of all smooth vector fields on the circle S^1. A smooth vector field amounts precisely to a derivation in the algebra of smooth functions on S^1, $C^\infty(S^1)$. It is convenient to treat $C^\infty(S^1)$ as a dense subalgebra of the Abelian C^*-algebra $C(S^1)$ of all continuous functions on S^1. In Section 8.5, we contrasted the two C^*-algebras $C(S^1)$ and $A(R_\theta)$ where $A(R_\theta)$ denotes the irrational rotation C^*-algebra constructed from a given irrational rotation angle θ. There are important differences between the two algebras, and similarities as well.

Recently, an impressive number of differential geometric,

and K-theoretic, results have been developed for $A(R_\theta)$. The reader is referred to the pioneering paper [Con 3], and to [Ri 4] and [Jo 32] for a list of references and for historical remarks. The study of $A(R_\theta)$ is often referred to under the heading "noncommutative differential geometry". The C^*-algebra $A(R_\theta)$ is simple, and seems to provide a test case for more general results on simple C^*-algebras. The aim is to study curvature, connections, etc. for such noncommutative algebras. It was proved in [CR] that Yang-Mills functions are defined in terms of differentiable structures for $A(R_\theta)$, or rather for the dense subalgebra $A_\infty(R_\theta)$ consisting of smooth elements in $A(R_\theta)$. More specifically, the so-called *"constant curvature connections"* on $A_\infty(R_\theta)$ are identified in [CR] with the extremal functionals for the Yang-Mills problem.

An important similarity between the two algebras $C^\infty(S^1)$ and $A_\infty(R_\theta)$ stems from the definition of the product. The product in each of the two algebras may be defined in terms of Fourier coefficients as follows: Let $f(e^{it}) \sim \sum_n a_n e^{in2\pi t}$, and $h(e^{it}) \sim \sum_n b_n e^{in2\pi t}$, be the Fourier series for the respective elements f and h in $C^\infty(S^1)$. Then the Fourier coefficients $\{c_n\}$ of the pointwise product, $(fh)(e^{it}) := f(e^{it})h(e^{it})$, is given by the familiar Cauchy formula

$$c_n = \sum_m a_m b_{n-m}, \qquad (8.6.1)$$

and the summation in (8.6.1) is automatically absolutely convergent. The convergence properties result from the Paley-Wiener theorem [Katn]: The condition that a function f on S^1 be C^∞ is equivalent to the statement that the corresponding Fourier series $\{a_n\}$ is rapidly decreasing, i.e., for all $k \in \mathbb{N}$ there is a constant M such that

(8.6.2) $$|a_n| \le M(1+|n|)^{-k}, \qquad n \in \mathbb{Z}.$$

For the algebra $A_\infty(R_\theta)$, we have Fourier series indexed by the integer lattice $\mathbb{Z}^2$, rather than by $\mathbb{Z}$. For given $a \in A_\infty(R_\theta)$, the Fourier series $\{a_{m,n}\}_{(m,n)\in\mathbb{Z}^2}$ is determined by

(8.6.3) $$a_{m,n} = \text{trace}(u(m,n)^* a)$$

where "trace" refers to the unique trace state on $A(R_\theta)$. It is a simple matter to check that elements a in $A_\infty(R_\theta)$ are characterized by the two-dimensional Schwartz condition (8.6.2), i.e.: for all $k \in \mathbb{N}$ there is a constant M such that

(8.6.4) $$|a_{m,n}| \le M(1+|m|+|n|)^{-k}, \qquad (m,n) \in \mathbb{Z}^2.$$

Note that this latter condition is θ-independent.

For a pair of elements a,b, the product ab may be expressed in terms of the associated pair of Fourier series, but the resulting Cauchy product, corresponding to (8.6.1) above, has a twist tacked onto it. This is the point where the irrational angle θ enters the picture. The formula for the Fourier series $\{c_{m,n}\}$ for the product ab in $A_\infty(R_\theta)$ is

(8.6.5) $$c_{m,n} = \sum_{(k,\ell)\in\mathbb{Z}^2} e^{i2\pi\ell(m-k)\theta}\, a_{k,\ell}\, b_{m-k,n-\ell}.$$

Although a small amount of computation is involved in the verification of formula (8.6.5) for the product in $A_\infty(R_\theta)$, it is an immediate consequence of the definition of the generators of $A(R_\theta)$ in terms of a projective representation of $\mathbb{Z}^2$. Formulas (8.1.14) and (8.2.8-9) from Sections 8.1-8.2 are relevant in this connection.

In Section 8.2, we introduced a diophantine condition for bicharacters. We showed that every irrational number θ, $0 < \theta < 1$, defines a nondegenerate antisymmetric bicharacter on $\mathbb{Z}^2$. The corresponding C^*-algebra is simple, and it is called the irrational rotation algebra. The *diophantine condition* on the bicharacter amounts to the following estimate for θ: The sequence $(e^{i2\pi n\theta}-1)^{-1}$ grows at most polynomially in n. It follows from Roth's theorem [Sch] that this holds for all algebraic irrational numbers, and from Khintchine's theorem [Sch, p. 60] that it holds for all θ in a subset of $(0,1)$ of Lebesgue measure one.

We define cocycles B on the Abelian group $\Gamma = \mathbb{Z}^2$. Recall from Section 8.1, that if B is assumed to satisfy the cocycle identity (8.1.1), then a Lie bracket $[\ ,\ ]$ may be defined on Γ by the formula (8.1.2). In other words,

$$[\xi,\xi'] := B(\xi,\xi')(\xi+\xi').$$

The cocycle (8.5.2) on $\mathbb{Z}$ generalizes to $\mathbb{Z}^2$ if we define

$$B((m,n),(m',n')) = m'-m + n'-n, \quad \text{for} \quad (m,n) \in \mathbb{Z}^2,\ (m',n') \in \mathbb{Z}^2.$$

We have studied two different Lie algebras built on $\mathbb{Z}^2$ in Sections 8.2 and 8.5. In this section, we shall point out an important difference between representations of, on the one hand, the irrational rotation Lie algebras constructed in Section 8.2, and, on the other hand, the Virasoro type algebras studied in Section 8.5.

In Section 8.5, we considered the algebra of operators $\mathfrak{A}_\rho = \rho(\mathfrak{U}(\tilde{\mathfrak{v}}))$ generated by a given highest weight representation ρ of the algebra $\mathfrak{U}(\tilde{\mathfrak{v}})$ where $\tilde{\mathfrak{v}}$ is the Virasoro Lie algebra. The operator algebra $\mathfrak{A}_\rho$ is important for string theory since it carries a covariant representation of the group G_1 of all orientation preserving diffeomorphisms of S^1. More specifically, we constructed in Theorem 8.5.5 a unitary representation W of G_1 such that

$$W_g\,\rho(x)W_g^* = \rho(x^g), \qquad g \in G_1,\ x \in \tilde{\mathfrak{v}}, \tag{8.6.6}$$

where x^g is defined by (8.5.22) and (8.5.25).

In Section 8.2, we constructed a Lie algebra L_θ as a central extension of $\mathbb{Z}^2$ associated to a cocycle parametrized by an irrational rotation angle θ. The corresponding universal enveloping algebra $\mathfrak{U}_{\mathbb{C}}(L_\theta)$ is naturally represented on Hilbert space. The representations are analogous to the highest weight representations. Let ρ_θ be the representation of $\mathfrak{U}_{\mathbb{C}}(L_\theta)$ on the trace Hilbert space $\mathcal{H}_2$ considered in Section 8.2A, and let $\mathfrak{A}_\theta = \rho_\theta(\mathfrak{U}_{\mathbb{C}}(L_\theta))$ be the corresponding operator algebra.

We shall describe below the symmetries of $\mathfrak{A}_\theta$ and compare the result with the analogous algebra $\mathfrak{A}_\rho$ studied in Section 8.5.

We will note, among other things, that there is no nontrivial action of G_1 on $\mathfrak{A}_\theta$. It follows from (8.6.6), however, that $\mathfrak{A}_\rho$ carries a natural automorphic action of $SL_2(\mathbb{R})$. This is because $SL_2(\mathbb{R})$ is embedded as a subgroup of G_1. At the Lie algebra level, this embedding is obtained by mapping the generators (5.4.9) for $s\ell_2(\mathbb{R})$ into the following three vector fields:

$$(\cos t)\frac{d}{dt}\,,\quad (\sin t)\frac{d}{dt}\,,\quad \frac{d}{dt}.$$

Since

$$[(\cos t)\frac{d}{dt}\,,\ (\sin t)\frac{d}{dt}] = \frac{d}{dt},$$

$$[\frac{d}{dt}\,,\ (\cos t)\frac{d}{dt}] = -(\sin t)\frac{d}{dt},$$

and

$$[\frac{d}{dt}\,,\ (\sin t)\frac{d}{dt}] = (\cos t)\frac{d}{dt}.$$

This is indeed a copy of $s\ell_2(\mathbb{R})$. The automorphic action on $\mathfrak{A}_\rho$ is obtained by just restricting the action (8.5.24) to the $SL_2(\mathbb{R})$-subgroup inside G_1. Alternatively, the action of $SL_2(\mathbb{R})$ on the circle S^1 may be identified *globally*, rather than in terms of the vector field *generators*. The homomorphism from $s\ell_2(\mathbb{R})$ into G_1 may be defined simply as the composite:

$$SL_2(\mathbb{R}) \xrightarrow{\ \varphi\ } SU(1,1) \xrightarrow{\ \psi\ } \mathrm{Aut}(S^1).$$

The first arrow φ is the familiar isomorphism $g \longrightarrow \Phi^{-1}g\Phi$ where $\Phi = \begin{bmatrix} 1 & -1 \\ -i & -i \end{bmatrix}$, and the second arrow is simply the fractional linear representation of matrices

$$\begin{bmatrix} \alpha & \bar{\beta} \\ \beta & \bar{\alpha} \end{bmatrix}, \qquad \alpha,\beta \in \mathbb{C}, \quad |\alpha|^2 - |\beta|^2 = 1,$$

on S^1, given by

$$z \longrightarrow \frac{\alpha z + \bar{\beta}}{\beta z + \bar{\alpha}}, \qquad z \in S^1.$$

Not only is there no automorphic action of G_1 on $\mathfrak{A}_\theta$, but there is also not any faithful action of $SL_2(\mathbb{R})$ on $\mathfrak{A}_\theta$. We shall determine below precisely what Lie group actions can be realized on the irrational rotation algebra $\mathfrak{A}_\theta$ when the irrational number θ is assumed to satisfy the generic *diophantine* condition.

This will involve very recent research, more specifically, the paper [BEℓGJ] entitled, "*Smooth Lie group actions on non-commutative tori.*" The classification results from this paper are valid in a much more general setting than $\mathfrak{A}_\theta$ but we shall restrict attention here to the important special case.

When classifying automorphic actions of Lie groups G on the algebra $\mathfrak{A}_\theta$, it is convenient to reduce the study to faithful representations. We shall further restrict discussion to smooth actions. Let $\mathfrak{A}_\theta$ be the simple algebra of elements with Fourier series satisfying (8.6.4). Then $\mathfrak{A}_\theta$ is norm-dense in the C^*-algebra A_θ. Recall that the simple C^*-algebra A_θ is generated by two unitaries u_1, u_2 satisfying

(8.6.7a) $$u_2 u_1 = e^{i2\pi\theta} u_1 u_2, \qquad 0 < \theta < 1.$$

We also recall that A_θ carries a canonical representation

(8.6.7b) $$\alpha : \mathbb{T}^2 \longrightarrow \mathrm{Aut}(A_\theta)$$

given by

$$(8.6.8) \qquad \alpha_{\left[e^{it_1}, e^{it_2}\right]}(u_1^m u_2^n) = e^{i(mt_1+nt_2)} u_1^m u_2^n, \qquad (m,n) \in \mathbb{Z}^2.$$

This action identifies a differentiable structure on the C^*-algebra A_θ, and the set of C^∞-elements for the representation α coincides with the algebra $\mathfrak{A}_\theta$.

The action α is further *ergodic* and *faithful*. Recall that *ergodicity* amounts to the assertion that the fixed-point algebra

$$A_\theta^\alpha = \{a \in A_\theta : \alpha_g(a) = a, \ g \in \mathbb{T}^2\}$$

is one-dimensional. The kernel of the homomorphism (8.6.7b) is trivial. A representation with trivial kernel is said to be *faithful*.

The study of arbitrary continuous actions (8.6.9) may easily be reduced to that of the faithful ones. Simply mod out by the kernel: The kernel of a continuous action is a closed normal subgroup N of the given Lie group G. It follows then from [Hel, Thm. 3.2] that the quotient G/N is again a Lie group, and that the given representation passes to the quotient.

The smoothness requirement for a representation

$$(8.6.9) \qquad \pi : G \longrightarrow \mathrm{Aut}(A_\theta)$$

is a *relative* notion. We make the precise

Definition 8.6.1. An action $\pi : G \longrightarrow \mathrm{Aut}(A_\theta)$ of a given Lie group on the irrational rotation C^*-algebra is said to be *smooth* if for all $a \in \mathfrak{A}_\theta$ the mapping

$$(k,g) \longrightarrow \alpha_k \, \pi_g(a)$$

is smooth from $\mathbb{T}^2 \times G$ to A_θ.

Note that the defining smoothness condition on π implies in particular that:

(i) π_g maps $\mathfrak{A}_\theta$ into $\mathfrak{A}_\theta$ for all $g \in G$, and

(ii) the smooth algebra $\mathfrak{A}_\theta = C^\infty(\alpha)$ is contained in

$C^\infty(\pi)$ = the C^∞-elements for the representation π.

As a first step in solving the classification problem for smooth actions on A_θ, we show that the problem may be *completely linearized.*

In this connection we need:

Lemma 8.6.2. Let

$$\pi : G \longrightarrow \mathrm{Aut}(A_\theta)$$

be a strongly continuous representation of the Lie group G and let g be the corresponding Lie algebra.

Assume that

(a) $\mathfrak{A}_\theta \subset C^\infty(\pi)$,

(b) $d\pi(x)$ maps $\mathfrak{A}_\theta$ into $\mathfrak{A}_\theta$ for all $x \in g$, and

(c) π_g maps $\mathfrak{A}_\theta$ onto $\mathfrak{A}_\theta$ for all $g \in G$.

Then it follows that π is a smooth action.

If conversely π is assumed smooth, then properties (a)-(c) hold, and moreover $\mathfrak{A}_\theta$ is a *core* for $d\pi(x)$, $x \in g$.

Proof. The proof of this result uses vector calculus techniques from Section 5.3, as well as Baire category arguments. It is worked out in all detail in Sectin 2 of [BEℓGJ]. Since it is somewhat technical, it will be omitted here.

It follows that every smooth action π of a Lie group gives rise to a finite-dimensional Lie subalgebra $\mathcal{L} := d\pi(g)$ of the infinite-dimensional Lie algebra of all derivations in $\mathfrak{A}_\theta$, i.e.,

$$\mathcal{L} \subset \mathrm{Der}(\mathfrak{A}_\theta). \tag{8.6.10}$$

In most cases, the smooth Lie group actions on A_θ are completely parametrized by finite-dimensional Lie subalgebras (8.6.10). We proved in [BEℓJG, Prop. 2.2] that this is indeed the case when the irrational number θ is further assumed to satisfy the generic *diophantine* condition. This amounts to the statement that *every* finite-dimensional Lie algebra (8.6.10) automatically exponentiates to a smooth action of the corresponding simply connected Lie group.

The proofs in [BEℓGJ] are based on perturbation theory for Lie group representations, developed in [JM, Ch. 9]. We sketch some of the details below. The diophantine assumption on θ will be in force throughout the rest of the section.

In Section 8.3 we proved that the Lie algebra $\mathrm{Der}(\mathfrak{A}_\theta)$ is the semidirect sum of the two-dimensional canonical Lie algebra

$$\mathcal{L}_0 = d\alpha(\mathbb{R}^2), \tag{8.6.11}$$

and

$$\mathcal{P} = \{\mathrm{ad}(h) : h \in \mathfrak{A}_\theta,\ h^* = -h\}. \tag{8.6.12}$$

(Lie algebras will be taken over the real numbers.) Recall that

$$\mathcal{L}_0 \cap \mathcal{P} = \{0\}, \tag{8.6.13}$$

and that $\mathcal{L}_0$ is spanned by the two derivations δ_1, δ_2 which are given by

$$\delta_i(u_j) = \begin{cases} \sqrt{-1}\ 2\pi\ u_j, & i = j \\ 0, & i \neq j \end{cases}$$

where u_1, u_2 are the two unitary generators satisfying the commutation relation (8.6.7).

The perturbation result from [JM, Thm. 9.9(c)] is about finite-dimensional Lie algebras which are obtained as bounded perturbations of a given fixed base-point Lie algebra. We may take as base-point Lie algebra $\mathcal{L}_0 = d\alpha(\mathbb{R}^2)$. Since

$$\mathrm{Der}(\mathfrak{A}_\theta) = \mathcal{L}_0 + \mathcal{P}, \tag{8.6.14}$$

it follows from [JM, loc.cit.] that *every* finite-dimensional Lie subalgebra of $\mathrm{Der}(\mathfrak{A}_\theta)$ *exponentiates*.

Let $\mathcal{L} \subset \mathrm{Der}(\mathfrak{A}_\theta)$ be a given Lie algebra, and let $\mathcal{L}_b$ denote the ideal of bounded elements in $\mathcal{L}$. It follows from

(8.6.13) and (8.6.14) that

(8.6.15) $$\mathcal{L}_b = \mathcal{L} \cap \mathcal{B}.$$

Every finite-dimensional Lie algebra $\mathcal{L}$ has a *Levi decomposition* ([Ja] or [Va]) into a direct sum of a *semisimple* subalgebra and a *solvable radical*. We shall denote by $\mathcal{S}$ the semisimple component, and by $\mathcal{R}$ the solvable radical.

Theorem 8.6.3 [BEℓGJ]. Let $\mathcal{L}$ be a finite-dimensional real Lie subalgebra of $\mathrm{Der}(\mathfrak{A}_\theta)$, and let $\mathcal{L} = \mathcal{S}+\mathcal{R}$ be the Levi decomposition.

Then

(a) $\mathcal{S} \subset \mathcal{B}$, and $\mathcal{S}$ has negative definite Killing form, i.e., is compact.

(b) $\mathcal{S}$ and $\mathcal{R}$ are *commuting* ideals of $\mathcal{L}$.

(c) $\mathcal{R}_b$ is Abelian, and $[\mathcal{R},\mathcal{R}] \subset \mathcal{R}_b$; thus the derived series of $\mathcal{R}$ has only two steps: $(0) \subset [\mathcal{R},\mathcal{R}] \subset \mathcal{R}$.

(d) $\mathcal{R}_b = \mathcal{R}_b(0) \oplus [\mathcal{R},\mathcal{R}_b]$ where $\mathcal{R}_b(0)$ denotes the centralizer of $\mathcal{R}$ in $\mathcal{R}_b$. Moreover, $[\mathcal{R},\mathcal{R}_b]$ has even dimension and is the direct sum of minimal two-dimensional ideals.

(e) The adjoint representation of $\mathcal{R}$ on the complexification $\mathcal{R}_b^{\mathbb{C}}$ is diagonalizable with purely imaginary weights.

(f) The quotient $\mathcal{R}/\mathcal{R}_b$ is isomorphic to a Lie subalgebra of $\mathcal{L}_0$.

Proof. The proof of this particular result is outside the scope of the present monograph. The reader is referred to [BEℓGJ, Thm. 2.5]. We shall note however that the essential ingredient in the proof is the Singer-Jorgensen-Moore theorem [JM, Thm. G.3, p. 472]. The latter result identifies the bounded operators in the infinitesimal operator Lie algebra of a given uniformly bounded representation of a Lie group in a Banach space.

Theorem 8.6.3 is therefore not particularly special to the

irrational relation algebra. It merely serves to show that the classification problem breaks up into two separate parts: (i) the *compact* case, and (ii) the *solvable* case.

We show in [BEℓGJ] that every compact Lie algebra may be realized. The compact actions are inner. We shall restrict attention here to the smooth actions of solvable Lie groups. This amounts to identifying the solvable Lie subalgebras of $\mathrm{Der}(\mathfrak{A}_\theta)$. The following result reveals a further *dichotomy*:

Theorem 8.6.4. Let $\mathfrak{R} \subset \mathrm{Der}(\mathfrak{A}_G)$ be a finite-dimensional solvable Lie subalgebra. Then either:

(a) $\mathfrak{R}$ contains a Lie subalgebra isomorphic to the three-dimensional Heisenberg Lie algebra, or else

(b) $\mathfrak{R}$ contains an Abelian Lie subalgebra $\mathfrak{T}$ such that

$$\mathfrak{R} = \mathfrak{T} + \mathfrak{R}_b \quad \text{(semidirect sum)}.$$

Moreover, the two possibilities are mutually exclusive.

Proof. The proof of this result is also beyond the scope of this text. The details are in [BEℓGJ]. Here we shall just add a note on the distinction between the two cases (a) and (b).

By virtue of Theorem 8.6.3, $\mathfrak{R}$ must be of dimension at least 3. We shall choose terminology such that the first solvable Lie algebra is the two-dimensional ax+b algebra. But this Lie algebra must be excluded by virtue of Theorem 8.6.3. It has a real weight; *not* purely imaginary.

Part (f) of Theorem 8.6.3 asserts that $\mathfrak{R}/\mathfrak{R}_b \subsetneqq \mathcal{L}_0$, and so, in particular, the quotient is Abelian. The embedding of $\mathfrak{R}/\mathfrak{R}_b$ into $\mathcal{L}_0$ is given explicitly as follows: Elements δ in $\mathfrak{R}$ decompose uniquely in the form:

$$\delta = \delta_0 + \mathrm{ad}h, \qquad \delta_0 \in \mathcal{L}_0, \quad h \in \mathfrak{A}_0. \tag{8.6.16}$$

The mapping $\delta \longrightarrow \delta_0$ is a homomorphism since

$$[\delta_0 + \mathrm{ad}h, \delta_0' + \mathrm{ad}h'] = \mathrm{ad}(\delta_0(h') - \delta_0'(h) + [h, h']). \tag{8.6.17}$$

The kernel of the homomorphism, $\delta \longrightarrow \delta_0 : \mathfrak{K} \longrightarrow \mathcal{L}_0$ is

$$\mathfrak{K} \cap \mathfrak{P} = \mathfrak{K}_b.$$

It follows that $\mathfrak{K}/\mathfrak{K}_b$ is naturally embedded as a Lie subalgebra of $\mathcal{L}_0$ as asserted.

We claim that case (b) occurs if and only if the short exact sequence

$$0 \longrightarrow \mathfrak{K}_b \longrightarrow \mathfrak{K} \longrightarrow \mathfrak{K}/\mathfrak{K}_b \longrightarrow 0$$

splits in the category of Lie algebras. It clearly splits if the quotient $\mathfrak{K}/\mathfrak{K}_b$ is one-dimensional. It follows that, in this case, there is no copy of the 3-dimensional Heisenberg Lie algebra inside $\mathfrak{K}$.

We shall restrict attention to the case $\dim(\mathfrak{K}/\mathfrak{K}_b) = 2$; then

(8.6.18) $$0 \longrightarrow \mathfrak{K}_b \longrightarrow \mathfrak{K} \longrightarrow \mathcal{L}_0 \longrightarrow 0$$

is a short exact sequence. Let $\psi : \mathfrak{K} \longrightarrow \mathcal{L}_0$ be the canonical homomorphism defined by (8.6.16). If (8.6.18) splits, there is a homomorphism of Lie algebras $\varphi : \mathcal{L}_0 \longrightarrow \mathfrak{K}$ such that

(8.6.19) $$\psi \circ \varphi = \mathrm{id}_{\mathcal{L}_0}.$$

Let $\mathcal{T} = \varphi(\mathcal{L}_0)$. Then

(8.6.20) $$\mathfrak{K} = \mathcal{T} + \mathfrak{K}_b \quad \text{(semidirect sum)}$$

and $\mathcal{T}$ is Abelian. Indeed, for $\xi, \eta \in \mathcal{L}_0$, we have

$$[\varphi(\xi), \varphi(\eta)] = \varphi([\xi, \eta]) = \varphi(0) = 0,$$

so $\mathcal{T}$ is Abelian since $\mathcal{L}_0$ is.

The decomposition (8.6.20) is a consequence of (8.6.19). Indeed, if $\xi \in \mathfrak{K}$ is given, then the element $\xi - \varphi(\psi(\xi))$ is in the kernel of ψ since $\psi(\varphi(\psi(\xi))) = \psi(\xi)$. It is therefore in $\mathfrak{K}_b$, which is the assertion (8.6.20) with $\mathcal{T} = \varphi(\mathcal{L}_0)$. (It would appear natural to conjecture that (8.6.18) always splits, but it does not!)

The other half of the assertion is trivial: If (8.6.20) holds with $\mathcal{T}$ Abelian, then (8.6.18) splits.

There are trivial examples of solvable Lie algebras $\mathfrak{R}$ such that $\mathfrak{R} \longrightarrow \mathfrak{R}/\mathfrak{R}_b$ splits;—both with $\mathfrak{R}/\mathfrak{R}_b$ one-dimensional, as well as two-dimensional. Take, e.g., $\mathfrak{R}$ to be spanned by the three derivations δ_1, and the pair of inner derivations $\sqrt{-1}$ Re u_1, $\sqrt{-1}$ Im u_1. Similarly, we could make the example four-dimensional by including δ_2.

In the remainder of the section, we shall concentrate on case (a). This is the case treated in [BEℓGJ, Section 5]. In the case when $\mathfrak{R}$ is the 3-dimensional Heisenberg Lie algebra, we give a complete classification of the smooth actions up to unitary equivalence.

It is not at all obvious that the 3-dimensional Heisenberg group can be made to act on the irrational rotation algebra. (Our results about the Heisenberg actions hold for arbitrary irrational θ.)

To give a flavor of the ideas involved in the proof of Theorem 8.6.4, we show that possibilities (a) and (b) are mutually exclusive. Assume (b), and let ξ_1, ξ_2, ζ be elements in $\mathfrak{R}$ such that

$$[\xi_1, \xi_2] = \zeta, \quad \text{and} \quad [\zeta, \xi_k] = 0, \qquad k = 1,2.$$

We claim that $\zeta = 0$.

Consider the decomposition $\mathfrak{R} = \mathcal{T} \oplus \mathfrak{R}_b$ with $\mathcal{T}$ Abelian. There are weights (i.e., linear functionals) on $\mathcal{T}$ such that

$$\mathfrak{R}^{\mathbb{C}} = M(0) \oplus \sum_{\psi \neq 0}^{\oplus} M(\psi) \tag{8.6.21}$$

where the weights ψ are nonzero linear functionals on $\mathcal{T}$, and

$$M(0) = \{\eta \in \mathfrak{R}^{\mathbb{C}} : \exists k \text{ s.t. } (\text{ad } \tau)^k(\eta) = 0, \ \tau \in \mathcal{T}\},$$

and

$$M(\psi) = \{\eta \in \mathfrak{R}^{\mathbb{C}} : \exists k \text{ s.t. } ((\mathrm{ad}\ \tau)-\psi(\tau))^k(\eta) = 0,\ \tau \in \mathcal{T}\}.$$

We expand the formula:

$$\begin{aligned} 0 &= (\mathrm{ad}\ \tau-\psi(\tau))^k(\eta) \\ &= \sum_i \begin{bmatrix} k \\ i \end{bmatrix} (-\psi(\tau))^i (\mathrm{ad}\ \tau)^{k-i}(\eta) \\ &= (-\psi(\tau))^k \eta + k(-\psi(\tau))^{k-1}[\tau,\eta] + \cdots. \end{aligned}$$

If $\psi(\tau) \neq 0$, it follows that $\eta \in [\mathfrak{R}, \mathfrak{R}^{\mathbb{C}}] \subset \mathfrak{R}_b^{\mathbb{C}}$. Thus $M(\psi) \subset \mathfrak{R}_b^{\mathbb{C}}$ if $\psi \neq 0$. It follows from Theorem 8.6.3 that

$$[\tau,\eta] = \psi(\tau)\eta, \qquad \tau \in \mathcal{T}.$$

Consider the decomposition

$$\xi_k = \delta_k + \eta_0^k + \sum_\psi \eta_\psi^k, \qquad k = 1,2,$$

with $\delta_k \in \mathcal{T}$, $\eta_0^k \in M(0)$, and $\eta_\psi^k \in M(\psi)$, $k = 1,2$, according to (8.6.21). For the commutator, we have

$$\begin{aligned} \zeta = [\xi_1,\xi_2] &= [\delta_1, \eta_0^2 + \sum_\psi \eta_\psi^2] + [\eta_0^1 + \sum_\psi \eta_\psi^1, \delta_2] \\ &= \sum_\psi (\psi(\delta_1)\eta_\psi^2 - \psi(\delta_2)\eta_\psi^1). \end{aligned} \tag{8.6.22}$$

A further substitution yields:

$$0 = [\xi_k,\zeta] = \sum_\psi \psi(\delta_k)\{\psi(\delta_1)\eta_\psi^2 - \psi(\delta_2)\eta_\psi^1\}.$$

It follows that, for all ψ, $\psi(\delta_1)\eta_\psi^2 - \psi(\delta_2)\eta_\psi^1 = 0$, or $\psi(\delta_1) = \psi(\delta_2) = 0$. But if $\zeta \neq 0$, then δ_1 and δ_2 must be linearly independent. So we cannot have $\psi(\delta_1) = \psi(\delta_2) = 0$ for $\psi \neq 0$. It follows that $\psi(\delta_1)\eta_\psi^2 - \psi(\delta_2)\eta_\psi^1 = 0$ must hold. Substitution back into (8.6.22) yields the desired conclusion $\zeta = 0$.

The construction of representations of the 3-dimensional

Heisenberg Lie algebra relies on, among other things, known facts about K-theory of the C^*-algebra A_θ, and the Fréchet algebra $\mathfrak{A}_\theta \subset A_\theta$. Rieffel (and Powers) proved in [Ri 4] the existence of a projection e in $\mathfrak{A}_\theta$ satisfying $\text{trace}(e) = \theta$. He further showed that $K_0(A_\theta) = K_0(\mathfrak{A}_\theta)$. It follows that

$$K_0(\mathfrak{A}_\theta) = \mathbb{Z}[1] + \mathbb{Z}[e]$$

where $[1]$ is the equivalence class of the trivial projection, and $[e]$ is the equivalence class of the Rieffel projection. It was proved in [PiVo] that there is a unital embedding of A_θ into an approximately finite-dimensional (AF for short) C^*-algebra B_θ, and the Elliott group ([Ell] or [Eff]) of B_θ is $\mathbb{Z} + \theta\mathbb{Z}$. Moreover, the trace functional is 1-1 from $K_0(A_\theta)$ onto $(\mathbb{Z}+\theta\mathbb{Z}) \cap [0,1]$. It follows, in particular, that there is a large supply of nontrivial projections in $\mathfrak{A}_\theta$.

Let $f,g \in C(\mathbb{R}/\mathbb{Z})$ satisfy:

$$f(s) = \begin{cases} 1 & , \quad 1-\theta \le s \le \theta \\ 1-f(s-\theta) & , \quad \theta \le s \le 1 \end{cases}$$

and

$$g(s) = \begin{cases} 0 & , \quad 0 \le s \le \theta \\ (f(s)-f(s)^2)^{1/2} & , \quad \theta \le s \le 1 \end{cases} .$$

Then it can be verified directly that the element

$$e = g(u_2)u_1 + f(u_2) + (g(u_2)u_1)^*$$

satisfies $e = e^* = e^2$. If the two functions f,g are picked in $C^\infty(S^1)$, $S^1 \simeq \mathbb{R}/\mathbb{Z}$, then $e \in \mathfrak{A}_\theta$.

Let M be a right module over a ring A, and let L be a Lie algebra of derivations of A. A connection of (M,A,L) is a linear mapping

$$\nabla : L \longrightarrow \text{End}_{\mathbb{C}}(M)$$

satisfying

(8.6.23) $$\nabla_x(\xi a) = \nabla_x(\xi)a + \xi x(a), \qquad x \in L, \quad a \in A, \quad \xi \in M.$$

In the beginning of this section, we studied highest weight representations ρ of the Virasoro algebra. We noted (and proved in Section 8.5) that ρ exponentiates to a projective representation W of G_1 on the Hilbert space $\mathcal{H}$ obtained from the GNS-representation applied to the algebra $\mathfrak{U}_{\mathbb{C}}(\tilde{\mathfrak{v}})$. We noted that $SL_2(\mathbb{R})$ is naturally embedded as a subgroup of G_1. The covariance formula (8.6.6) holds in particular for $g \in SL_2(\mathbb{R})$. Since W is a projective representation of $SL_2(\mathbb{R})$, it is a representation of a central extension of $SL_2(\mathbb{R})$ by a cocycle. We shall denote the group obtained from $SL_2(\mathbb{R})$ by this central extension $CSL_2(\mathbb{R})$, or $C_2(\mathbb{R})$ for short. The covariance formula gives rise to a module and a connection as follows: For M we take the space of C^∞-vectors for the representation W of $C_2(\mathbb{R})$ on $\mathcal{H}$, and for A we take $\mathfrak{U}_{\mathbb{C}}(\tilde{\mathfrak{v}})$ with the action of A on M given by

$$\xi a := \rho(a^*)\xi, \qquad a \in A, \quad \xi \in M.$$

The calculation below shows that the study of connections, and covariant systems amounts to one and the same thing. It is easy to check that this turns M into a right-module over A.

We claim that the representation W of $C_2(\mathbb{R})$ is determined by a connection ∇ of (M,A) defined over the Lie algebra of $C_2(\mathbb{R})$. For x in the Lie algebra, we define

$$g(t) = \exp(tx), \qquad t \in \mathbb{R},$$

$$x(a) = \frac{d}{dt}\bigg|_{t=0} a^{g(t)}, \qquad a \in \mathfrak{U}_{\mathbb{C}}(\tilde{\mathfrak{v}}),$$

and

$$\nabla_x \xi = dW(x)\xi, \qquad \xi \in M.$$

It follows from (8.6.6) that $\{\nabla_x\}$, defined this way, is indeed a connection of (M,A) as asserted.

We verify that ∇_x satisfies (8.6.23). For $a \in A$, $\xi \in M$, we have

$$\begin{aligned}\nabla_x(\xi a) &= dW(x)\rho(a^*)\xi \\ &= \rho(a^*)dW(x)\xi + \rho(x(a)^*)\xi \\ &= (\nabla_x\xi)a + \xi x(a)\end{aligned}$$

where we utilized the differentiated version of formula (8.6.6).

We now turn to the study of connections on finitely generated projective modules over the ring $\mathfrak{A}_\theta$ where θ is a given irrational number, $0 < \theta < 1$.

The *curvature* of a given connection ∇ is a bilinear function on the Lie algebra L with values in $\mathrm{End}_A(M)$, defined by

(8.6.24) $$C(x,y) = [\nabla_x,\nabla_y] - \nabla_{[x,y]}, \qquad x,y \in L.$$

We shall consider the case where L is the canonical action of the Lie algebra $\mathbb{R}^2$ on $\mathfrak{A}_\theta$, given by (8.6.11). Connes [Con 3] defines the *Chern character* of ∇, and, for the module $e\mathfrak{A}_\theta$, it may be identified with the number

$$c_1(e) = \frac{1}{2\pi i}\,\mathrm{trace}(e[\delta_1(e),\delta_2(e)]).$$

Let $\mathcal{S}$ be the Schwartz space on $\mathbb{R}$, i.e., the space of C^∞-vectors for the Schrödinger representation on $L^2(\mathbb{R})$. It is well known that $\mathcal{S}$ is also an algebra. Let

$$\{h_n : n = 0,1,2,\cdots\}$$

be the orthonormal basis for $L^2(\mathbb{R})$ formed from the Hermite functions, cf. (6.2.17), and let

$$f = \sum_{n=0}^{\infty} \langle f,h_n\rangle h_n$$

be the orthogonal Hermite-expansion, with

$$a_n = \langle f, h_n \rangle = \int_{\mathbb{R}} f(v) h_n(v) dv.$$

The following characterization of the Schwartz space can easily be verified using elementary properties of the Schrödinger representation:

$$\mathcal{S} = \{f \in L^2(\mathbb{R}) : \forall k, \exists M \text{ s.t. } : |a_n| \le M(1+n)^{-k}\}.$$

It is interesting to compare this with (8.6.2) above.

Let $p, q \in \mathbb{N}$, and let $M = \mathbb{C}^q \otimes \mathcal{S}(\mathbb{R})$. We shall regard M as a module. Let $L = \mathbb{R}^2$ with basis x_1, x_2. Define

(8.6.25)
$$(\nabla_{x_1} \xi)(s) = \frac{d}{ds} \xi(s),$$

$$\nabla_{x_2} \xi(s) = \frac{2\pi i s}{\epsilon} \xi(s), \qquad s \in \mathbb{R},$$

where $\epsilon = \frac{p}{q} - \theta$.

The *curvature tensor* of this connection is constant. Since $L = \mathbb{R}^2$ is Abelian, we may identify the curvature with

$$C(x_1, x_2) = \nabla_{x_1} \nabla_{x_2} - \nabla_{x_2} \nabla_{x_1}.$$

Following [Co 3], [CR] and [Ri 4], we turn $M = \mathbb{C}^q \otimes \mathcal{S}(\mathbb{R})$ into a module over $\mathfrak{A}_\theta$ by defining an action of $\mathfrak{A}_\theta$ on M as follows: Let w_1, w_2 be a pair of unitaries on $\mathbb{C}^q$ satisfying

$$w_2 w_1 = \exp(-i2\pi/q) w_1 w_2,$$

and define the action of $\mathfrak{A}_\theta$ on M, on generators, by

$$(\xi u_1)(s) = w_1 \xi(s+\epsilon), \qquad s \in \mathbb{R},$$

$$(\xi u_2)(s) = \exp(i2\pi s) w_2^p \xi(s), \qquad s \in \mathbb{R}.$$

Connes showed [Co 3, Thm. 7] that M, defined this way, is a projective module over $\mathfrak{A}_\theta$ of dimension $|p-q\theta|$, and that the connection ∇, defined by (8.6.25), has constant curvature equal $(\theta-p/q)^{-1}$.

Moreover the two modules $(M, \mathfrak{A}_\theta)$ and $(e\mathfrak{A}_\theta, \mathfrak{A}_\theta)$ are isomorphic. When the isomorphism is applied, it follows that the module $e\mathfrak{A}_\theta$ has a *canonical constant curvature connection*. The same result applies to the module $(1-e)\mathfrak{A}_\theta$ with the only modification that θ is replaced by $1-\theta$. Let $\nabla^{\pm}$ be the constant curvature connections on the respective modules $e\mathfrak{A}_\theta$ and $(1-e)\mathfrak{A}_\theta$. For $x \in L = \mathbb{R}^2$, define

$$a(x) = \nabla_x^+(e) + \nabla_x^-(1-e). \tag{8.6.26}$$

We have

Proposition 8.6.5 [BEℓGJ]. The mapping

$$x \longrightarrow \delta_x + \mathrm{ad}(a(x)) \tag{8.6.27}$$

defines a projective representation of $\mathbb{R}^2$, and a faithful representation of the 3-dimensional Heisenberg Lie algebra.

Proof. The result in [BEℓGJ] is in fact more general. It shows that representations are associated with arbitrary partitions of unity in $\mathfrak{A}_\theta$, i.e., a finite family of nontrivial projections in $\mathfrak{A}_\theta$ with sum one, and conversely. We have restricted attention here to the case $\{e,1-e\}$.

We define the two derivations

$$\xi_k = \delta_k + \mathrm{ad}\ a(x_k), \qquad k = 1,2.$$

Since

$$[\xi_1,\xi_2] = \mathrm{ad}(\delta_1(a(x_2))-\delta_2(a(x_1))+[a(x_1),a(x_2)])$$

by virtue of (8.6.17), we must verify that the element

$$c := \delta_1(a(x_2)) - \delta_2(a(x_1)) + [a(x_1),a(x_2)]$$

satisfies

$$\xi_k(c) = 0, \qquad k = 1,2,$$

or equivalently,

$$\delta_k(c) = -[a(x_k),c], \qquad k = 1,2. \tag{8.6.28}$$

The connection ∇^+ on $e\mathfrak{A}_\theta$ was picked so that the curvature

$$C^+(x_1,x_2)(eb) = [\nabla^+_{x_1},\nabla^+_{x_2}](eb)$$
$$= \lambda\ eb$$

where λ is a numerical constant. In fact, $\lambda = (\theta-p/q)^{-1}$.

Using this, the connection formula (8.6.23), and Leibniz' rule, we show in [BEℓGJ, Prop. 4.2] that:

$$ece = \lambda\ e$$

and

$$(1-e)c(1-e) = \frac{\theta\lambda}{\theta-1}(1-e).$$

It follows that

$$c = \lambda\ e + \frac{\theta\lambda}{\theta-1}(1-e),$$

and in particular that the commutator $[\xi_1,\xi_2] = \mathrm{ad}(c)$ is a nonnzero derivation.

To verify (8.6.28) we need only check that

$$\delta_x(e) = -[a(x),e], \qquad x \in L.$$

We do this in two steps: Set $a^0(x) = e\delta_x(e)+(1-e)\delta_x(1-e)$. Using Leibniz' rule for δ_x, we get:

$$\delta_x(e) = -[a^0(x),e] \quad \text{and} \quad ea^0(x)e = 0.$$

It follows that

$$\begin{aligned}
-\delta_x(e) &= [a^0(x),e] \\
&= [a^0(x)+e\nabla^+_x(e)e+(1-e)\nabla^-_x(1-e)(1-e),e] \\
&= [\nabla^+_x(e)e+e\delta_x(e),e] + [\nabla^-_x(1-e)(1-e)+(1-e)\delta_x(1-e),e] \\
&= [\nabla^+_x(e)+\nabla^-_x(1-e),e] \\
&= [a(x),e], \qquad x \in L,
\end{aligned}$$

as asserted.

Corollary 8.6.6 [BEℓGJ]. If ∇ is a connection on $e\mathfrak{A}_\theta$

with constant curvature $C(x,y)$, then the curvature tensor is a Lie cocycle on L, i.e., identity (2.5.3) holds:

$$C([x,y],z) + C([y,z],x) + C([z,x],y) = 0, \qquad x,y,z \in L.$$

We also note that the representations of the 3-dimensional Heisenberg Lie algebra $\mathfrak{h}$, constructed above, all exponentiate to smooth automorphic representations of the 3-dimensional Heisenberg group H, i.e., group actions $\pi : H \longrightarrow \mathrm{Aut}(A_\theta)$ such that $d\pi(\mathfrak{h})$ is identical to the original Lie algebra from Proposition 8.6.5 above. This follows from [JM, Thm. 9.9(c)].

There are two additional results in [BEℓGJ] regarding Lie algebra representations of $\mathfrak{h}$, or equivalently Lie actions of the group H on A_θ.

The first one is that every smooth faithful action

$$\pi : H \longrightarrow \mathrm{Aut}(A_\theta)$$

arises this way from a finite partition of unity in $\mathfrak{A}_\theta$, and a corresponding field of constant curvature connections. More specifically, a partition of unity is a finite orthogonal system $\mathcal{E} = \{e\}$ of projections in $\mathfrak{A}_\theta$ with sum 1. A field of connections is a family of connections ∇^e indexed by $\mathcal{E}$. We say that the field is constant curvature if each connection ∇^e on $e\mathfrak{A}_\theta$ is constant curvature.

The cocycles and the Heisenberg representation is defined from this field as above, *mutatis mutandis*.

Let $\mathcal{H}_2$ be the trace Hilbert space with trace vector Ω, and define a unitary representation W of H on $\mathcal{H}_2$ as follows. Let $\xi_k = \delta_k + \mathrm{ad}\, a(x_k)$, $k = 1,2$, be the generators for a representation of $\mathfrak{h}$, and let $\zeta = [\xi_1,\xi_2]$. Define

$$\text{(8.6.29)} \qquad \begin{aligned} dW(x_k)(b\Omega) &= (\delta_k(b)+a(x_k)b)\Omega, \\ dW([x_1,x_2]) &= cb\Omega, \end{aligned}$$

where $c = \delta_1(a(x_2)) - \delta_2(a(x_1)) + [a(x_1),a(x_2)]$, and let W

be the corresponding representation of H which exists by [JM, Thm. 9.9(c)]. Then W is a finite direct sum of Schrödinger representations $(S_\lambda, H, \mathcal{L}^2(\mathbb{R}))$ corresponding to Planck's constant λ, and for a finite set of λ's in $\mathbb{R}\setminus\{0\}$. The multiplicity $m(\lambda)$ is finite for each λ in the spectrum.

We refer to [BEℓGJ, Thm. 5.5] for details.

Note that the noncommutative Schrödinger system (8.6.29) is obtained as a bounded perturbation of a unitary representation U of $\mathbb{R}^2$ generated by the two commuting operators

$$dU(x_k)(bU) = \delta_k(b)\Omega, \qquad k = 1,2, \quad b\Omega \in \mathcal{H}_2.$$

This is somewhat surprising since the two operators $dW(x_k)$, $k = 1,2$, have Lebesgue spectrum (in particular, absolutely continuous), while the two commuting operators $dU(x_k)$ have purely discrete specturm, even periodic, despite the fact that one pair is obtained from the other by a system of smooth perturbations. The perturbing operators are given explicitly by equations (8.6.29).

In Section 5.9 of [BEℓGJ], we finally show that, for a given partition of unity $\mathcal{E} = \{e\}$ in $\mathfrak{A}_\theta$, the parameter space, for unitary equivalence classes of smooth Heisenberg actions based on $\mathcal{E}$, is the space

$$\prod_e (\mathbb{T}^2)^{d(e)-1} / S_{d(e)}$$

where $S_{d(e)}$ is the symmetric group on $d(e)$ letters, and the action of $S_{d(e)}$ on $(\mathbb{T}^2)^{d(e)-1}$ is the quotient of the action on $(\mathbb{R}^2)^{d(e)}$ by permutation of coordinates. Finally the function, $e \longrightarrow d(e) : \mathcal{E} \longrightarrow \mathbb{N}$, is determined as follows: Rieffel [Ri 4] showed that, for each e, there is a θ' and $d(e) \in \mathbb{N}$ such that

$$e\mathfrak{A}_\theta e \simeq \mathfrak{A}_{\theta'} \otimes M_{d(e)}(\mathbb{C})$$

where θ' is obtained from θ by a $SL_2(\mathbb{Z})$ fractional linear transformation, and $M_{d(e)}(\mathbb{C})$ is the ring of all $d(e)$ by $d(e)$ complex matrices.

The details on these last two results are, unfortunately, beyond the scope of this book.

APPENDIX: INTEGRABILITY OF LIE ALGEBRAS

In this short appendix, we summarize the results on integrability of Lie algebras of operators which are used in the book, and we include some recent developments in the subject. Although proofs will not be given, some implications relating the different results will be pointed out with varying amounts of detail.

For finite-dimensional representations, the integrability of the Lie algebra representation to the Lie group is quite easy. Only *very basic* facts on the exponential mapping from Lie theory are needed.

For the purpose of setting down the notation, we shall review this finite-dimensional problem first.

Let $\mathfrak{g}$ be a finite-dimensional Lie algebra over $\mathbb{R}$, and let V be a finite-dimensional vector space over $\mathbb{C}$. Let $\rho : \mathfrak{g} \longrightarrow \text{End } V$ be a representation, i.e., a linear mapping preserving the respective Lie brackets:

(A1) $$\rho([x,y]) = [\rho(x),\rho(y)] := \rho(x)\rho(y)-\rho(y)\rho(x), \quad x,y \in \mathfrak{g}.$$

By Ado's theorem ([Ado], [HS]), $\mathfrak{g}$ is the Lie algebra of some Lie group. By going to the universal *simply connected* covering group, if necessary, we may assume that the Lie group, say G, is simply connected. The integrability results listed below will only produce representations of this group. If G covers some different group, say H, then a representation π of G may not, in general, pass to a representation of H, although G and H have the same Lie algebra. But it is easy to decide when it does: Suppose that $d\pi = \rho$, i.e.,

(A2) $$\rho(x)v = \frac{d}{dt}\Big|_{t=0} \pi(\exp_G tx)v, \qquad x \in \mathfrak{g}, \quad v \in V.$$

Let $\varphi : G \longrightarrow H$ be the universal covering homomorphism. Then

$$G/\text{kernel}(\varphi) \simeq H,$$

and it follows that π passes to a representation of H if and only if $\pi(g) = I$ for all $g \in \text{kernel}(\varphi)$.

In Section 6.2, we studied this problem for $g = s\ell_2(\mathbb{R})$. In the latter case, V was infinite-dimensional, but the condition is the same. We checked that, if G is the universal simply connected covering group of $SL_2(\mathbb{R})$, and if H is the two-sheeted metaplectic cover of $SL_2(\mathbb{R})$, then the representation π obtained by exponentiating the Lie algebra representation ρ, given by (A1), is indeed trivial on the kernel for the covering mapping $G \longrightarrow H$. But it is *not* trivial on the covering homomorphism $H \longrightarrow SL_2(\mathbb{R})$;—hence the metaplectic group H.

The construction of the group representation π, from ρ, is by exponentiation, and this is where the finite-dimensionality of V reduces the problem to linear algebra. If V is instead infinite-dimensional, it becomes an intricate problem in operator theory.

Assume first $\dim V < \infty$. Since G is connected, it is generated, as a group, by

$$\{\exp x : x \in g\},$$

and moreover, there is a neighborhood $\mathcal{U}$ of 0 in g such that

$$\{\exp x : x \in \mathcal{U}\}$$

is a neighborhood of the origin e in G. This follows from the Implicit Function Theorem since it is known that the exponential mapping, $\exp : g \longrightarrow G$ is nonsingular at the origin 0 in g. In fact, an explicit formula for the differential of the exponential mapping, $\exp : g \longrightarrow G$, is known [He 3, I, Thm. 6.5, and II, Thm. 1.7]: Let $x \in g$, and identify the tangent space g_x with g itself. The differential $d\exp_x$ maps from the tangent space $g_x \simeq g$ to the tangent space $G_{\exp x}$. Let U_x denote left multiplication by $\exp x$, and let dU_x be the differential of U_x at the origin e in G.

Then

$$\text{(A3)} \qquad d\exp_x = dU_x \circ \frac{I-e^{-adx}}{adx}$$

where the fraction is defined by functional calculus of the entire function $\sum_{n=0}^{\infty} (-z)^n/(n+1)!$ applied to the linear endomorphism, $adx : g \longrightarrow g$.

To exponentiate a given representation $\rho : g \longrightarrow \text{End } V$ on a finite-dimensional vector space V, we use the same functional calculus argument, now applied to the exponential function $\sum_{n=0}^{\infty} z^n/n!$, $z \in \mathbb{C}$.

For every $x \in g$, the linear mapping $\rho(x)$ may be substituted. The resulting power series $\sum_{n=0}^{\infty} \rho(x)^n/n!$ is convergent, and defines an invertible linear mapping from V to V.

If we pick a neighborhood $\mathcal{U}$ of 0 in g such that $\exp \mathcal{U}$ is a neighborhood of e in G, then a function π may be defined on $\exp \mathcal{U}$ as follows:

$$\text{(A4)} \qquad \pi(\exp x) = \exp \rho(x) := \sum_{n=0}^{\infty} \rho(x)^n/n!.$$

Defined this way, π will be a local representation on neighborhood $\mathcal{W} = \exp \mathcal{U}$.

We may take $\mathcal{U}$ to be a convex neighborhood. We claim

(i) if g_1, g_2, and $g_1 \cdot g_2 \in \mathcal{W}$, then

$$\pi(g_1 \cdot g_2) = \pi(g_1)\pi(g_2),$$

and

(ii) $$\pi(g^{-1}) = \pi(g)^{-1},$$

The condition in (i) amounts to the assertion

$$\text{(A5)} \qquad \pi(e^x e^{ty})\pi(e^{-ty})\pi(e^{-x}) = I$$

for all $x, y \in \mathcal{U}$ such that $e^x e^{ty} \in \mathcal{W}$ for all $t \in [0,1]$.

The expression (A5) is C^∞, even analytic, as a function of t, and (A5) holds for $t = 0$. The latter amounts to condition (ii) which is straightforward. We shall check, using (A1) and (A4) above, that

$$\frac{d}{dt}\{\pi(e^x e^{ty})\pi(e^{-ty})\pi(e^{-x})\} = 0.$$

The desired formula (A5) follows from this.

We claim that

$$\text{(A6)} \qquad \frac{d}{dt}\{\pi(e^x e^{ty})\} = \pi(e^x e^{ty})\rho(y).$$

The desired conclusion follows from this and the product rule for the derivative:

$$\frac{d}{dt}\{\pi(e^x e^{ty})\pi(e^{-ty})\pi(e^{-x})\}$$

$$= (\pi(e^x e^{ty})\rho(y))\pi(e^{-ty})\pi(e^{-x}) - \pi(e^x e^{ty})(\rho(y)\pi(e^{-ty}))\pi(e^{-x})$$

$$\equiv 0.$$

In the verification of (A6), it is convenient to choose a matrix representation for $\mathfrak{g}$. We shall also choose a basis $x_1, \cdots, x_n$ for $\mathfrak{g}$, and introduce coordinate functions $\{f_k(t) : 1 \leq k \leq n\}$ for $e^x e^{ty}$ relative to that basis. As a result we get the expansion

$$e^x e^{ty} = e^{f_1(t)x_1} e^{f_2(t)x_2} \cdots e^{f_n(t)x_n},$$

and the formula

$$\frac{d}{dt} e^x e^{ty} = \sum_{n=0}^{\infty} f_k'(t) e^{f_1(t)x_1} \cdots e^{f_{k-1}(t)x_{k-1}} x_k e^{f_k(t)x_k} \cdots e^{f_n(t)x_n}$$

$$= \sum_{n=0}^{\infty} f_k'(t) e^{f_1(t)x_1} \cdots e^{f_n(t)x_n} \operatorname{Ad}(g_k(t)^{-1})(x_k)$$

where $g_k(t) = e^{f_{k+1}(t)x_{k+1}} \cdots e^{f_n(t)x_n}$.

After introducing the terminology

$$x(t) = \sum_{k=1}^{\infty} f'_k(t)\ \mathrm{Ad}(g_k(t)^{-1})(x_k) \in \mathfrak{g},$$

the above formula takes the form

$$\frac{d}{dt}\, e^x e^{ty} = e^x e^{ty}\, x(t).$$

It follows that $x(t) = y$ for all $t \in [0,1]$.

Using the analytic power series definition (A4), we get

$$\rho(\mathrm{Ad}(g_k(t)^{-1})(x_k)) = \pi(g_k(t)^{-1})\rho(x_k)\pi(g_k(t)),$$

and therefore

$$\begin{aligned}\frac{d}{dt}\,\pi(e^x e^{ty}) &= \pi(e^x e^{ty}) \sum_{k=1}^{n} f'_k(t)\pi(g_k(t)^{-1})\rho(x_k)\pi(g_k(t)) \\ &= \pi(e^x e^{ty}) \sum_{k=1}^{n} f'_k(t)\rho(\mathrm{Ad}(g_k(t)^{-1})x_k) \\ &= \pi(e^x e^{ty})\rho\Big(\sum_{k=1}^{n} f'_k(t)\mathrm{Ad}(g_k(t)^{-1})x_k\Big) \\ &= \pi(e^x e^{ty})\rho(y),\end{aligned}$$

which is the desired conclusion (A6).

As already noted, formula (A5) follows from this.

To summarize, we have defined a local representation π by the exponential formula (A4) applied to the given Lie algebra representation ρ. We have worked with the assumption $\dim V < \infty$, and, to be sure, there are shorter proofs of (A5) in this special case. But the present proof has the advantage that it can be extended and generalized to the most general case when V is an infinite-dimensional topological linear space with a Fréchet linear topology. The reader is referred to [JM, Chs. 8-9] for the details of these more substantial considerations.

Since G is assumed simply connected, it is enough to have π as a local representation. Extension of a local homomorphism from a simply connected Lie group is always possible by general theory, see, e.g., [Hoc 1, Thm. 3.1, Ch. IV, p. 54].

The setting of our integrability theorems will be that of linear operators in a given Banach space $\mathfrak{X}$. Representations ρ of the given finite-dimensional real Lie algebra $\mathfrak{g}$ will be considered. The domain of ρ will be denoted $\mathfrak{D}$. Then ρ is a homeomorphism of Lie algebras $\rho : \mathfrak{g} \longrightarrow \text{End } \mathfrak{D}$, and we shall always assume that $\mathfrak{D}$ is dense in $\mathfrak{X}$. We may assume without loss of generality that ρ is faithful. If not, just pass to the quotient $\mathfrak{g}/\text{kernel}(\rho)$ to get a faithful representation. As a result, it is possible to identify the given Lie algebra $\mathfrak{g}$ with the Lie algebra of operators $L = \rho(\mathfrak{g})$. It is nonetheless convenient to keep the two distinct since $\mathfrak{g}$ is often obtained as the Lie algebra of a given Lie group G, and the representation ρ may not be readily available.

We are now just in the setting of Banach space and the representation ρ is not assumed to be Hermitian in any sense. (Anyway, there is no inner product!) Consequently, there is no guarantee that the operator $\rho(x)$, $x \in \mathfrak{g}$, is even *closable*, let alone closed. Moreover, we are *not* placing assumptions on all the operators in $L = \{\rho(x) : x \in \mathfrak{g}\}$.

We are, however, placing an assumption on the elements in a subset of L, but *not* a basis, in the general case. A subset $M \subset L$ is said to *generate* L as a Lie algebra if L is spanned over $\mathbb{R}$ by elements in M along with all possible iterated commutators formed from elements in M. Equivalently, L is the smallest Lie subalgebra of $\text{End } \mathfrak{D}$ which contains M.

We shall assume that there is a generating set M such that each ξ in M is closable as a linear operator in ξ, and moreover that the closed operator $\bar{\xi}$ is the infinitesimal generator of a strongly continuous one-parameter group of bounded operators in $\mathfrak{X}$.

A subset $M \subset L$ with this property is said to be *generic* in L.

The given representation ρ is said to be exact if there is a strongly continuous representation U of some Lie group G with $\mathfrak{g}$ as Lie algebra such that $\rho = dU$. The interpretation of this formula requires some degree of specificity. If U is a strongly continuous representation of G, and $x \in \mathfrak{g}$, then the operator $dU(x)$ is the infinitesimal generator of the one-parameter group,

$$t \longrightarrow U(\exp tx),$$

and it is closed. The Gårding space for U is a core for each of the operators $dU(x)$. The formula

$$\rho = dU \tag{A7a}$$

is meant to summarize that each operator $\rho(x)$, $x \in \mathfrak{g}$, is closable, and that

$$\overline{\rho(x)} = dU(x), \qquad x \in \mathfrak{g}. \tag{A7b}$$

It will follow from the three results below that, from assumptions on only a generic subset of L, we will get conclusions about all the operators in L.

Theorem A1. Let ρ be a representation of a Lie algebra $\mathfrak{g}$. Let $\mathfrak{D}$ be the domain of ρ, Assume that the Lie algebra $L = \rho(\mathfrak{g}) \subset \text{End}\ \mathfrak{D}$ contains a generic set M. For $\xi \in M$, let $\{U_\xi(t) : t \in \mathbb{R}\}$ denote the corresponding one-parameter subgroup of $\mathfrak{B}(\mathfrak{X})$ generated by $\overline{\xi}$. Assume conditions (i)-(ii) below:

(i) For all $\xi \in M$,

$$U_\xi(t)\mathfrak{D} = \mathfrak{D}, \qquad t \in \mathbb{R}, \tag{A8}$$

holds.

(ii) For all $\xi, \eta \in M$ and $a \in \mathfrak{D}$, the function $t \longrightarrow \eta U_\xi(t)a$ is bounded in $\mathfrak{X}$ for at least one nontrivial open subinterval of $\mathbb{R}$.

Then it follows that L is integrable, i.e., that the representation ρ is exact.

Theorem A2. Let $(\rho, \mathfrak{g}, \mathfrak{D}, \mathfrak{X})$ be a representation, i.e., $\mathfrak{g}$ is a finite-dimensional Lie algebra over $\mathbb{R}$, and ρ is a representation of $\mathfrak{g}$ with dense domain $\mathfrak{D}$ in a given Banach space $\mathfrak{X}$.

Assume that $\mathfrak{D}$ is a Fréchet space in a linear topology which is stronger than the norm topology induced from $\mathfrak{X}$.

If (A8) is satisfied for all ξ in a generic subset of $L = \rho(g)$, then it follows that L is integrable, i.e., that ρ is exact.

Let L be a finite-dimensional Lie algebra of operators on a normed vector space $\mathfrak{D}$, and let $\eta_1, \cdots, \eta_n$ be a basis for L. Then we may define a graph norm $\|\cdot\|_1$ relative to the basis operators as follows:

$$a \longrightarrow \|a\| + \sum_{i=1}^{n} \|\eta_i(a)\|,$$

where $\|\cdot\|$ denotes the given norm on $\mathfrak{D}$. If a different basis for L is chosen, then the corresponding two graph norms will be equivalent.

Theorem A3. Let $(\rho, g, \mathfrak{D}, \mathfrak{X})$ be a Lie algebra representation with dense domain in a Banach space. Let $M \subset L = \rho(g)$ be a generic set. The one-parameter groups $\{U_\xi(t)\}_{t \in \mathbb{R}}$ satisfy exponential estimates

$$\text{(A9)} \qquad \|U_\xi(t)\| \leq A \exp(B|t|), \qquad t \in \mathbb{R},$$

for constants A and B depending on ξ.

Assume, for each $\xi \in M$, that there are complex numbers $z_\pm$ satisfying

(i) $\qquad \operatorname{Re} z_+ > B(\xi), \quad \operatorname{Re} z_- < -B(\xi),$

and

(ii) each of the spaces

$$\{z_\pm a - \xi(a) : a \in \mathfrak{D}\}$$

is dense in $\mathfrak{D}$ relative to the graph norm defined by a basis for L. (We refer to this condition as Graph Density (GD).)

Then it follows that L is integrable.

The above three theorems represent joint work with R. T. Moore [JM, Chs. 8-9], and we refer to [JM] for details

The following result may be obtained as a corollary of Theorem A3 above. It is from [GJ2, Thm. 3.1]. It was also proved, independently, in an unpublished preprint by Rusinek.

Corollary A4. Let $(\rho, \mathfrak{g}, \mathcal{D}, \mathcal{X})$ be a Lie algebra representation with dense domain in a Banach space, and assume that $L = \rho(\mathfrak{g})$ contains a generic set M. Assume further that every $a \in \mathcal{D}$ is an analytic vector for each $\xi \in M$.

Then it follows that L is integrable.

Proof of A4. Let $\|\cdot\|$ denote the norm of $\mathcal{X}$. For $\xi \in M$, we have

$$\text{(A10)} \qquad \|U_\xi(t)\| \leq A\, e^{B|t|}, \qquad t \in \mathbb{R},$$

with A and B depending on ξ. But there is an equivalent norm on $\mathcal{X}$, also denoted $\|\cdot\|$, such that, in the new norm, we have

$$\text{(A11)} \qquad \|U_\xi(t)\| \leq e^{B|t|}, \qquad t \in \mathbb{R}, \quad \xi \in M.$$

The new equivalent norm may be defined as follows:

$$a \longrightarrow \max_{\xi \in M} \sup_{t \in \mathbb{R}} e^{-B(\xi)|t|} \|U_\xi(t)a\|,$$

in terms of the original norm. It follows from (A9) that this new norm is equivalent to the original one, and that (A11) holds relative to the new norm.

Let $\|\cdot\|_1$ denote the graph norm defined relative to a basis $\eta_1, \cdots, \eta_n$ for L. Give L the norm $\|\sum_{i=1}^{n} c_i \eta_i\| = \sum_{i=1}^{n} |c_i|$. If T denotes the operator norm of $\mathrm{ad}\xi : L \longrightarrow L$, and if

$$\text{(i)} \qquad \|\xi^m a\| \leq CD^m\, m!$$

(ii) $$\|\xi^m \eta_i a\| \le CD^m\, m!, \qquad 1 \le i \le n,$$

for constants C,D, then it follows that

$$\|\xi^m a\|_1 \le C \sum_{k=0}^{m} \binom{m}{k} T^{m-k} D^k\, k!\, \|a\|_1.$$

It follows from this that every $a \in \mathfrak{D}$ is analytic for each $\xi \in M$ relative to the graph norm $\|\cdot\|_1$.

Indeed, if η is one of the basis elements, we have the combined estimate:

$$\begin{aligned}
\left\|\eta\xi^m a\right\| &\le \sum_{k=0}^{m} \binom{m}{k} \left\|\xi^k (\text{ad } \xi)^{m-k}(\eta)(a)\right\| \\
&\le \sum_{k=0}^{m} \binom{m}{k} CD^k\, k!\, \|(\text{ad } \xi)^{m-k}(\eta)\|\, \|a\|_1 \\
&\le \sum_{k=0}^{m} \binom{m}{k} CD^k\, k!\, \|\text{ad } \xi\|^{m-k} \|\eta\|\, \|a\|_1 \\
&\le \sum_{k=0}^{m} \binom{m}{k} CD^k\, k!\ T^{m-k} \|a\|_1,
\end{aligned}$$

as asserted.

By virtue of (A11), we have the estimate

$$\|za - \xi a\| \ge (|\text{Re } z| - B)\|a\|, \quad a \in \mathfrak{D}, \quad z \in \mathbb{C}, \quad |\text{Re } z| > B.$$

But

$$\begin{aligned}
\|\eta(za - \xi a)\| &= \|z\eta a - \xi\eta a + [\xi, \eta]a\| \\
&\ge (|\text{Re } z| - B)\|\eta a\| - T\|a\|_1.
\end{aligned}$$

It follows that

$$\|za - \xi a\|_1 \ge (|\text{Re } z| - B - nT)\|a\|_1$$

for $z \in \mathbb{C}$, $|\text{Re } z| > B + nT$.

Let $\mathfrak{X}_1$ be the completion of $\mathfrak{D}$ relative to the norm

$\|\cdot\|_1$. This completion may be realized concretely as the completion in the Cartesian product $\mathcal{X}^{n+1}$ of the multivariable graph $\{(a,\eta_1 a,\cdots,\eta_n a) : a \in \mathcal{D}\}$. Since we do not know, *a priori*, that the basis operators η_i are closable, the completion $\mathcal{X}_1$ may not naturally embed as a subspace of $\mathcal{X}$. But the mapping $J : \mathcal{D} \longrightarrow \mathcal{X}_1$ defined by

$$Ja := (a,\eta_1 a,\cdots,\eta_n a)$$

identifies $\mathcal{D}$ as a dense subspace of $\mathcal{X}_1$. For every $\xi \in L$, we define the operator $\xi^{(1)}$ in $\mathcal{X}^1$ by

$$\xi^{(1)} Ja = J\xi a, \qquad a \in \mathcal{D}.$$

We shall apply a variant of the Hille-Yosida Theorem [BR 2, vol. I, Thm. 3.2.22] to each of the operators

$$\{\xi^{(1)} : \xi \in M\}.$$

It follows that each $\xi^{(1)}$ is closable, and that $\overline{\xi^{(1)}}$ generates a strongly continuous group of bounded operators in $\mathcal{X}_1$, satisfying

$$\|U_\xi^{(1)}(t)\| \leq \exp((B(\xi)+nT)|t|), \qquad t \in \mathbb{R}.$$

We now apply the converse implication in the classical Hille-Yosida Theorem [HP] to $\xi^{(1)}$, and conclude that

$$(z-\xi^{(1)})J\mathcal{D}$$

is dense in $\mathcal{X}_1$ relative to the $\|\cdot\|_1$-norm for $z \in \mathbb{C}$, $|\mathrm{Re}\, z| > B(\xi)+nT$. Since

$$J(za-\xi a) = zJa - \xi^{(1)} Ja, \qquad a \in \mathcal{D},$$

it follows that condition (GD) in Theorem A3 above is verified. The integrability of L therefore follows from A3 as asserted.

The following perturbation result was used on numerous occasions in Section 8.6.

Let $L \subset \mathrm{End}\,\mathcal{D}$ be a finite-dimensional Lie algebra over $\mathbb{R}$

of operators in a normed linear space, and let $P \subset \text{End}\,\mathfrak{D}$ be a Lie subalgebra consisting of bounded operators. We shall assume that P is normalized by L, i.e., that $[L,P] \subset P$, or more precisely, $[\xi,\eta] = \xi\eta - \eta\xi \in P$ for all $\xi \in L$ and $\eta \in P$.

Let $L +) P$ denote the corresponding semidirect sum Lie algebra. Recall that $L +) P$ is isomorphic to the extension

$$E(P) = \{(\xi,\eta) : \xi \in L,\ \eta \in P\}$$

with bracket:

$$[(\xi,\eta),(\xi',\eta')] = ([\xi,\xi'],[\xi,\eta'] + [\xi',\eta] + [\eta,\eta']),$$

and we have a short exact sequence

$$0 \longrightarrow P \longrightarrow E(P) \longrightarrow L \longrightarrow 0.$$

Let K be a finite-dimensional Lie subalgebra of the (generally) infinite-dimensional Lie algebra $L +) P \simeq E(P)$. We say that K is a *semidirect product perturbation* of L relative to the perturbation class P.

We have

Theorem A5. Let K be a semidirect product perturbation of a given finite-dimensional Lie algebra $L \subset \text{End}\,\mathfrak{D}$ and let $\mathfrak{X}$ be the norm-completion of $\mathfrak{D}$.

Suppose that L is integrable in $\mathfrak{X}$, and that K contains a generic subset. Then it follows that K is also integrable, and that, for each m, the space of C^m vectors for L is contained in that of K.

If the perturbation class $P \subset \text{End}\,\mathfrak{D}$ is restricted further, then the assumption in Theorem A5 that K contain a generic subset may be dropped.

A linear operator η in a normed linear space $\mathfrak{D}$ is said to be *conservative* if each of the two operators $\pm\eta$ is dissipative.

This is know [BR 2, vol.I, new edn.] to be equivalent to the *a priori* estimates:

$$\|a+t\eta(a)\| \geq \|a\|, \qquad a \in \mathfrak{D}, \quad t \in \mathbb{R}.$$

We have

Theorem A6. Let K be a semidirect product perturbation of an integrable Lie algebra $L \subset \text{End}\,\mathfrak{D}$ relative to a perturbation class P which consists of conservative and bounded operators. Then it folows that K is also integrable.

The results of Theorems A5-A6 may be obtained from Theorem A3, and we refer to [JM, Thm. 9] for details.

We conclude with a very recent integrability result from [BGJR].

Theorem A7. Let $(\rho, \mathfrak{g}, \mathfrak{D}, \mathfrak{X})$ be a Lie algebra representation with dense domain in a Banach space. Assume $L = \rho(\mathfrak{g})$ has a basis $\eta_i = \rho(x_i)$, $i = 1, \cdots, n$, satisfying $\|a+t\eta_i a\| \geq m\|a\|$, $a \in \mathfrak{D}$, $t \in \mathbb{R}$, $1 \leq i \leq n$, where m is some constant, $0 < m \leq 1$.

Assume further that the closure of the operator $\sum_{i=1}^{n} \eta_i^2$ generates a strongly continuous semigroup S_t of bounded operators on $\mathfrak{X}$ satisfying

(i) $$S_t\mathfrak{X} \subset \mathfrak{D}, \qquad 0 < t \leq 1,$$

and

(ii) $$\|\eta_i S_t a\| \leq \text{const.}\|a\| t^{-\frac{1}{2}}, \qquad a \in \mathfrak{X},\ 0 < t \leq 1,\ 1 \leq i \leq n.$$

Then it follows that L is integrable.

Remark. Note that condition (ii) states that the operator S_t is bounded from $\mathfrak{X}$ into $\mathfrak{X}_1$ with norm of type $O(t^{-\frac{1}{2}})$.

We finally note that the integrability theorems A1-A4, and

Theorem A7 also are valid in the reverse implication.

The other implications are interesting and nontrivial, and the reader is referred to [JM] and [BGJR] for details.

REFERENCES

[Aa] J.F. Aarnes, Fröbenius reciprocity of differentiable representations, Bull. Amer. Math.Soc. 80(1974), 337-340.

[AM] R. Abraham and J.E. Marsden, Foundation of Mechanics 2nd edn., The Benjamin Cummings Publ. Co., Reading, Mass., 1978.

[Ad] J.F. Adams, Lectures in Lie Groups, Benjamin, New York, 1969.

[Ado] I.D. Ado, Lie groups, Amer. Math. Soc. Transl., Series I, 9(1962), 308.

[AP] C.A. Akemann and G.K. Pedersen, Central sequences and inner derivations of separable C^*-algebras, Amer. J. Math. 101(1979), 1047-1061.

[AG] N.I. Akhiezer and I.M. Glazman, Theory of Linear Operators in Hilbert Spaces I, F. Ungar Publ. Co., New York,, 1961, pp. 147 (translated from the Russian).

[Am] W. Ambrose, Spectral resolution of groups of unitary operators, Duke Math. J. 11(1944), 589-595.

[AR] R.L. Anderson and R. Raczka, Coupling Problem for U(p,q) Ladder Representations, II. Limitation on the Coupling of U(p,q) Ladders, Proc. Roy. Soc. London A302(1968), 501-508.

[An] E. Angelopoulos, On unitary irreducible representations of semidirect products with nilpotent normal subgroup, Ann. Inst. H. Poincaré, Sect. A (N.S.) 18 (1973), 39-55.

[ABF] E. Angelopoulos, F. Bayen and M. Flato, On the localizability of massless particles, Phys. Ser. 9(1974), 173-183.

[Ar 1] H. Araki, Bogoliubov automorphisms and Fock representations of canonical anticommutation relations, in Operator Algebras and Mathematical Physics, Contemporary Mathematics, 62 Amer. Math. Soc., Providence, R.I.,1986.

[Ar 2] __________, On quasifree states of CAR and Bogoliubov automorphisms, Publications of RIMS, Kyoto Univ. 6 (1970), 385-442.

[Ar 3] __________, On representations of the canonical commutation relations, Comm. Math. Phys. 20(1971), 9-25.

[AW] H. Araki and E.J. Woods, Representation of canonical commutation relations describing a nonrelativistic free Bose gas, J. Math. Phys. no. 5, 4(1963), 637-641.

[Arv 1] W.B. Arveson, Subalgebras of C^*-algebras, Acta Math. 123(1969), 141-224.

[Arv 2] ________, On groups of automorphisms of operator algebras, J. Funct. Anal. 15(1974), 217-243.

[Arv 3] ________, Notes on extensions of C^*-algebras, Duke Math. J. 44(1977), 329-355.

[Arv 4] ________, An Invitation to C^*-algebras, Springer-Verlag, New York, 1976.

[At 1] M.F. Atiyah, K-theory and reality, Quart. J. Math., Oxford 17(1966), 367-386.

[At 2] ________, K-theory, Benjamin, New York, 1967.

[At 3] ________, Global Theory of Elliptic Operators, Proc. Int. Conf. on Func. Anal. and Related Topics, Tokyo, 1969.

[At 4] ________, Instantons in two and four dimensions, Commun. Math. Phys. 93(1984), 437-451.

[At 5] ________ & R. Bott, A Lefschetz fixed point formula for elliptic complexes: II. Applications, Ann. of Math. 88(1968), 451-491.

[At 6] ________ & R. Bott, The Yang-Mills equations over Riemann surfaces, Philos. Trans. Roy. Soc. London 308A(1982), 523-615.

[At 7] ________ & A.N. Pressley, Convexity and loop groups In: Arithmetic and Geometry; papers dedicated to I.R. Shafarevich on the occasion of his sixtieth birthday, Vol. II: Geometry, Birkhäuser, 1983.

[AK] L. Auslander & B. Kostant, Polarization and unitary representations of solvable Lie groups, Inv. Math. 14(1971), 255-354.

[Barg] V. Bargmann, Irreducible unitary representations of the Lorentz group, Ann. Math. 48(1947), 568-640.

[BB] A.O. Barut & W. Brittin, Lectures in Theoretical Physics 10: de Sitter and Conformal Groups and Their Applications, Colorado Associated Univ. Press, 1971.

[B Rac] A.O. Barut & R. Raczka, Theory of Group Representations and Applications, Polish Scientific Publ., Warsaw, 1977.

[Bat 1] C.J.K. Batty, Derivations on compact spaces, Proc. London Math. Soc. (3) 42(1981), 299-330.

[Bat 2] ________, Local operators and derivations on C^*-algebras, Trans. Amer. Math. Soc. 287(1985), 343-352.

[BG] R.W. Beals & P.C. Greiner, Calculus on Heisenberg manifolds, Princeton, 1986 (Ann. Math. Studies) (to appear).

[BGV] R.W. Beals, P.C. Greiner & J. Vauthier, The Laguerre calculus on the Heisenberg group in R.A. Askey, *et al.*, "Special functions", Group theoretical aspects and applications", Reidel, Dordrecht, 1984, 189-216.

[BGGV] R.W. Beals, B. Gaveau, P.C. Greiner & J. Vauthier, The Laguerre calculus on the Heisenberg group, Bull. Sci. Math. (2e série) 110(1986), 225-288.

[Be] E. Beggs, De Rham's theorem for infinite dimensional manifolds (to be publilshed in Quart J. Math., Oxford).

[BF] C. Berg & G. Forst, Potential Theory on Locally Compact Abelian Groups, Springer-Verlag, Berlin-Heidelberg-New York, 1975.

[Berez] F.A. Berezin, Some remarks on the associative envelope of a Lie algebra (in Russian), Funksion. Anal. Priloz. 1(1967), 1-14.

[Bern] P. Bernat, Sur le corps des quotients de l'algèbre enveloppante d'une algèbre de Lie, C.R. Acad. Sci. Paris 254(1962), 1712-1714.

[BGG 1] I.N. Bernstein, I.M. Gelfand & S.I. Gelfand, Differential oprators on the base affine space and a study of g-modules, 21-64. In: Lie groups and their representations, Summer School of the Bolyai Janos Math. Soc., ed., I.M. Gelfand, Wiley, New York, 1975.

[BGG 2] ________, The structure of representations generated by vectors of largest weight (in Russian), Funksion. Anal. Priloz. 5(1971), 1-19.

[BGG 3] ________, Differential operators on the base affine space and a study of g-modules. In: Publ. of 1971 Summer School in Math., ed., I.M. Gelfand, Janos Bolyai Math. Soc., Budapest, 21-64.

[Bi 1] G.D. Birkhoff, Singular points of ordinary differential equations, Trans. Amer. Math. Soc. 10(1909), 436-470.

[Bi 2] __________, Equivalent singular points of ordinary linear differential equations, Math. Ann. 74(1913), 34-139.

[Bla] B. Blackadar, K-theory for Operator Algebras, Springer Verlag, New York, 1986.

[Blat 1] R.J. Blattner, On induced representations, Amer. J. Math. 83(1961), 79-98.

[Blat 2] __________, On induced representations, II. Infinitesimal induction, Amer. J. Math. 83(1961), 499-512.

[Blat 3] __________, Positive definite measures, Proc. Amer. Math. Soc. 14(1963), 423-448.

[Blat 4] __________, Induced and produced representations of Lie algebras, Trans. Amer. Math. Soc. 144(1969), 457-474.

[BO'R] A. Böhm & L. O'Raifeartaigh, A mass-splitting theorem for general definitions of mass, Phys. Rev. 171(1968), 1698-1701.

[Bony] J.M. Bony, Principe du maximum, inégalité de Harnack et unicité du problème de Cauchy pour les opérateurs elliptiques dégénérés, Ann. Inst. Fourier, Grenoble, 19(1969), 277-304.

[BM] A. Borel & G.D. Mostov, On semisimple automorphisms of Lie algebras, Ann. of Math. 61(1955), 389-405.

[BW] A. Borel & N. Wallach, Continuous cohomology, discrete subgroups, and representation of reductive groups, Ann. of Math. Studies, 94, Princeton Univ. Press, Princeton, 1980.

[Bott 1] R. Bott, An application of Morse theory to the topology of Lie groups, Bull. Soc. Math. France 84(1956), 251-281.

[Bott 2] __________, Homogeneous vector bundles, Ann. of Math. 66(1957), 203-248.

[Bott 3] __________, The space of loops on a Lie group, Michigan Math. J. 5(1958), 35-61.

[Bott 4] __________, The stable homotopy of the classical groups, Ann. of Math. 70(1959), 313-317.

[Bott 5] __________ & L.W. Tu, Differential Forms in Algebraic Topology, Springer-Verlag, New York, 1982.

[Bou 1] N. Bourbaki, Espaces Vectoriels Topologiques, Hermann, Paris, 1964.

[Bou 2] __________, Groupes et algèbres de Lie, Ch.1, Hermann, Paris, 1960; Ch. 2 et 3, Hermann, Paris, 1972; Ch. 4, 5 et 6, Hermann, Paris, 1968; Ch. 7 et 8, Hermann, Paris, 1975; Ch. 9, Masson, Paris, 1982.

[Bou 3] __________, Integration (rev. edn.), Hermann, Paris 1965.

[Bouw] I.Z. Bouwer, Standard representations of simple Lie algebras, Canad. J. Math. 20(1968), 344-361.

[Bra 1] O. Bratteli, Inductive limits of finite-dimensional C^*-algebras, Trans. Amer. Math. Soc. 171(1972), 195-234.

[Bra 2] __________, Structure spaces of approximately finite-dimensional C^*-algebras, J. Funct. Anal. 16(1974), 192-204.

[Bra 3] __________, On dynamical semigroups and compact group actions. In: Springer Lecture Notes in Mathematics 1055 (1984), 46-61.

[Bra 4] __________, A remark on extensions of quasifree derivations on the CAR-algebra, Letters Math. Phys. 6 (1982), 499-504.

[Bra 5] __________, Derivations and free group actions, Proc. of Conf. on Operator Algebras and Mathematical Physics, Iowa, 17-21/6-85, P.E.T. Jorgensen and P. Muhly, eds., Contemporary Mathematics vol., 62, AMS, Providence, 1986.

[Bra 6] __________, Unbounded derivations and C^*-dynamics, Proc. Fifth Symposium of Pure and Applied Mathematics, Kangneung College, Kangwon, Korea, 22-27/7-85, Kun Soo Chang and Yong Mooon Park, eds.

[Bra 7] __________, Derivations, dissipations, and group actions on C^*-algebras, Springer-Verlag, Heidelberg, New York, 1987 (LNM 1229).

[BEE] O. Bratteli, G.A. Elliott & D.E. Evans, Locality and differential operators on C^*-algebras, J. Diff. Equations 64(1986), 221-273.

[BEll] O. Bratteli & G.A. Elliott, Structure spaces of approximately finite-dimensional C^*-algebras II, J. Func. Anal. 30(1978), 74-82.

[BER 1] O. Bratteli, G.A. Elliott & D.W. Robinson, The characterization of differential operators by locality: Classical flows, Compositio Math. 58(1986), 279-319.

[BER 2] __________, Strong topological transitivity and C^*-dynamical systems, J. Math. Soc. Japan 37(1985), 115-133.

[BER 3] __________, The characterization of differential operators by locality: Dissipations and ellipticity, Publ. RIMS Kyoto Univ. 21(1985), 1031-1049.

[BER 4] __________, The characterization of differential operators by locality: C^*-algebras of type I, J. Operator Theory (to appear).

[BEv 1] O. Bratteli & D.E. Evans, Dynamical semigroups commuting with compact abelian actions, Ergod. Thy. & Dynam. Sys. 3(1983), 187-217.

[BEv 2] __________, Derivations tangential to compact groups: The nonabelian case, Proc. London Math. Soc. 52 (1986), 369-384.

[BGoo] O. Bratteli & F.M. Goodman, Derivations tangential to compact group actions: Spectral conditions in the weak closure, Canad. J. Math. 37(1985), 160-192.

[BElGJ] O. Bratteli, G.A. Elliott, F.M. Goodman & P.E.T. Jorgensen, Smooth Lie group actions on non-commutative tori, Preprint 1986.

[BElGJ] __________, Smooth Lie group actions on the irrational rotation algebra, Preprint 1987, Univ. of Iowa (submitted).

[BEvGJ] O. Bratteli, D.E. Evans, F.M. Goodman & P.E.T. Jorgensen, A dichotomy for derivations on $\mathcal{O}_n$, Publ. of the RIMS, Kyoto U. 22(1986), 103-117.

[BEJ] O. Bratteli, G.A. Elliott & P.E.T. Jorgensen, Decomposition of unbounded derivations into invariant and approximately inner parts, J. reine angew. Math. (Crelle's J.) 346(1984), 166-193.

[BGJ] O. Bratteli, F. M. Goodman & P.E.T. Jorgensen, Unbounded derivations tangential to compact groups of automorphisms, II, J. Func. Anal. 61(1985), 247-289.

[BGJR] O. Bratteli, F.M. Goodman, P.E.T. Jorgensen & D.W. Robinson, The heat semigroup and integrability of Lie algebras, Preprint 1987, Austral. Natl. Univ. (submitted).

[BKJR] O. Bratteli, A. Kishimoto, P.E.T. Jorgensen & D.W. Robinson, A C^*-algebraic Shoenberg theorem, Ann. Inst. Fourier Grenoble 33(1984), 155-187.

[BJ 1] O. Bratteli & P.E.T. Jorgensen, Unbounded *-derivations and infinitesimal generators on operator algebras. In: Proceedings of the AMS Summer Institute on Operator Algebras, Kingston, Ontario, 1980. Symposia in Pure Math. (38) 2(1982), 353-365.

[BJ 2] __________, Unbounded derivations tangential to compact groups of automorphisms, J. Func. Anal. 48(1982), 107-133.

[BJ 3] __________, Derivations commuting with abelian gauge actions on lattice systems, Commun. Math. Phys. 87 (1982), 353-364.

[BK 1] O. Bratteli & A. Kishimoto, Automatic continuity of derivations on eigenspaces, Proc. Conf. on Operator Algebras and Mathematical Physics, Iowa 1985, P.E.T. Jorgensen and P. Muhly, eds., Contemporary Math. vol. 62, AMS, Providence (1986).

[BK 2] __________, Derivations and free group actions on C^*-algebras, J. Operator Theory 15(1986), 377-410.

[BKR] O. Bratteli, A. Kishimoto & D.W. Robinson, Embedding product type actions into C^*-dynamical systems, J. Func. Anal. (to appear).

[BR 1] O. Bratteli & D.W. Robinson, Unbounded derivations on C^*-algebras, I and II, Comm. Math. Phys. 42(1975), 253-268; 46(1976), 11-30.

[BR 2] __________, Operator Algebras and Quantum Statistical Mechanics, I and II, Springer-Verlag, New York, 1979/81 (vol. I in new edn. 1987).

[BDF] L.G. Brown, R.G. Douglas & P.A. Fillmore, Extensions of C^*-algebras and K-homology, Ann. of Math. 105 (1977), 265-324.

[CE] A.L. Carey & D.E. Evans, On an automorphic action of U(n,1) on $\mathcal{O}_n$, J. Func. Anal (to appear).

[CRu] A.L. Carey & S.N.M. Ruijsenaars, On fermion gauge groups, current algebras, and Kac-Moody algebras, Preprint, A.N.U., Canberra, 1985.

[C 1] E. Cartan, Complément au mémoire sur la géométire des groupes simples, Ann. Math. Pura Appl. 5(1928), 253-260.

[C 2] __________, Sur la détermination d'un système orthogonal complet dans un espace de Riemann symétrique clos, Rend. Circ. Mat. Palermo 53(1929), 217-252.

[C 3] __________, Groupes simples clos et ouverts et géométrie Riemannienne, J. Math. Pures Appl. 8(1929), 1-33.

[C 4] __________, Leçons sur la géométrie des espaces de Riemann, 2d edn., Gauthier-Villars, Paris (1951), pp. 378.

[C 5] __________, The Theory of Spinors, Hermann, Paris (1966), pp. 157.

[Ca 1] P. Cartier, On H. Weyl's character formula, Bull. Amer. Math. Soc. 67(1961), 228-230.

[Ca 2] __________, Quantum mechanical commutation relations and theta functions, Proc. Symp. Pure Math. 9(1966), 361-383.

[CD] P. Cartier & J. Dixmier, Vecteurs analytique dans les représentations de groupes de Lie, Amer. J. Math. 80 (1958), 131-145.

[Cas] H. Casimir, Über die Konstruktion Einer zu den Irreduziblen Darstellungen Halbeinnfacher Kontinuierlicher Gruppen gehörigen Differentialgleichung, Proc. Roy. Acad. Amsterd. 34(1931), 844-846.

[Che] P. Chernoff, Note on product formulas for operator semigroups, J. Funct. Anal. 2(1968), 238-242.

[CM] P. Chernoff & J.E. Marsden, On continuity and smoothness of group actions, Bull. Amer. Math. Soc. 76 (1970), 1044.

[Ch] C. Chevalley, Theory of Lie Groups, vol. I, Princeton Univ. Press, Princeton, N.J., 1946, pp. 217.

[CT] A. Chodos & C. Thorn, Making the massless string massive, Nuclear Phys. B72(1974), 509-522.

[Cob] L. Coburn, The C^*-algebra generated by an isometry, Bull. Amer. Math. Soc. 73(1967), 722-726.

[CL] E.A. Coddington & N. Levinson, Theory of ordinary differential equations, McGraw-Hill, New York, 1955.

[Co] P.M. Cohn, Lie Groups, Cambridge Univ. Press, Cambridge, 1967, pp. 164.

[Con 1] A. Connes, Sur la théorie non-commutative de l'integration, Lect. Notes in Math. 725, 19-143, Springer-Verlag, Berlin, N.Y., 1979.

[Con 2] __________, A survey of foliations and operator algebras, In: Operator Algebras and Applications, Proc. Symp. Pure Math. (38) vol. 1, Amer. Math. Soc., Providence, R.I., 1980, 521-632.

[Con 3] __________, C^*-algèbres et géométrie différentielle, C.R. Acad. Sci. Paris 290(1980), 599-604.

[Con 4] __________, Spectral sequences and homology of currents for operator algebras, Tagungsbericht 42/81, Oberwolfach 1981, 4-5.

[Con 5] __________, Cohomologie cyclique et foncteurs Ext^n, C.R. Acad. Sci. Sér. A, Paris 296(1983), 953-958.

[Con 6] __________, Cyclic cohomology and the transverse fundamental class of a foliation, Preprint I.H.E.S. M/84/7 (1984).

[Con 7] __________, Non-commutative differential geometri, I. The Chern character in K-homology: II. De Rham homology and non-commutative algebra, Publ. Math. IHES 62 (1985), 257-360.

[CR] A. Connes & M.A. Rieffel, Yang-Mills for noncommutative two-tori, In: "Operator Algebras and Mathematical Physics," Contemporary Math., vol. 62 (eds. P.E.T. Jorgensen & P.S. Muhly), Amer. Math. Soc., Providence, R.I., 1986.

[CV] N. Conze & M. Vergne, Idéaux primitifs des algèbres enveloppantes des algèbres de Lie résolubles, C.R. Acad. Sci. Paris (A) 272(1971), 985-988.

[Cu 1] J. Cuntz, Simple C^*-algebras generated by isometries, Commun. Math. Phys. 57(1977), 173-185.

[Cu 2] __________, Automorphisms of certain simple C^*-algebras, In: Quantum Fields—Algebras, Processes, L. Streit, ed., Springer-Verlag, Wien-New York, 1980, 187-196.

[CEℓGJ] J. Cuntz, G.A. Elliott, F.M. Goodman & P.E.T. Jorgensen, On the classification of non-commutative tori. II, C.R. Math. Rep. Acad. Sci. Canada 7(1985), 189-194.

[Dav] E.B. Davies, A generation theorem for operators commuting with group actions, Math. Proc. Cambridge Phil. Soc. 96(1984), 315-322.

[Di 1] P.A.M. Dirac, The electron wave equation in de Sitter space, Ann. of Math. 36(1935), 657-669.

[Di 2] __________, The Principles of Quantum Mechanics, 3rd edn., Clarendon Press, Oxford, 1947.

[Dix 1] J. Dixmier, Sur la relation i(PQ-QP) = I, Compositio Math. 13(1958), 263-270

[Dix 2] __________, Sur les représentations unitaires des groupes de Lie nilpotents, II, Bull. Soc. Math. France 85(1957), 325-388.

[Dix 3] __________, Sur l'algébre enveloppante d'une algèbre de Lie nilpotente, Arch. Math. 10(1959), 321-326.

[Dix 4] __________, Sur les représentations des groupes de Lie nilpotents, IV, Canad. J. Math. 11(1959), 321-344.

[Dix 5] __________, Sur les représentations unitaires des groupes de Lie nilpotents, V, Bull. Soc. Math. France 87(1959), 65-79.

[Dix 6] __________, Sur les représentations unitaires des groupes de Lie nilpotents, VI, Canad. J. Math. 12 (1960), 324-352.

[Dix 7] __________, Représentations intégrables du groupe de Sitter, Bull. Soc. Math. France 89(1961), 9-41.

[Dix 8] __________, Les C^*-algébres et leurs représentations Gauthier-Villars, Paris, 1964, pp. 382.

[Dix 9] __________, Représentations irréductibles des algébres de Lie nilpotents, An. Acad. Brasil. Ci. 35 (1963), 491-519.

[Dix 10] __________, Représentations irréductibles des algébres de Lie résolubles, J. Math. Pures Appl. 45 (1966), 1-66.

[Dix 11] __________, Sur le dual d'un groupe de Lie nilpotent, Bull. Sci. Math. 90(1966), 113-118.

[Dix 12] __________, Sur le centre de l'algèbre enveloppante d'une algèbre de Lie, C.R. Acad. Sci. Paris (A) 265 (1967), 408-410.

[Dix 13] __________, Sur les représentations unitaires des groupes de Lie nilpotents, I, Amer. J. Math. 81 (1969), 160-170.

[Dix 14] __________, Les algébres d'operateurs dans l'espace hilbertien, Gauthier-Villars, Paris, 1969, pp. 367.

[Dix 15] __________, Représentations induites des algébres de Lie, L'Enseignement Math. 16(1970), 1699-175.

[Dix 16] __________, Sur les algébres de Weyl, II, Bull. Sci. Math. 94(1970), 2899-301.

[Dix 17] __________, Sur les représentations induites des algèbres de Lie, J. Math. Pures Appl. 50(1971), 1-24.

[Dix 18] __________, Polarisations dans les algèbres de Lie, Ann. Sci. Ecole Norm. Sup. 4(1971), 321-336.

[Dix 19] __________, Idéaux primitifs dans l'algèbre enveloppante d'une algèbre de Lie semi-simple complexe; (a) C.R. Acad. Sci. Paris (A) 271(1970), 134-136; (b) *Ibid.* (A) 272(1971), 1628-1630.

[Dix 20] __________, Idéaux maximaux dans l'algèbre enveloppante d'une algèbre de Lie semi-simple complexe, C.R. Acad. Sci. Paris (A) 274(1972), 228-230.

[Dix 21] __________, Sur les homomorphismes d'Harish-Chandra, Inv. Math. 17(1972), 167-176.

[Dix 22] __________, Quotients simples de l'algèbre envelop-pange de sl_2, J. Algebra 24(1973), 551-564.

[Dix 23] __________, Algèbres Enveloppantes, Gauthier-Villars, Paris, (1974), pp. 349.

[Dix 24] __________, Enveloping Algebras, North-Holland, Amsterdam, 1974. (Translation of [Dix 23].)

[DM] J. Dixmier & P. Malliavin, Factorisations de functions et de vecteurs indefiniment differentiables, Bull. Sci. Math. 2e série 102(1978), 305-330.

[DoMe] H.D. Doebner & O. Melsheimer, On representations of Lie algebras with unbounded generators: I. Physical consequences, Il Nuvo Cimento, v. ILA, N.1 (1967), 73-97.

[D] L. Dolan, Kac-Moody algebra is hidden symmetry of chiral models, Phys. Rev. Lett. 47(1981), 1371-1374.

[DR] S. Doplicher & J.E. Roberts, Compact Lie groups associated with endomorphisms of C^*-algebras, Bull. Amer. Math. Soc. 11(1984), 333-338.

[Do] Dorfmeister, G. & J., Classifications of certain pairs of operators (P,Q) satisfying [P,Q] = -i, J. Funct. Anal.57(1984), 301-328.

[Dou] R.G. Douglas, Banach Algebra Techniques in Operator Theory, Academic Press, New York, 1972.

[Du 1] M. Duflo, Sur les représentations irréductibles des algèbres de Lie contenant un idéal nilpotent, C.R. Acad. Sci. Paris (A) 270(1970), 504-506.

[Du 2] __________, Sur les extensions des représentations irréductibles des groupes de Lie nilpotents, Ann. Sci. Ecole Norm. Sup. 5(1972), 71-120.

[Du 3] __________, Certaines algèbres de type fini sont des algèbres de Jacobson, J. Algebra 27(1973), 358-365.

[DV] M. Duflo & M. Vergne, Une propriété de la représentation coadjointe d'une algèbre de Lie, C.R. Acad. Sci. Paris (A) 268(1969), 583-585.

[DS 1] N. Dunford & J.T. Schwartz, Linear Operators, Part I, Interscience, New York, 1967.

[DS 2] __________, Linear Operators, Part II, Interscience, New York, 1963.

[Eff] E.G. Effros, Dimensions and C^*-algebras, C.B.M.S. Lecture Notes, No. 46, Amer. Math. Soc., Providence, R.I., 1981.

[EH] E. Effros & F. Hahn, Locally compact transformation groups and C^*-algebras, Mem. Amer. Math. Soc. 75 (1967).

[E] Y.V. Egorov, Subelliptic operators, Russian Math. Survey (2) 30(1975), 59-118; (3) 30(1975), 55-105.

[Ell 1] G.A. Elliott, Some C^*-algebras with outer derivations, III, Ann. of Math. 106(1977), 121-143.

[Ell 2] __________, On the K-theory of the C^*-algebra generated by a projective representation of a torsion-free discrete abelian group, Operator Algebras and Group Representation. I, New York, 1983, 157-184.

[Ev 1] __________, D.E. Evans, Irreducible quantum dynamical semigroups, Commun. Math. Phys. 54(1977), 293--297.

[Ev 2] __________, On $\mathcal{O}_n$, Publ. RIMS Kyoto Univ. 16(1980), 915-927.

[Ev 3] __________, The C^*-algebras of topological Markov chains, Tokyo Metropolitan Univ. Lecture Notes, 1983.

[EK] D.E. Evans & A. Kishimoto, Duality for automorphisms on a compact C^*-dynamical system, Preprint, 1986.

[FP] C. Feffermann & D. Phong, Pseudo-differential operators with positive symbols, Séminaire Goulaouic-Meyer-Schwartz, 1980-81.

[F 1] J.M.G. Fell, Non-unitary dual spaces of groups, Acta Math. 114(1965), 267-310.

[F 2] __________, An extension of Mackey's method to Banach *-algebraic bundles, Mem. Amer. Math. Soc. 90(1969), pp. 168.

[Fe] W. Feller, On the generation of unbounded semigroups of bounded linear operators, Ann. Math. (2) 58(1953), 166-174.

[Flato] M. Flato, Theory of analytic vectors and applications, In: Mathematical Physics and Physical Mathematics, K. Maurin & R. Raczka, eds., Reidel—PWN, Dordrecht-Warsaw, 1976, 231-250.

[FS 1] M. Flato & D. Sternheimer, Remarks on the connection between external and internal symmetries, Phys. Rev. Letters 15(1965), 934-935.

[FS 2] __________, On the connection between external and internal symmetries of strongly interacting particles, J. Math. Phys. 7(1966), 1932-1958.

[FS 3] __________, Local representations and mass spectrum, Phys. Rev. Lett. 16(1966), 1185-1186.

[FS 4] __________, Poincaré partially integrable local representations and mass spectrum, Commun. Math. Phys. 12(1969),296-303

[FS 5] __________, On an infinite-dimensional group, Comm. Math. Phys. 14(1969), 5-12.

[FS 6] M. Flato & J. Simon, Separate and joint analyticity in Lie group representations, J. Funct. Anal. 13 (1973), 268-276.

[FSS 1] M. Flato, J. Simon & D. Sternheimer, Conformal covariance of field equations, Ann. of Phys. (N.Y.) 61(1973), 78-97.

[FSS 2] __________, Sur l'integrabilité des représentations antisymétriques des algèbres de Lie compactes, C.R. Acad. Sc. Paris 277(1973), 939-942.

[FSSS] M. Flato, J. Simon, H. Snellman & D. Sternheimer, Simple facts about analytic vectors and integrability, Ann. Scient. de l'École Norm. Sup. 5(1972), 423-434.

[Fo 1] G.B. Folland, A fundamental solution for a subelliptic operator, Bull. Amer. Math. Soc. 79(1973), 373-376.

[Fo 2] __________, Subelliptic estimates and function spaces on nilpotent Lie groups, Ark. mat. 13(1975), 161-208.

[Fo 3] __________, On the Rothschild-Stein lifting theorem, Comm. in P.D.E. 212(1977), 165-191.

[FS 1] G.B. Folland & E.M. Stein, Estimates for the $\bar{\partial}_b$ complex and analysis on the Heisenberg group, Commun. Pure Appl. Math. 27(1974), 429-522.

[FS 2] __________, Hardy Spaces on Homogeneous Groups, Princeton University Press, Princeton, 1982.

[Fr 1] I.B. Frenkel, Two constructions of affine Lie algebra representations and boson-fermion correspondence in quantum field theory, J. Funct. Anal. 44(1981), 259-327.

[Fr 2] __________, Representations of affine Lie algebras, Hecke modular forms and Korteweg-de Vries type equations, pp. 71-110. In: Lie algebras and related topics, Proceednigs of a Conference held at New Brunswick, N.J., May 29-31, 1981. Lecture Notes in Math. Vol. 933, Springer-Verlag, N.Y., 1982.

[Fr 3] __________, Orbital theory for affine Lie algebras, Invent. Math. 77(1984), 301-352.

[Fr 4] __________, Representations of Kac-Moody algebras and dual resonance models, Lectures in Appl. Math. Vol. 21, 325-353, American Mathematical Society, 1985.

[FK] I.B. Frenkel & V.G. Kac, Basic representations of affine Lie algebras and dual resonance models, Invent. Math. 62(1981), 23-66.

[FV] H. Freudenthal & H. de Vries, Linear Lie Groups, Academic Press, New York, 1969.

[Frö] J. Fröhlich, Application of commutator theorems to the integration of representations of Lie algebras and commutation relations, Comm. Math. Phys. 54 (1977), 135-150.

[Fu 1] B. Fuglede, On the relation PQ-QP = -iI, Math. Scand., 20(1967), 79-88.

[Fu 2] __________, Conditions for two self-adjoint operators to commute or to satisfy the Weyl relation, Math. Scand. 51(1982), 163-178.

[Ga] P. Gabriel, Représentations des algèbres de Lie résolubles, Sém. Bourbaki 347(1968-1969), 1-22, Lecture Notes in Math., Springer-Verlag, Berlin.

[Gan] R. Gangolli, On the Plancherel formula and the Paley-Wiener theorem for spherical functions on semi-simple Lie groups, Ann. of Math. 93(1971), 150-155.

[Gå 1] L. Gårding, Note on continuous representations of Lie groups, Proc. Nat. Acad. Sci. USA 33(1947), 331-332.

[Gå 2] __________, Vecteurs analytiques dans les représentations des groupes de Lie, Bull. Soc. Math. France 88 (1960), 73-93.

[Gar 1] H. Garland, The arithmetic theory of loop algebras, J. Algebra 53(1978), 480-551.

[Gar 2] __________, The arithmetic theory of loop groups, Pub. Math. I.H.E.S. 52(1980), 5-136.

[Gav 1] B. Gaveau, Solutions fondamentales, représentations, et estimées sous-elliptiques pour les groupes nilpotent d'ordre 2, C.R. Acad. Sc. Paris 282(1976), A 563-566.

[Gav 2] __________, Principe de moindre action, propagation de la chaleur et estimées sous elliptiques sur certaines groupes nilpotents, Acta Math. 139(1977), 95-153.

[GGV] B. Gaveau, P.C. Greiner & J. Vauthier, Intégrals de Fourier quadratiques et calcul symbolique exact sur le groupe d'Heisenberg, J. Funct. Anal. 1986 (to appear).

[Ge] S. Gelbart, Fourier analysis on Matrix space, Mem. Amer. Math. Soc. 108, Providence, R.I., 1971, pp. 71.

[G 1] I.M. Gelfand, On one-parametrical groups of operators in a normed space, Dokl. Akad. Nauk SSSR 25(1939), 713-718 (Russian).

[G 2] __________, Spherical functions in symmetric Riemann spaces, Dokl, Akad. Nauk SSSR 70(1950), 5-8 (Russian).

[G 3] __________, The center of an infinitesimal group ring, Math. Sb. 26(1950), 103-112 (Russian).

[G 4] __________, Automorphic functions and the theory of representations, Trudy Moskov. Mat. Obsc. 12(1963), 389-412 (Russian).

[G 5] __________, (ed.) Lie groups and their representations, Summer School of the Bolyai János Math. Soc. Budapest, Aug. 16/Sept. 3, 1970, Academiai Kiadó, Budapest and A. Hilger Ltd., London, pp. 726.

[G 6] __________, (ed.) Representation Theory, London Math. Soc. Lecture Note Series 69, Cambridge Univ. Press, Cambridge, 1982.

[GGVe] I.M. Gelfand, M.I. Graev & A.M. Vershik, Representations of the group of smooth mappings of a manifold into a compact Lie group, Compositio Math. 35(1977), 299-334.

[GK 1] I.M. Gelfand & A.A. Kirillov, Sur les corps liés aux algèbres enveloppantes des algèbres de Lie, Publ. Inst. Hautes Etudes Sci. 31(1966), 5-19.

[GK 2] __________, On the structure of the field of quotients of the enveloping algebra of a semi-simple Lie algebra (in Russian), Dokl. Akad. Nauk SSSR 180 (1968), 775-777.

[GK 3] __________, The structure of the enveloping field of a semi-simple Lie algebra (in Russian), Funktion. Anal. Prilo. 3(1969), 7-26.

[GMS] I.M. Gelfand, R.A. Minlos & Z.Ya. Sapiro, Representations of the Rotation Group and of the Lorentz Group and Their Applications, MacMillan, New York, 1963, pp. 368.

[GN 1] I.M. Gelfand & M.A. Naimark, Unitary representations of the Lorentz group, Izv. Akad. Nauk SSSR, Math. Ser. 11(1947), 411-504 (Russian).

[GN 2] __________, Unitary representations of the classical groups, Trudy Mat. Inst. Steklov 36(1960), Moskov-Leningrad, pp. 288 (Russian).

[GN 3] __________, Unitary representation of an unimodular group containing the identity representation of the unitary subgroup, Trudy Moskov. Mat. Obsc. 1(1952), 423-474 (Russian).

[GN 4] __________, Unitäre Darstellungen der klassischen Gruppen, Akademie Verlag, Berlin, 1957, pp. 333.

[GP 1] I.M. Gelfand & V.Ya. Ponomarev, Indecomposable representations of the Lorentz group, Usp. Mat. Nauk. 23(1968), 3-60.

[GP 2] __________, Remarks on the classification of a pair of commuting linear transformation in a finite-dimensional space, Function. Anal. Prilozen. 3(1969), 81-82.

[GV] I.M. Gelfand & N.Ya. Vilenkin, Generalised Functions, vol. 4, Academic Press, New York, 1964.

[GGVi] I.M. Gelfand, M.I. Graev & N.Ya. Vilenkin, Generalized functions, vol. 5, Integral Geometry and Representation Theory, Academic Press, New York, 1966, pp. 449.

[Gel 1] D. Geller, Local solvability and homogeneous distributions on the Heisenberg group, Comm. Partial Diff. Equa. (5) 5(1980), 475-560.

[Gel 2] __________, Liouville's theorem for homogeneous groups, Comm. in P.D.E. (to appear).

[Go] R. Godement, A theory of spherical functions, I, Trans. Amer. Math. Soc. 73(1952), 496-556.

[Goo 1] R.W. Goodman, Analytic and entire vectors for representations of Lie groups, Trans. Amer. Math. Soc. 143 (1969), 55-76.

[Goo 2] __________, Analytic domination by fractional powers of a positive operator, J. Funct. Anal. 3(1969), 246-264.

[Goo 3] __________, One-parameter groups generated by operators in an enveloping algebra, J. Func. Anal. 6 (1970), 218-236.

[Goo 4] __________, Complex Fourier analysis on nilpotent Lie groups, Trans. Amer. Math. Soc. 160(1971), 373-391.

[Goo 5] __________, Differential operators of infinite order on a Lie group, I, J. Math. Mech. 19(1970), 879-894.

[Goo 6] __________, Differential operators of infinite order on a Lie group, II, Indiana Univ. Math. J. 21(1971), 383-409.

[Goo 7] __________, Some regularity theorems for operators in an enveloping algebra, J. Differential Equations 10(1971), 448-470.

[Goo 8] __________, On the boundedness and unboundedness of certain convolution operators on nilpotent Lie groups, Proc. Amer. Math. Soc. 39(1973), 409-413.

[Goo 9] __________, Nilpotent Lie groups, Lecture Notes in Mathematics vol. 562, Springer-Verlag, Heidelberg, 1976.

[Goo 10] __________, Singular integral operators on nilpotent Lie groups, Ark. Mat. (1) 18(1980), 1-11.

[GW 1] R.W. Goodman & N.R. Wallach, Structure and unitary cocycle representations of loop groups and the group of diffeomorphisms of the circle, J. Reine Angew. Math. 347(1984), 69-133.

[GW 2] __________, Projective unitary positive-energy representations of Diff(S^1), J. Funct. Anal. 63(1985), 299-321.

[GJ 1] F. Goodman & P.E.T. Jorgensen, Unbounded derivations commuting with compact group actions, Comm. Math. Phys. 82(1981), 399-405.

[GJ 2] __________, Lie algebras of unbounded derivations, J. Funct. Anal. 52(1983), 369-384.

[Gℓ] J. Glimm, Locally compact transformation groups, Trans. Amer. Math. Soc. 101(1961), 124-138.

[GℓJa] J. Glimm & A. Jaffe, Quantum Physics, Springer-Verlag, New York, 1981.

[GSW] M.B. Green, J.H. Schwarz & E. Witten, Superstring Theory, vols. 1-2, Cambridge Univ. Press, Cambridge, 1987.

[Gre 1] P.C. Greiner, A fundamental solution for a non-elliptic partial differential operator, Canad. J. Math. (5) XXXI(1979), 1107-1120.

[Gre 2] __________, On the Laguerre calculus of left-invariant convolution (pseudo-differential) operators on the Heisenberg group, "Séminaire Goulaouic-Meyer-Schwartz: Equatins aux dérivées partielles", 1980/1981, exp. no. 11, Pallaiseau, École Polytechnique, Centre de Mathématiques, 1981.

[GS 1] P.C. Greiner & E.M. Stein, On the solvability of some differential operators of type $\square_b$, "Several complex variables [1976/1977. Cortona]", 106-165, Pisa, Scuola normale superiore, 1978.

[GS 2] __________, Estimates for the $\bar{\partial}$-Neumann problem, Mathematical Notes, 19, Princeton University Press, Princeton, 1977.

[GK] B. Gruber & A.U. Klimyk, Properties of linear representations with a highest weight for the semisimple Lie algebras, J. Math. Phys. 16(1975), 1816-1832.

[GR] B. Gruber & L. O'Raifeartaigh, S theorem and construction of the invariants of the semisimple compact Lie algebras, J. Math. Phys. 5(1964), 1796-1804.

[Gru 1] V.V. Grusin, On a class of hypoelliptic operators, Mat. Sb. (125) 83(1970), 456-473 (Math. USSR Sb. 12 (1970), 458-476).

[Gru 2] __________, Hypoelliptic differential equations and pseudodifferential operators with operator-valued symbols, Mat. Sb. (130) 88(1972), 504-521 (Math. USSR Sb. 17(1972), 497-514.

[HHW] R. Haag, N.M.Hugenholtz & M. Winnink, On the equilibrium states in quantum statistical mechanics, Commun. Math. Phys. 5(1967), 215-236.

[Ham] M. Hamermesh, Group Theory and Its Application to Physical Problems, Addison-Wesley, 1962, pp. 509.

[Ha] R.S. Hamilton, The inverse function theorem of Nash and Moser, Bull. Amer. Math. Soc. (New ser.) 7(1982), 65-222.

[HC 1] Harish-Chandra, On some applications of the universal enveloping algebra of a semisimple Lie algebra, Trans. Amer. Math. Soc. 70(1951), 28-96.

[HC 2] __________, Plancherel formula for the 2×2 real unimodular group, Proc. Nat. Acad. Sci. USA 38(1952), 337-342.

[HC 3] __________, Representations of a semisimple Lie group on a Banach space, I, Trans. Amer. Math. Soc. 75(1953), 185-243.

[HC 4] __________, Representations of semisimple Lie groups, II, Trans. Amer. Math. Soc. 76(1954), 26-65.

[HC 5] __________, Representations of semisimple Lie groups, III, Trans. Amer. Math. Soc. 76(1954), 234-253.

[HC 6] __________, The Plancherel formula for complex semisimple Lie groups, Trans. Amer. Math. Soc. 76(1954), 485-528.

[HC 7] __________, Representations of semisimple Lie groups, IV, Amer. J. Math. 77(1955), 743-777.

[HC 8] __________, Representations of semisimple Lie groups, V, Amer. J. Math. 78(1956), 1-41.

[HC 9] __________, Representations of semisimple Lie groups, VI, Amer. J. Math. 78(1956), 564-628.

[HC 10] __________, Spherical functions on a semisimple Lie group, I, Amer. J. Math. 80(1958), 241-310.

[HC 11] __________, Spherical functions on a semisimple Lie group, II, Amer. J. Math. 80(1958), 553-613.

[HC 12] __________, Invariant distributions on Lie algebras, Amer. J. Math. 86(1964), 271-309.

[HC 13] __________, Invariant differential operators and distributions on a semisimple algebra, Amer. J. Math. 86 (1964), 534-564.

[HC 14] __________, Invariant eigendistributions on a semisimple Lie algebra, Inst. Hautes Études Sci. Publ. Math. 27(1965), 5-54.

[HC 15] __________, Invariant eigendistributions on a semisimple Lie group, Trans. Amer. Math. Soc. 119(1965), 457-508.

[HC 16] __________, Discrete series for semisimple Lie groups, I, Acta Math. 113(1965), 241-318.

[HC 17] __________, Discrete series for semisimple Lie groups, II, Acta Math. 116(1966), 1-111.

[HC 18] __________, Two theorems on semisimple Lie groups, Ann. of Math. 83(1966), 74-128.

[HC 19] __________, Some applications of the Schwartz space of a semisimple Lie group, In: Lectures in Modern Analysis and Applications II, Lecture Notes in Mathematics 140, Springer-Verlag, Berlin-Heidelberg-New York, 1970, 1-7.

[HC 20] __________, Harmonic analysis on semisimple Lie groups, Bull. Amer. Math. Soc. 76(1970), 529-551.

[HS] M. Hausner & J. Schwartz, Lie Groups; Lie Algebras, Gordon and Breach, New York-London-Paris, 1968, pp 229.

[Haz 1] M. Hazewinkel, On deformation, approximations, and nonlinear filtering, Systems and Control Letters, 1 (1981), 32-36.

[Haz 2] __________, Formal Groups and Applications, Academic Press, New York, 1978.

[HM] M. Hazewinkel & S.I. Marcus, On Lie algebras and finite-dimensional filtering, Stochastics, 7(1982), 29-62.

[HW] M. Hazewinkel & J.C. Willems (eds.), Stochastic Systems: The Mathematics of Filtering and Identification and Applications, D. Reidel Publ. Co., Dordrecht, 1980.

[He 1] B. Helffer, Hypoellipticité microlocale: présentation historique, Bull. Soc. Math. France Mem. 51-52(1977), 5-12.

[He 2] __________, Hypoellipticité pour des opérateurs différentiels sur des groupes nilpotents, Pseudo-differential Operators with Applications, Cours du C.I.M.E. (1977), Liguori Editor.

[He 3] __________, Hypoellipticité des opérateurs quasi-homogénes á coefficients polynomiaux II, Preprint de l'Ecole Polytechnique, Mai 1979.

[He 4] __________, (D'après B. Helffer, G. Métivier & J. Nourrigat), On the hypoellipticity of the operators of the type $\sum_{j=1}^{n} Y_j^2+\frac{1}{2} \Sigma C_{j,k}[Y_j,Y_k]$, Séminaire de l'Université de Nantes, 1981-82.

[He 5] __________, Partial differential equations on nilpotent groups, Lecture Notes in Mathematics, no. 1077: Lie Group Representations III. Springer-Verlag, 1986.

[He 6] __________, On a conjecture of Jorgensen and Klink, Publ. Res. Inst. Math. Sci., Kyoto University (to appear).

[HN 1] B. Helffer & J. Nourrigat, Hypoellipticité pour des groupes de rang 3, Comm. P.D.E. (8) 3(1978), 693-743.

[HN 2] __________, Caractérisation des opérateurs hypoelliptiques homogènes invariants à gauche sur un groupe nilpotent gradué, Comm. P.D.E. (9) 4(1979), 899-958.

[HN 3] __________, Hypoellipticité pour des opérateurs quasi-homogènes à coefficients polynomiaux, Preprint de l'Ecole Polytechnique, Mars 1979.

[HN 4] __________, Approximation d'un système de champs de vecteurs et applications à l'hypoellipticité, Ark. Mat. (2) (1979), 237-254.

[HN 5] __________, Hypoellipticité maximale pour des opérateurs polynömes de champs de vecteurs, (Une nouvelle démonstration d'un th. de L.P. Rothschild), Preprint, 1980.

[HN 6] __________, Hypoellipticité par des opérateurs quasi-homogènes à coefficients polynomiaux, Actes du Colloque de St. Cast, Juin 1979.

[HN 7] __________, Hypoellipticité maximale pour des opérateurs polynömes de champs de vecteurs, Note aux CRAS, Oct. 1979. Manuscript et actes du Colloque de St. Jean de Monts, Juin 1980.

[HN 8] __________, Hypoellipticité maximale pour des opérateurs polynömes de champs de vecteurs, Rend. Sem. Mat. Univ. Polit. Torino (1984), 115-134.

[HN 9] __________, Théorèmes d'indice pour des opérateurs différentiels à coefficients polynomiaux, Preprint, Février 1979; et Partial Differential Equations Banach Center Publ. 10(1983).

[HN 10] __________, Hypoellipticité Maximale pour des Opérateurs Polynömes de Champs de Vectuers, Birkhäuser, Boston-Basel 1985.

[HH] G.C. Hegerfeldt & J. Henning, Coupling of Space-Time and internal symmetry, Fortschr. Physik 16(1968), 491-544.

[HKW] G.C. Hegerfeldt, K. Kraus & E.P. Wigner, Proof of the Fermion superselection rule without the assumption of time-reversal invariance, J. Math. Phys. 9,(1968), 2029-2031.

[Hel 1] S. Helgason, Differential Geometry and Symmetric Spaces, Academic Press, New York, 1962.

[Hel 2] __________, A duality for symmetric spaces with applications to group representations, Advances in Math. 5(1970), 1-154.

[Hel 3] __________, Differential Geometry, Lie Groups, and Symmetric Spaces, Academic Press, New York, 1978.

[HeH] J.W. Helton & R.E. Howe, Integral operators: commutators, traces, index and homology. In: Proceedings of a Conference on Operator Theory (Dalhousie Univ., Halifax, N.S., 1973), Lecture Notes in Mathematics vol. 345, 141-209, Springer-Verlag, Berlin, 1973.

[Her] M.-R. Herman, Simplicité du groupe des difféomorphismes de classe C^{∞}, isotopes a l'identité, du tore de dimension n, C.R. Acad. Sci. Paris, Sér. A, 273 (1971), 232-234.

[HR 1] E. Hewitt & K.A. Ross, Abstract Harmonic Analysis I, Springer-Verlag, Berlin-Göttingen-Heidelberg, 1963.

[HR 2] __________, Abstract Harmonic Analysis II, Springer-Verlag, Berlin-Heidelberg, New York, 1970.

[HP] E. Hille & R.S. Phillips, Functional analysis and semigroups, AMS Colloquium Publ. Providence, R.I., 1957.

[Hoc 1] G. Hochschild, The Structure of Lie Groups, Holden-Day, San Francisco, 1965.

[Hoc 2] __________, Algebraic groups and Hopf algebras, Illinois J. Math. 14(1970), 52-65.

[HM] G. Hochschild & G.D. Mostow, Cohomology of groups, Illinois J. Math. 6(1962), 367-401.

[Hö 1] L. Hörmander, Hypoelliptic second order differential equations, Acta Math. 119(1968), 147-171.

[Hö 2] __________, Linear Partial Differential Operators, 3rd ed., Springer-Verlag, Berlin, 1969.

[Hö 3] __________, The Analysis of Linear Partial Differential Operators I, Springer-Verlag, Berlin, 1983.

[Hö 4] __________, A class of hypoelliptic pseudo-differential operators with double characteristics, Math. Ann. (2) 217(1975).

[Hö 5] __________, Subelliptic operators, Seminar on singularities of solutions of linear partial differential equatins, Ann. of Math. Studies 91, Princeton U., 1979.

[Hö 6] __________, The Weyl calculus, Comm. Pure Appl. Math. 32(1979), 359-443.

[Hö 7] __________, On the subelliptic test estimates, Comm. Pure Appl. Math. 33(1980), 330-363.

[HM] L. Hörmander & A. Melin, Free systems of vector fields, Ark. Mat. (1) 16(1978).

[Hul 1] A. Hulanicki, The distribution of energy in the Brownian motion in Gaussian fields and analytic hypoellipticity of certain subelliptic operators on the Heisenberg group, Studia Math. 56(1976), 165-173.

[Hul 2] __________, Commutative subalgebra of $L^1(G)$ associated with a subelliptic operator on a Lie group, Bull. Amer. Math. Soc. 81(1975), 121-124.

[Ja] N. Jacobson, Lie Algebras, Interscience, New York, 1962.

[Je] J. Jenkins, Dilations, gauges, and growth in nilpotent Lie groups, Preprint, June 1975.

[Joh] R.W. Johnson, Homogeneous Lie algebras and expanding automorphisms, Preprint, 1974.

[Jo 1] P.E.T. Jorgensen, Representations of differential operators on a Lie group, J. Functional Anal. 20 (1975), 105-135.

[Jo 2] __________, Perturbation and analytic continuation of group representations, Bull. Amer. Math. Soc. 82 (1976), 921-924.

[Jo 3] __________, Approximately reducing subspaces for unbounded linear operators, J. Functional Anal. 23 (1976), 392-414.

[Jo 4] __________, Approximately invariant subspaces for unbounded linear operators II, Math. Ann. 227(1977), 177-182.

[Jo 5] __________, Essential self-adjointness of semibounded operators, Math. Ann. 237(1978), 187-196.

[Jo 6] __________, Trace states and KMS states for approximately inner dynamical one-parameter groups of *-automorphisms, Commun. Math. Phys. 53(1977), 135-142.

[Jo 7] __________, Self-adjoint operator extensions satisfying the Weyl commutation relations, Bull. Amer. Math. Soc. 1 (New ser.)(1979), 266-269.

[Jo 8] __________, Commutation properties for automorphism groups of von Neumann algebras and exponentiation of derivations, J. Functional Anal. 34(1979), 138-145.

[Jo 9] __________, Commutators of Hamiltonian operators and non-Abelian algebras (Extensions of symmetric operators and unbounded derivations), J. Math. Anal. Appl. 73(1980), 151-133.

[Jo 10] __________, Self-adjoint extension operators commuting with an algebra, Math. Zeit. 169(1979), 41-62.

[Jo 11] __________, Spectral theory for domains in $\mathbb{R}^n$ of finite measure, Proc. Nat. Acad. Sci. USA 77(1980), 5050-5051.

[Jo 12] __________, Partial differential operators and discrete subgroups of a Lie group, Math. Ann. 247(1980), 101-110.

[Jo 13] __________, Unbounded operators: Perturbation and commutativity problems, J. Functional Anal. 39(1980), 281-307.

[Jo 14] __________, A uniqueness theorem for the Heisenberg-Weyl commutation relations with non-self-adjoint position operator, Amer. J. Math. 103(1981), 273-287.

[Jo 15] __________, Spectral theory of finite-volume domains in $\mathbb{R}^n$, Advances in Math. 44(1982), 105-120.

[Jo 16] __________, Monotone convergence of operator semi-groups and the dynamics of infinite particle systems, J. Approximation Theory 43(1985), 205-230.

[Jo 17] __________, Spectral theory for infinitesimal generators of one-parameter groups of isometries: The Min-Max principle, and compact perturbations, J. Math. Anal. Appl. 90(1982), 243-370.

[Jo 18] __________, Ergodic properties of one-parameter automorphism groups on operator algebras, J. Math. Anal. Appl. 87(1982), 354-372.

[Jo 19] __________, Extensions of unbounded $*$-derivations in UHF C^*-algebras, J. Functional Anal. 45(1982), 341-356.

[Jo 20] __________, An optimal spectral estimator for multi-dimensional time series with an infinite number of sample points, Math. Zeit. 183(1983), 381-398.

[Jo 21] __________, Spectral representations of unbounded nonlinear operators on Hilbert-space, Pacific J. Math. 111(1984), 93-104.

[Jo 22] __________, The integrability problem for infinite-dimensional representations of finite-dimensional Lie algebras, Expositiones Math. 4(1983), 289-306.

[Jo 23] __________, A structure theorem for Lie algebras of unbounded derivations in C^*-algebras & Appendix, Compositio Math. 52(1984), 85-98.

[Jo 24] __________, New results on unbounded derivations and ergodic groups of automorphisms, Expositiones Math. 2(1984), 3-24.

[Jo 25] __________, Representations of differential operators on a Lie group, and conditions for a Lie algebra of operators to generate a representation of the group, J. d'Analyse Math. 43(1983/84), 251-288.

[Jo 26] __________, Second order right-invariant partial differential equations on a Lie group, Comm. P.D.E. (accepted).

[Jo 27] __________, Nilpotent ordinary differential operators with polynomial coefficients, J. Diff. Equa. 65(1986), 1-18.

[Jo 28] __________, Analytic continuation of local representations of Lie groups, Pacific J. Math. (2) 124(1986), 397-408.

[Jo 29] __________, Commutative algebras of unbounded operators, J. Math. Anal. Appl. 122(1987).

[Jo 30] __________, Unitary dilations and the C^*-algebra $\mathcal{O}_2$, Israel J. Math. (8) 20(1986).

[Jo 31] __________, Ergodicity of the automorphic action of U(1,H) on $C^*(\mathfrak{L}(H))$, Abstracts Amer. Math. Soc. 5 (1984), 212.

[Jo 32] __________, Book review of "Derivations, dissipations and group actions on C^*-algebras" (O. Bratteli) for the Amer. Math. Soc. Bulletin 17, no. 1, July 1987.

[JK 1] P.E.T. Jorgensen & W.H. Klink, Quantum mechanics and nilpotent groups, I: The curved magnetic field, Publ. RIMS, Kyoto Univ. 21(1985), 969-999.

[JK 2] __________, Spectral transform for the sub-Laplacian on the Heisenberg group, Preprint, Univ. of Iowa, 1986 (submitted).

[JM] P.E.T. Jorgensen & R.T. Moore, "Operator Commutation Relations," D. Reidel Publ. Co., Mathematics and Its Applications, Dordrecht-Boston-Lancaster, 1984.

[JMu] P.E.T. Jorgensen & P.S. Muhly, Self-adjoint extensions satisfying the Weyl operator commutation relations, J. d'Anal. Math. 37(1980), 46-99.

[JPe] P.E.T. Jorgensen & S. Pedersen, Harmonic analysis on tori, Acta Appl. Math. (to appear).

[JPr] P.E.T. Jorgensen & G.L. Price, Extending quasi-free derivations on the CAR-algebra, J. Operator Theory 16(1986), 147-155.

[Jos 1] A. Joseph, Proof of the Gelfand-Kirillov conjecture for solvable Lie algebras, Proc. Amer. Math. Soc. 45(1974), 1-10.

[Jos 2] __________, A generalization of the Gelfand-Kirillov conjecture (to appear).

[Jt] R. Jost, Eine Bemerkung zu einem Letter von L. O'Raifeartaigh und einer Entgegnung von M. Flato und D. Sternheimer, Helv. Phys. Acta 39(1966), 369-375.

[JB] A. Jucys & A.A. Bandzaitis, The Theory of Angular Momentum in Quantum Mechanics, Mintis, Vilnius, pp. 463 (Russian), 1978.

[Ka 1] V.G. Kac, Simple irreducible graded Lie algebras of finite growth, Math. USSR Izv. 2(1968), 1271-1311.

[Ka 2] __________, Infinite-dimensional algebras, Dedekind's η-function, classical Möbius function and the very strange formula, Advances in Math. 30(1978), 85-136.

[Ka 3] __________, Highest weight representations of infinite dimensional Lie algebras. In: Proceedings of ICM," pp. 299-304, Helsinki, 1978.

[Ka 4] __________, An elucidation of "Infinite dimensional algebras...and the very strange formula," $E_8^{(1)}$ and the cube root of the modular invariant j, Adv. in Math. 35(1980), 264-273.

[Ka 5] __________, Infinite root systems, representations of graphs and invariant theory, I, Invent. Math. 56 (1980), 57-92; II: J. Algebra 78(1982), 141-162.

[Ka 6] __________, Infinite Dimensional Lie Algebras, Birkhäuser, Boston, 1983.

[Ka 7] __________, Some problems on infinite-dimensional Lie algebras, In: "Lie Algebras and Related Topics," Lecture Notes in Mathematics, vol. 933, Springer-Verlag, Berlin-Heidelberg-New York, 1982.

[Kad 1] R.V. Kadison, Transformation of states in operator theory and dynamics, Topology 3(1965), 177-198.

[Kad 2] __________, Derivations on operator algebras, Ann. Math. 83(1966), 280-293.

[Kad 3] __________ (ed.), Operator algebras and applications, Proc. Symposia Pure Math., vol. 38, parts 1 and 2, Amer. Math. Soc., Providence, RI, 1982.

[KR] R.V. Kadison & J.R. Ringrose, Fundamentals of the Theory of Operator Algebras, Vols. I and II, Academic Press, 1983/86.

[Km] R.R. Kallman, Unitary groups and automorphisms of operator algebras, Amer. J. Math. 91(1969), 785-806.

[Kap] I. Kaplansky, Modules over operator algebras, Amer. J. Math. 75(1953), 839-859.

[Kas] G.G. Kasparov, The operator K-functor and extensions of C^*algebras, Math. USSR Izv. 44(1981), 513-572.

[KMLM] D. Kastler, M. Mebkkout, G. Loupias & L. Michel, Central decomposition of invariant states applications to the groups of time translations and of Euclidean trans formations in algebraic field theory, Commun. Math. Phys. 27(1972), 1995-222.

[Kat 1] T. Kato, On the commutation relation AB-Ba = c, Arch. Rat. Mech. Anal. 10(1962), 273-275.

[Kat 2] __________, Perturbation Theory for Linear Operators, Springer-Verlag, New York, 1980.

[Katn] Y. Katznelson, An Introduction to Harmonic Analysis, Dover Publications, Inc., New York, 1976.

[Kir 1] A.A. Kirillov, Unitary representations of nilpotent groups, Russian Math. Survey 17(1952), 53-104.

[Kir 2] __________, The method of orbits in the theory of unitary representations of Lie groups, Funkcional. Anal. i Prilozen. 2(1968), 96-98.

[Kir 3] __________, Characters of unitary representations of Lie groups, Funccional. Anal. i Prilozen. 2(1968), 40-55 (Russian).

[Kir 4] __________, Representations of infinite dimensional unitary group, Dokl. Akad. Nauk SSSR 212(1973), 288-290 (Russian).

[Kir 5] __________, Unitary representations of diffeomorphism group and some of its subgroups, Preprint, Moscow Univ., 1976 (Russian).

[Kir 6] __________, Elements of the Theory of Representations, Springer-Verlag, Berlin, 1976.

[Kis] A. Kishimoto, Automorphisms and covariant irreducible representations, Yokohama Math. J. 31(1983), 159-168.

[KiRo 1] A. Kishimoto & D.W. Robinson, Derivations, dynamical systems and spectral restrictions, Math. Scand. 56 (1985), 159-168.

[KiRo 2] __________, On unbounded derivations commuting with a compact group of *-automorphisms, Publ. RIMS, Kyoto Univ. 18(1982), 1121-1136.

[KiRo 3] __________, Dissipations, derivations, dynamical systems, and asymptotic abelianness, J. Operator Theory 13(1985), 237-253.

[Ki 1] J. Kisynski, On the Integrability of Banach Space Representations of Lie Algebras, Preprint, Univ. of Warsaw, 1973, pp. 43.

[Ki 2] __________, An integration of Lie algebra representations in Banach space, International Atomic Energy Agency, Tireste, Internal Report No. 130, (1974).

[Kl] A. Kleppner & R.L. Lipsman, The Plancherel formula for group extensions, I, Ann. Sci. École Norm. Sup. 5 (1972), 459-516; II, 6(1973), 103-132.

[Kn] A.W. Knapp, Representation Theory of Semisimple Groups, An Overview Based on Exampless, Princeton Univ. Press, Princeton Univ., 1986.

[KS 1] A.W. Knapp & E.M. Stein, Intertwining operators for semisimple groups, Ann. of Math. (2) 93(1971), 489-578.

[KS 2] __________, Singular integrals and the principal series III, Proc. Nat. Acad. Sci. USA 71(1974), 4622-4624.

[KS 3] __________, Singular integrals and the principal series IV, Proc. Nat. Acad. Sci. USA 72(1975), 2459-2461.

[KV] A. Koranyi & S. Vagi, Singular integrals on homogeneous spaces and some problems of classical analysis, Ann. Sci. Norm. Sup. Pisa, Sci. fis. e mat., Ser. 3, 25(1971), 575-648.

[Ko 1] B. Kostant, A formula for the multiplicity of a weight, Trans. Amer. Math. Soc. 93(1959), 53-73.

[Ko 2] __________, The principal three-dimensional subgroup and the Betti numbers of a complex simple Lie group, Amer. J. Math. 81(1959),

[Ko 3] __________, Lie group representations on polynomial rings, Amer. J. Math. 85(1963), 327-404.

[Ko 4] __________, On the existence and irreducibility of certain series of representations, Bull. Amer. Math. Soc. 75(1969), 627-642.

[Ko 5] __________, Quantization and unitary representations, Part I, Prequantization Lectures in Modern Analysis and Applications, Lecture Notes in Math. 170, Springer Verlag, Berlin-Heidelberg-New York, 1970, 87-208.

[Ko 6] __________, On convexity, the Weyl group and the Iwasawa decomposition, Ann. Sci. École Norm. Sup. 6 (1973), 413-455.

[Ko 7] __________, ,On the existence and irreducibility of certain series of representations, In: Lie Groups and Their Representations, I. Gelfand, ed., Akad. Kiadó, Budapest-Hilger-London, 1975, 231-329.

[Ko 8] On the tensor product of a finite and an infinite dimensional representation, J. Funct. Anal. 20(1975), 227-285.

[KØ] B. Kostant & B. Ørsted, K-finite vectors for representations of u(p,q), In preparation.

[Kr] S.G. Krein, Linejnye differentialnyue uravnenija v Banachovom prostranstve, Preprint, 1967.

[Ku] N.H. Kuiper, The homotopy type of the unitary group of Hilbert space, Topology 3(1965), 19-30.

[KS 1] R.A. Kunze & E.M. Stein, Uniformly bounded representations and harmonic analysis of the real 2×2 unimodular group, Amer. J. Math. 82(1960), 1-62.

[KS 2] __________, Uniformly bounded representations, III. Intertwining Operators for the Principal Series on Semisimple Groups, Amer. J. Math. 89(1967), 385-442.

[L 1] S. Lang, Introduction to Differentiable Manifolds, Interscience, New York, 1962.

[L 2] __________, "SL(2,ℝ)," Addison-Wesley, Reading, MA, 1975.

[La 1] R.P. Langlands, Semigroups and Representations of Lie Groups, Thesis, Yale Univ., 1960.

[La 2] __________, Some holomorphic semigroups, Proc. Nat. Acad. Sci. Ser. USA 46(1960), 361-363.

[Las] R.K. Lashof, Lie algebras of locally compact groups, Pacific J. Math. 7(1957), 1145-1162.

[Lep] J. Lepowsky, Generalized Verma modules, loop space cohomology and Macdonald type identities, Ann. Sci. École Norm. Sup. (4) 12(1979), 169-234.

[LW] J. Lepowsky & N.R. Wallach, Finite and infinite dimensional representations of linear semisimple groups, Trans. Amer. Math. Soc. 184(1973), 223-246.

[Le] H. Lewy, An example of a smooth linear partial differential equation without solution, Ann. of Math. Ser. 2, T. 66(1957), 155-158.

[Lip 1] R. Lipsman, Group Representations, A Survey of Some Current Topics, Lecture Notes in Math. 388, Springer-Verlag, Berlin-New York, 1974, pp. x+166.

[Lip 2] __________, An explicit realization of Kostant's complementary series with applications to uniformly bounded representations (to appear).

[Lo] R. Longo, Automatic relative boundedness of derivations in C^*-algebras, J. Functional Anal. 34(1979), 21-28.

[Loo 1] L.H. Loomis, An Introduction to Abstract Harmonic Analysis, Van Nostrand, New York, 1953, pp. 190.

[Loo 2] __________, Positive definite functions and induced representations, Duke Math. J. 27(1960), 569-579.

[Lu] G. Lumer, Spectral operators, Hermitian operators, and bounded groups, Acta Sci. Math. 25(1964), 75-85.

[Lus] M. Luscher, Analytic representations of semisimple Lie groups and their continuation to contractive representations of holomorphic Lie semigroups, Deutsches Elektronen Synchroton, 75/51, 1975.

[Mac] I.G. Macdonald, Kac-Moody algebras, In: Lie Algebras and Related Topics, R.V. Moody, ed., Conference Proceedings of the Canadian Mathematical Society, vol. 5, Amer. Math. Soc., 1986.

[Ma 1] G.W. Mackey, On induced representation of groups, Amer. J. Math. 73(1951), 576-592.

[Ma 2] __________, Induced representations of locally compact groups, I, Ann. of Math. 55(1952), 101-139.

[Ma 3] __________, Induced representations of locally compact groups, II, Ann. of Math. 58(1953), 193-221.

[Ma 4] __________, The Theory of Group Representations, Lecture Notes (Summer, 1955), Dept. of Math., Univ. of Chicago, vol. 1-3, 1956, pp. 182, (Mimeographed Notes, Univ. of Chicago).

[Ma 5] __________, Unitary representations of group extensions, I, Acta Math. 99(1958), 265-311.

[Ma 6] __________, Infinite-dimensional group representations, Bull. Amer. Math. Soc. 69(1963), 628-686.

[Ma 7] __________, Group representations and noncommutative harmonic analysis, Mimeographed Notes, Univ. of California, Berkeley, 1965.

[Ma 8] __________, Ergodic theory and virtual groups, Math. Ann. 166(1966), 187-207.

[Ma 9] __________, Induced Representations of Groups and Quantum Mechanics, Benjamin, New York-Amsterdam, 1968, pp. 167.

[Ma 10] __________, Induced representations of locally compact groups and applications, In: Functional Analysis and Related Fields, F.E. Browder, ed., Springer-Verlag, Berlin-Heidelberg, 1970, 132-171.

[Ma 11] __________, Ergodicity in the theory of group representations, Actess Cong. Int. Math. Nice, Vol. 2, Gauthier-Villars, Paris, 1970, 401-417.

[Ma 12] __________, Products of subgroups and projective multipliers, In: Coll. Math. Societatis J. Bolya 5, Hilbert Space Operators, Tihany, Hungary, North-Holland Publ., 1970, 401-413.

[Ma 13] __________, On the analogy between semisimple Lie groups and certain related semidirect product groups, In: Lie Groups and Their Representations, I, I. Gelfand, ed., Hilgar, London, 1975, 339-363.

[MMc] D. Masson & W.K. McClary, Classes of C^{∞}-vectors and essential self-adjointness, J. Funct. Anal. 10(1972), 19-32.

[Mau 1] F.I. Mautner, On the decomposition of unitary representations of Lie group, Proc. Amer. Math. Soc. 2 (1951), 490-496.

[Mau 2] __________, Fourier analysis and symmetric spaces, Proc. Nat. Acad. Sci. USA 37(1951), 529-533.

[Mau 3] __________, Induced representations, Amer. J. Math. 74(1952), 737-758.

[McC] J. McCully, The Laguerre transform, SIAM Rev. 2(1960), 185-191.

[Mel 1] A. Melin, Parametrix constructions for some classes of right invariant differential operators on the Heisenberg group, Comm. Partial Differential Equa. (12) 6(1981), 1363-1405.

[Mel 2] __________, Parametrix constructions for some classes of right invariant differential operators on nilpotent groups, Preprint, 1981 (to appear in Ann. Global Anal. and Geometry).

[Mel 3] __________, Lie filtrations and pseudo-differential operators, Preprint, 1982.

[Mel 4] __________, Lower bounds for pseudo-differential operators, Ark. Mat. 9(1971), 117-140.

[MS] A. Melin & J. Sjostrand, F.I.O. with complex valued phase function, Lecture Notes in Math. vol. 459(1975), 120-223.

[Mic] E.A. Michael, Locally multiplicatively-convex topological algebras, Mem. Amer. Math. Soc. 11(1952), 1-82.

[MZ] D. Montgomery & L. Zippin, Topological Transformation Groups, Interscience Publ. Inc., New York, 1955.

[M 1] R.V. Moody, A new class of Lie algebras, J. Algebra 10(1968), 211-230.

[M 2] __________, Euclidean Lie algebras, Canad. J. Math. 21(1969), 1432-1454.

[Mo 1] C.C. Moore, On the Frobenius reciprocity theorem for locally compact groups, Pacific J. Math. 12(1962), 359-365.

[Mo 2] __________, ed., Harmonic Analysis on Homogeneous Spaces, Proc. of Symposia in Pure Mathematics, Amer. Math. Soc. XXVI(1973), 467.

[Mo 3] __________ & J.A. Wolf, Square integrable representations of nilpotent groups, Trans. Amer. Math. Soc. 185(1976), 445-462.

[Mr 1] R.T. Moore, Lie algebras of operators and group representations on Banach spaces, Thesis, Princeton Univ., 1964, 1-164+ix.

[Mr 2] __________, Exponentiation of operator Lie algebras on Banach spaces, Bull. Amer. Math. Soc. 71(1965), 903-908.

[Mr 3] __________, Measurable, continuous and smooth vectors for semigroups and group representations, Mem. Amer. Math. Soc. 78(1968), 80.

[Mr 4] __________, Generation of equi-continuous semigroups by Hermitian and sectorial operatorss, I, II, Bull. Amer. Math. Soc. 77(1971), 224-229 and 368-373.

[Mr 5] __________, Core Fréchet spaces of C^∞-vectors for families of unbounded operators (in preparation).

[N 1] L. Nachbin, On the finite dimensionality of every irreducible unitary representation of a compact group, Proc. Amer. Math. Soc. 12(1961), 11-12.

[N 2] __________, The Haar Integral, Van Nostrand Co., Inc., Princeton, New York, 1965, pp. 156.

[Na 1] B. Sz.-Nagy, Spectraldarstellung linearer Transformationen des Hilbertschen Raumes, Springer-Verlag, Berlin, 1942.

[Na 2] __________, Prolongements de transformations de l'espace de Hilbert qui sortent de cet espace, Appendice au livre "Leçons d'analyse fonctionnelle" par F. Riesz et B. Sz. Nagy, Akademiai Kiadó, Budapest, 1955, 323-339.

[Nk] M.A. Naimark, Linear Representation of the Lorentz Group, Pergamon Press, London, 1964.

[Ne 1] E. Nelson, Kernel functions and eigenfunction expansions, Duke Math. J. 25(1958), 15-28.

[Ne 2] __________, An existence theorem for second order parabolic equations, Trans. Amer. Math. Soc. 88(1958), 414-429.

[Ne 3] __________, Analytic vectors, Ann. of Math. 70(1959), 572-615.

[Ne 4] __________, Time-ordered operator products of sharp-time quadratic forms, J. Funct. Anal. 11(1972), 211-219.

[Ne 5] __________, Topics in Dynamics, I: Flows, Math. Notes, Princeton Univ. Press, 1969.

[NS] E. Nelson & W.F. Stinespring, Representation of elliptic operators, Amer. J. Math. 81(1959), 547-560.

[vN 1] J. von Neumann, Die Eindeutigkeit der Schrödingerschen Operatoren, Math. Ann. 104(1931), 570-578.

[vN 2] __________, Mathematische Grundlagen der Quantenmechanik, Springer, Berlin; English translation by R.T. Beyer, Princeton Univ. Press, Princeton, NJ, 1955, pp. 262.

[No] Y. Nouazé, Remarques sur "Idéaux premiers de l'algèbre enveloppante d'une algèbre de Lie nilpotente," Bull. Sci. Math. 91(1967), 117-124.

[NG] Y. Nouazé & P. Gabriel, Idéaux premiers de l'algèbre enveloppante d'une algèbre de Lie nilpotente, J. Algebra 6(1967), 77-99.

[OPT] D. Olesen, G.K. Pedersen & M. Takesaki, Ergodic actions of compact abelian groups, J. Operator Theory 3(1980), 237-269.

[Ørs 1] B. Ørsted, A model for interacting quantum fields, J. Funct. Anal. 36(1980), 53-77.

[Ørs 2] __________, Composition series for analytic continuation of holomorphic discrete series, Trans. Amer. Math. Soc. 260(1980), 563-573.

[Ørs 3] __________, Induced representations and a new proof of the imprimitivity theorem, J. Funct. Anal. 31(1979), 355-359.

[O'R 1] L. O'Raifeartaigh, Mass differences and Lie algebras of finite order, Phys. Rev. Lett. 14(1965), 575-577.

[O'R 2] __________, Lorentz invariance and internal symmetry, Phys. Rev. 139B(1965), 1052-1062.

[O'R 3] __________, Broken symmetry, In: Group Theory and Its Applications, E.M. Loebl, ed., 469-539, 1968.

[O'R 4] __________, Unitary representations of Lie groups in quantum mechanics, Lecture Notes in Physics 6, Springer-Verlag, Berlin-Heidelberg-New York, 1970, 144-236.

[Pa 1] R. Palais, A global formulation of the Lie theory of transformation groups, Mem. Amer. Math. Soc. 22(1957).

[Pa 2] __________, On the homotopy type of certain groups of operators, Topology 3(1965), 271-279.

[Pan] S. Paneitz, Analysis in space-time, III, J. Funct. Anal. 54(1983), 18-112.

[PaS] W. Paschke & N. Salinas, Matrix algebras over $\mathcal{O}_n$, Michigan Math. J. 26(1979), 3-12.

[Ped] G.K. Pedersen, C^*-algebras and Their Automorphism Groups, Academic Press, London-New York, San Francisco, 1979.

[Pee] J. Peetre, The Weyl transform and Laguerre polynomials, Le Matematiche, Universita di Catania, Seminario 27(1972), 301-323.

[Pel] C. Peligrad, Derivations of C^*-algebras which are invariant under an automorphism group, In: Topics in Modern Operator Theory, OT Ser. 2, Birkhäuser-Verlag, 1981, 259-268.

[Pe] R. Penney, Abstract Plancherel theorems and Frobenius reciprocity theorem, J. Funct. Anal. 18(1975), 177-190.

[Ph] R.S. Phillips, Perturbation theory for semi-groups of linear operators, Trans. Amer. Math. Soc. 74(1954), 199-221.

[PiPo] M. Pimsner & S. Popa, The Ext groups of some C^*-algebras considered by J. Cuntz, Rev. Roum. Math. Pures Appl. 23(1978), 1069-1076.

[PiVo 1] M. Pimsner & D. Voiculescu, Imbedding the irrational rotation C^*-algebra into an AF-algebra, J. Operator Theory 4(1980), 201-210.

[PiVo 2] __________, Exact sequences for K-groups and Ext-groups of certain cross-product C^*-algebras, J. Operator Theory 4(1980), 93-118.

[Po 1] N.S. Poulsen, Regularity Aspects of the Theory of Infinite Dimensional Representations of Lie Groups, Ph.D. Thesis, Massachusetts Institute of Technology, 1970.

[Po 2] __________, Lecture notes on representaions of Lie groups, Aarhus, 1972 (unpublished).

[Po 3] __________, On C^∞-vectors and intertwining bilinear forms for representations of Lie groups, J. Funct. Anal. 9(1972), 87-120.

[Po 4] __________, On the canonical commutation relations, Math. Scand. 32(1973), 112-122.

[Pow 1] R.T. Powers, Self-adjoint algebras of unbounded operators, Commun. Math. Phys. 21(1971), 85-124.

[Pow 2] __________, Self-adjoint algebras of unbounded operators II, Trans. Amer. Math. Soc. 187(1974), 261-293.

[PSa 1] R.T. Powers & S. Sakai, Unbounded derivations in operator algebras, J. Funct. Anal. 19(1975), 91-95.

[PSa 2] __________, Existence of ground states and KMS states for approximately inner dynamics, Comm. Math. Phys. 39(1975), 273-288.

[PS] A. Pressley & G. Segal, Loop Groups, Oxford Math. Monographs, Oxford Univ. Press, Clarenden Press, Oxford, 1986.

[Pu 1] L. Pukánszky, The Plancherelformula for the universal covering group of SL(2,ℝ), Math. Ann. 156(1964), 96-143.

[Pu 2] __________, Leçons sur les représentations des groupes, Dunos, Paris, 1967, pp. 178.

[Pu 3] __________, On the characters and the Plancherel formula of nilpotent groups, J. Funct. Anal. 1(1967), 255-280.

[Put] C.R. Putnam, Commutation Properties of Hilbert Space Operators, Springer-Verlag, Berlin-New York, 1967.

[Rao] R.R. Rao, Unitary representations defined by boundary conditions—the case of sl(2,ℝ), Acta Math. 139(1977), 185-216.

[RS] M. Reed & B. Simon, Methods of Modern Mathematical Physics, Vols. 1 & 2, Academic Press, New York, 1972 and 1973.

[Ren] R. Rentschler, Propriétés fonctorielles de la bijection canonique entre Spec U(g) et $(\text{Spec } S(g))^g$ pour les algèbres dse Lie nilpotentes, C.R. Acad. Sci. Paris (A) 271(1970), 868-871.

[Ric] C.E. Rickart, General Theory of Banach Algebras, Van Nostrand, Princeton, NJ, 1960.

[Ri 1] M.A. Rieffel, Induced Banach representations of Banach algebras and locally compact groups, J. Funct. Anal. 1(1967), 443-491.

[Ri 2] __________, Square-integrable representations of Hilbert algebras, J. Funct. Anal. 3(1969), 265-300.

[Ri 3] __________, Induced representations of C^* algebras, Advances in Math. 13(1974), 176-257.

[Ri 4] __________, C^*-algebras associated with irrational rotations, Pacific J. Math. 93(1981), 415-429.

[RN] F. Riesz & B. Sz-Nagy, Functional Analysis, Blackie & Sons, Ltd., London, 1956, pp. 468.

[Rob 1] D.W. Robinson, Smooth derivations on abelian C^*-dynamical systems, J. Austral. Math. Soc. Ser. A (to appear).

[Rob 2] __________, Commutators and generators II, Quart. J. Math. (submitted).

[Rob 3] __________, Smooth cores of Lipschitz flows, Publ. RIMS Kyoto Univ. (to appear).

[Rob 4] __________, Differential operators on C^*-algebras, In: Operator Algebras and Mathematical Physics, P.E.T. Jorgensen & P. Muhly, eds., Contemporary Mathematics, vol. 62, Amer. Math. Soc., Providence, 1986.

[Rob 5] __________, The differential and integral structure of Lie groups, Preprint, 1986.

[Ro] C. Rockland, Hypoellipticity on the Heisenberg group, Representation theoretic criteria, Trans. Amer. Math. Soc. 240(1978), 1-52.

[Rot 1] L.P. Rothschild, A criterion for hypoellipticity of operators constructed of vector fields, Comm. in P.D.E. 4(1979), 645-699.

[Rot 2] __________, Nonexistence of optimal estimates for the boundary Laplacian operator on certain weakly pseudo-convex domains, Comm. in P.D.E. 5(1980), 897-912.

[RSt] L.P. Rothschild & E.M. Stein, Hypoelliptic differential operators and nilpotent groups, Acta Math. 37 (1977), 248-315.

[Ru 1] W. Rudin, The authomorphisms and the endomorphisms of the group algebra of the unit circle, Acta Math. 95 (1956), 39-55.

[Ru 2] __________, Fourier Analysis on Groups, Interscience Publ., 1967.

[Ru 3] __________, Functional Analysis, McGraw-Hill, New York, 1973.

[Rus 1] J. Rusinek, The integrability of a Lie algebra representation, Lett. Math. Phys. 2(1978), 367-371.

[Rus 2] __________, On the integrability of representations of real Lie algebras in a Banach space, Preprint (9), Warsaw Univ., 1981.

[Sa] P. Sally, Analytic continuation of the irreducible unitary representations of the universal covering group of SL(2,ℝ), Mem. Amer. Math. Soc. 69(1962).

[Sak 1] S. Sakai, On a conjecture of Kaplansky, Tôhoku Math. J. 12(1960), 31-33.

[Sak 2] __________, Derivations of W^*-algebras, Ann. Math. 83(1966), 273-279.

[Sak 3] __________, Derivations of simple C^*-algebras, J. Funct. Anal. 2(1968), 202-206.

[Sak 4] __________, C^*-algebras and W^*-algebras, Springer-Verlag, Berlin-Heidelberg-New York, 1971.

[Sak 5] __________, On one parameter groups of $*$-automorphisms on operator algebras and the corresponding unbounded derivations, Amer. J. Math. 98(1976), 427-440.

[Sak 6] __________, Theory of unbounded derivations on C^*-algebras, Lecture Notes, Copenhagen Univ. and Univ. of Newcastle upon Tyre, 1977.

[Sak 7] __________, Developments in the theory of unbounded derivations in C^*-algebras, Proc. of Symposia in Pure Mathematics II, 38(1982), 309-331.

[Sam] H. Samelson, Notes on Lie Algebras, Van Nostrand-Reinhold, New York, 1969.

[Sc] H.H. Schaefer, Topological Vector Spaces (3rd printing), Springer-Verlag, New York, 1971.

[Sch] W.M. Schmidt, Diophantine Approximation, Lecture Notes in Mathematics 785, Berlin-Heidelberg, New York, 1980.

[Scm] K. Schmüdgen, On a class of representations of the Heisenberg commutation relation PQ-QP = -iI, In: Operator Theory: Advances and Applications II, Birkhäuser, Basel, 1983.

[Schw 1] L. Schwartz, Théorie des distributions, vol I, Hermann Paris, 1957, pp. 162.

[Schw 2] __________, Théorie des distributions, vol. II, Hermann, Paris, 1957, pp. 173.

[GSe] G.B. Segal, Unitary representations of some infinite dimensional groups, Commun. Math. Phys 80(1981), 301-342.

[SW] G.B. Segal & G. Wilson, Loop groups and equations of K dV type, Pub. Math. I.H.E.S. 61(1985), 5-65.

[Se 1] I.E. Segal, The group ring of a locally compact group, I, Proc. Nat. Acad. Sci. USA 27(1941), 348-352.

[Se 2] __________, Irreducible representations of operator algebras, Bull. Amer. Math. Soc. 53(1947), 73-88.

[Se 3] __________, The group algebra of a locally compact group, Trans. Amer. Math. Soc. 61(1947), 69-105.

[Se 4] __________, The two-sided regular representation of a unimodular locally compact group, Ann. of Math. 51 (1950), 293-298.

[Se 5] __________, An extension of Plancherel's formula to separable unimodular groups, Ann. of Math. 52(1950), 272-292.

[Se 6] __________, Decomposition of operator algebras, I, Mem. Amer. Math. Soc. 9(1951), pp. 67; II, pp. 66.

[Se 7] __________, A class of operator algebras which are determined by groups, Duke Math. J. 18(1951), 221-265.

[Se 8] __________, Hypermaximality of certain operators on Lie groups, Proc. Amer. Math. Soc. 3(1952), 13-15.

[Se 9] __________, A noncommutative extension of abstract integration, Ann. of Math. 57(1953), 401-457.

[Se 10] __________, Infinite dimensional irreducible representations of compact semi-simple groups, Bull. Amer. Math. Soc. 70, 155-160.

[Se 11] __________, An extension of a theorem of L. O'Raifeartaigh, J. Funct. Anal. 1(1967), 1-21.

[Se 12] __________, Positive-energy particle models with mass splitting, Proc. Natl. Acad. Sci. USA 57(1967), 194-197.

[Se 13] __________, Notes towards the construction of nonlinear relativistic quantum fields, III. Properties of the C^*-dynamics for a certain class of interactions, Bull. Amer. Math. Soc. 75(1969), 1390-1395.

[Se 14] __________, Mathematical Cosmology and Extragalactic Astronomy, Academic Press, New York, 1976.

[SK] I.E. Segal & R.A. Kunze, Integrals and Operators, McGraw Hill, New York, 1968, pp. 308.

[SvN] I.E. Segal & J. von Neumann, A theorem on unitary representations of semisimple Lie groups, Ann. of Math. 52(1950), 509-517.

[Ser 1] J.P. Serre, Lie Algebras and Lie Groups, Benjamin, New York, 1964, pp. 247.

[Ser 2] __________, Algèbres de Lie semi-simple complex, Benjamin, New York, 1966, pp. 130.

[Sha] D. Shale, Linear symmetries of free Boson fields, Trans. Amer. Math. Soc. 103(1962), 149-167.

[Simm] D.J. Simms, Lie Groups and Quantum Mechanics, Lecture Notes in Mathematics 52, Springer-Verlag, Berlin-Heidelberg, 1968, pp. 90.

[Si 1] B. Simon, Functional Integration and Quantum Physics, Academic Press, New York, 1979.

[Si 2] __________, Trace ideals and their applications, London Math. Soc. Lecture Notes vol. 35, Cambridge Univ. Press, Cambridge, 1979.

[JSi] J. Simon, On the integrability of representations of finite dimensional real Lie algebras, Commun. Math. Phys. 28(1972), 39-46.

[SH] Ja.G. Sinai & A.Ja. Helmskii, A description of differentiations in algebras of the type of local observables of spin systems, Funk. Anal. Prilo. 6(1972), 99-100; Funct. Anal. Appl. 6(1973), 343-344 (English translation).

[Sing 1] I.M. Singer, Lie Algebras of Unbounded Operators, Thesis, Univ. of Chicago, 1950.

[Sing 2] __________, Uniformly continuous representations of Lie groups, Ann. Math. (2) 56(1952), 242-247.

[SiW] I.M. Singer & J. Wermer, Derivations on commutative normed algebras, Math. Ann. 129(1955), 260-264.

[Sj] J. Sjostrand, Parametrices for pseudo-differential operators with multiple characteristics, Ark. Mat. 12(1974), 85-130.

[Sℓ] J. Slawny, On factor representations and the C^*-algebra of canonical commutation relations, Comm. Math. Phys. 24(1971), 151-170.

[SoV] L. Solomon & D.N. Verma, Sur le corps des quotients de l'algèbre enveloppante d'une algèbre de Lie, C.R. Acad. Sci. Paris (A) 264(1967), 985-986.

[S] E.M. Stein, Topics in Harmonic Analysis, Princeton Univ. Press, 1970.

[Ste 1] D. Sternheimer, Extensions et unifications de algèbres de Lie, J. Math. Pures Appl. 47(1968), 249-289.

[Ste 2] __________, Propriétes spectrales dans les représentations de groupes des Lie, J. Math. Pures Appl. 47 (1968), 289-319.

[Sti 1] W.F. Stinespring, Positive functions on C^*-algebras, Proc. Amer. Math. Soc. 6(1955), 211-216.

[Sti 2] __________, A semi-simple matrix group is of Type I, Proc. Amer. Math. Soc. 9(1958), 965-967.

[Sti 3] __________, Integrability of Fourier transform for unimodular Lie groups, Duke Math. J. 9(1958), 123-131.

[St 1] M.H. Stone, Linear transformations in Hilbert space, III. Operational Methods and Group Theory, Proc. Natl. Acad. Sci. USA 16(1930), 172-175.

[St 2] __________, On one parameter unitary groups in Hilbert space, Ann. of Math. 33(1932), 643-648.

[St 3] __________, Linear Transformations in Hilbert Space, Colloq. Pub. 15, Amer. Math. Soc. Providence, RI, 1932.

[St 4] __________, On unbounded operators in Hilbert space, J. Indian Math. Soc. 15(1951), 155-192.

[Stφ] E. Stφrmer, Spectra of ergodic transformations, J. Funct. Anal. 15(1974), 202-215.

[Str] R.F. Streater, The representations of the oscillator group, Comm. Math. Phys. 4(1967), 217-236.

[Stri] R.S. Strichartz, Singular integrals on nilpotent Lie groups, Proc. Amer. Math. Soc. 53(1975), 367-374.

[Ta] H. Takai, On a problem of Sakai in unbounded derivations, J. Funct. Anal. 43(1981), 202-208.

[Tak] M. Takesaki, Fourier analysis of compact automorphism groups (An application of Tannaka duality theorem), in D. Kastler, ed., Algèbres d'opérateurs et leurs applications en physique mathématique, Éditions CNRS, Paris, 1979.

[Tay 1] M. Taylor, Noncommutative microlocal analysis (I), Preprint, 1983.

[Tay 2] __________, Noncommutative harmonic analysis, Amer. Math. Soc., Providence, RI, 1986.

[Th] L.H. Thomas, On unitary representations of the group of De Sitter space, Ann. of Math. 42(1941), 113-126.

[Tho 1] K. Thomsen, Dispersiveness and positive contractive semigroups, J. Funct. Anal. 56(1984), 348-359.

[Tho 2] __________, A note to the previous paper by Bratteli, Goodman and Jorgensen, J. Funct. Anal. 61(1985), 290-294.

[Til] H.G. Tillmann, Zur Eindeutigkeit der Lösungen der quantenmechanischen Vertauschungsrelationen, I, Acta Sci. Math. Szeged 24(1963), 258-270; II, Arch. Math. 15(1964), 332-334.

[Tit 1] J. Tits, Groups and group functors attached to Kac-Moody data, Arbeitstagung Bonn, 1984. Lecture notes in Mathematics, Vol. 1111, Springer-Verlag, Berlin, 1985.

[TW] J. Tits & L. Waelbroech, The integration of Lie algebra representations, Pacific J. Math. 26(1968), 595-600.

[Tom] J. Tomiyama, The Theory of Closed Derivations in the Algebra of Continuous Functions on the Unit Interval, Lecture Notes, Tsing Hua Univ., 1983.

[Tre 1] F. Trèves, Topological Vector Spaces, Distributions, and Kernels, Academic Press, New York, 1967.

[Tre 2] __________, On the existence and regularity of solutions of linear partial differential equations, Proc. Symp. Pure Math 23, Amer. Math. Soc., Providence, RI, 1973, 33-60.

[Us] T. Ushijima, On the generation and smoothness of semigroups of linear operators, J. Fac. Soc. Tokyo 19 (1972), 65-127.

[Va 1] V.S. Varadarajan, On the ring of invariant polynomials on a semisimple Lie algebra, Amer. J. Math. 90- (1968), 308-317.

[Va 2] __________, Geometry of Quantum Theory (2nd edn.), Springer-Verlag, Berlin-New York, 1985.

[Ve 1] M. Vergne, La structure de Poisson sur l'algèbre symétrique d'une algèbre de Lie nilpotente, C.R. Acad. Sci. Paris (A) 269(1969), 950-952.

[Ve 2] __________, La structure de Poisson sur l'algèbre de Lie Nilpotente, Bull. Soc. Math. France 100(1972), 301-335.

[Ve 3] __________, Cohomologie des algèbres dse Lie nilpotentes; application à l'étude de la variété des algèbre de Lie nilpotentes, Bull. Soc. Math. France 98 (1970), 81-116.

[Ve 4] __________, Seconde quantification et groupe symplectique, C.R. Acad. Sci. Sér. A, Paris 285(1977), 191-194.

[VR] M. Vergne & H. Rossi, Analytic continuation of the holomorphic discrete series of a semi-simple Lie group, Acta Math. 136(1976), 1-59.

[Ver 1] D.N. Verma, Structure of certain induced representations of complex semisimple Lie algebras, Dissertation, Yale Univ., 1966; Bull. Amer. Math. Soc. 74 (1968), 160-168, 628.

[Ver 2] __________, Möbius inversion for the Bruhat ordering on a Weyl group, Ann. Sci. École Norm. Sup. 4(1971), 393-398.

[VGG] A.M. Vershik, I.M. Gelfand & M.I. Graev, Representations of the group SL(2,ℝ) where ℝ is a ring of functions, Russ. Math. Surveys 28(1973), 87-132.

[Vi 1] I. Vidav, Spectra of perturbed semigroups with applications to transport theory, J. Math. Anal. Appl. 30 (1970), 264-279.

[Vi 2] __________, Perturbations of strongly continuous semigroups, Preprint, Univ. of Ljubljana, Yugoslavia, 1975.

[Vo] D.A. Vogan, Jr., Representations of Real Reductive Lie Groups, Birkhäuser, Boston-Basel, 1981.

[Voi 1] D. Voiculescu, Symmetries of some reduced free product C^*-algebras, Preprint, INCREST, 1983, in Springer Lecture Notes in Math. vol. 1132, pp. 556-588.

[Voi 2] __________, Dual algebraic structures on operator alge bras related to free products, J. Operator Theory 17 (1987), 85-98.

[Wa 1] N.R. Wallach, cyclic vectors and irreducibility for principal series representations, I, Trans. Amer. Math. Soc. 158(1971), 107-113; II, ibid. 164(1972), 389-396.

[Wa 2] __________, Harmonic Analysis on Homogeneous Spaces, Marcel Dekker, New York, 1973.

[War] G. Warner, Harmonic Analysis on Semi-simple Lie Groups, I and II, Springer-Verlag, Boston, 1972.

[We 1] A. Weil, L'integration dans les groupes topologiques et ses applications, Hermann, Paris, A.S.I., 1940, pp. 158.

[We 2] __________, On discrete subgroups of Lie groups, I, Ann. of Math. 72(1960), 369-384.

[Wey 1] H. Weyl, Theorie der Darstellung kontinuierlicher halbeinfacher Gruppen durch lineare Transformationen, I, Math. Z. 23(1925), 271-309; II, 24(1926), 328-376; III, 377-395, Nachtrag, 789-791.

[Wey 2] __________, Harmonics on homogeneous manifolds, Ann. of Math. 35(1934), 486-490.

[Wey 3] __________, The Structure and Representations of Continuous Groups, Lecture Notes, Princeton Univ. Press, Princeton, NJ, 1935.

[Wey 4] __________, The Classical Groups, Princeton Univ. Press, Princeton, NJ, 1939, pp. 302.

[Wey 5] __________, The Classical Groups. Their Invariants and Representations, Princeton Univ. Press, 1946, pp. 320.

[Wey 6] __________, The Theory of Groups and Quantum Mechanics, Dover, New York, 1950, pp. 442.

[Wm] A.S. Wightman, The problem of existence of solutions in quantum field theory, In: D. Feldman, ed., Proc. Fifth Annual Eastern Theoretical Phys. Conf., Benjamin, 1967.

[Wi 1] E.P. Wigner, Gruppentheorie und ihre Anwendung auf die Quantenmechanik der Atomspektren, Friedr. Vieweg, Braunschweig, 1931, pp. 332.

[Wi 2] __________, On unitary representations of the inhomogeneous Lorentz group, Ann. of Math. 40(1939), 149-204.

[Wi 3] __________, Relativistic invariance and quantum phenomena, Rev. Modern Phys. 29(1957), 255-268.

[Wi 4] __________, Group Theory and Its Applications to the Quantum Mechanics of Atomic Spectra, Academic Press, New York, 1959, pp. 372.

[Wi 5] __________, Unitary Representaions of the Inhomogeneous Lorentz Group Including Reflections, Lectures of the Istanbul Summer School of Theoretical Physics, F. Gürsey, ed., Gordon and Breach, 1962, 37-80.

[Win] A. Wintner, On the normalization of characteristic differential in continuous spectra, Phys. Rev. 72 (1947), 516-517.

[Wit 1] E. Witten, Non-abelian bosonization in two dimensions, Commun. Math. Phys. 92(1984), 455-472.

[Wit 2] __________, Noncommutative geometry and string field theory, Preprint, 1985, Princeton Univ. Press.

[Ya] T. Yao, Unitary irreducible representations of SU(2,2), I, II and III, J. Math. Phys.; I, 8(1967), 1931-1954; II, 9(1968), 1615-1626; III, 12(1971), 315-342.

[Yang] C.-T. Yang, Hilbert's fifth problem and related problems on transformation groups, in Proc. Sym-posia Pure Math. 28(1976), Amer. Math. Soc. Provi-dence, RI.

[Yo] K. Yosida, Functional Analysis (3rd edn.), Springer-Verlag, New York, 1971.

[Zy] A. Zygmund, Trigonometric Series, Second Edition, Cambridge Univ. Press, Cambridge, 1977.

INDEX